Applied Linear Algebra
The Decoupling Principle

LORENZO SADUN
University of Texas at Austin

PRENTICE HALL, *Upper Saddle River, New Jersey 07458*

Library of Congress Cataloging-in-Publication Data
Sadun, Lorenzo Adlai.
  Applied linear algebra : the decoupling principle / Lorenzo Sadun.
     p. cm.
  Includes index.
  ISBN 0-13-085645-2
  1. Algebras, Linear. I. Title.
  QA184.S23 2001
  512'.5—dc21   00-039991

Acquisition Editor: *George Lobell*
Assistant Vice President of Production and Manufacturing: *David W. Riccardi*
Executive Managing Editor: *Kathleen Schiaparelli*
Senior Managing Editor: *Linda Mihatov Behrens*
Production Editor: *Bob Walters*
Manufacturing Buyer: *Alan Fischer*
Manufacturing Manager: *Trudy Pisciotti*
Marketing Manager: *Angela Battle*
Marketing Assistant: *Vince Jansen*
Director of Marketing: *John Tweeddale*
Editorial Assistant: *Gale Epps*
Art Director: *Jayne Conte*
Cover Designer: *Bruce Kenselaar*
Cover Image: *Architecture Collection 1, The Images Publishing Group, PTY, Ltd.*

Prentice
Hall © 2001 by Prentice Hall, Inc.
       Upper Saddle River, New Jersey 07458

All right reserved. No part of this book may be reproduced, in any form or by any means, without permission in writing from the publisher.

Printed in the United States of America

10 9 8 7 6 5 4 3

ISBN: 0-13-085645-2

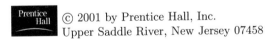

Prentice-Hall International (UK) Limited, *London*
Prentice-Hall of Australia Pty. Limited, *Sydney*
Prentice-Hall of Canada Inc., *Toronto*
Prentice-Hall Hispanoamericana, S.A., *Mexico*
Prentice-Hall of India Private Limited, *New Delhi*
Prentice-Hall of Japan, Inc., *Tokyo*
Pearson Education Asia Pte. Ltd.
Editora Prentice-Hall do Brasil, Ltda., *Rio de Janeiro*

For Anita, Rina, Allan and Jonathan

Contents

Preface .... **ix**

1 The Decoupling Principle **1**

2 Vector Spaces and Bases **9**
   2.1 Vector Spaces 9
   2.2 Linear Independence, Basis and Dimension 15
   2.3 Properties and Uses of a Basis 22
   2.4 Change of Basis 25
   2.5 Building New Vector Spaces from Old Ones 30

3 Linear Transformations and Operators **37**
   3.1 Definitions and Examples 37
   3.2 The Matrix of a Linear Transformation 42
   3.3 The Effect of a Change of Basis 46
   3.4 Infinite Dimensional Vector Spaces 49
   3.5 Kernels, Ranges, and Quotient Maps 52

4 An Introduction to Eigenvalues **57**
   4.1 Definitions and Examples 57
   4.2 Bases of Eigenvectors 59
   4.3 Eigenvalues and the Characteristic Polynomial 61
   4.4 The Need for Complex Eigenvalues 68
   4.5 When is an Operator Diagonalizable? 72
   4.6 Traces, Determinants, and Tricks of the Trade 77
   4.7 Simultaneous Diagonalization of Two Operators 82
   4.8 Exponentials of Complex Numbers and Matrices 86
   4.9 Power Vectors and Jordan Canonical Form 91

5 Some Crucial Applications — 97
- 5.1 Discrete-Time Evolution: $\mathbf{x}(n) = A\mathbf{x}(n-1)$ 97
- 5.2 First-Order Continuous-Time Evolution: $d\mathbf{x}/dt = A\mathbf{x}$. 103
- 5.3 Second-Order Continuous-Time Evolution: $d^2\mathbf{x}/dt^2 = A\mathbf{x}$ 108
- 5.4 Reducing Second-Order Problems to First-Order ... 117
- 5.5 Long-Time Behavior and Stability 120
- 5.6 Markov Chains and Probability Matrices 126
- 5.7 Linear Analysis near Fixed Points of Nonlinear Problems 135

6 Inner Products — 145
- 6.1 Real Inner Products: Definitions and Examples 145
- 6.2 Complex Inner Products 150
- 6.3 Bras, Kets, and Duality 153
- 6.4 Expansion in Orthonormal Bases: Finding Coefficients 159
- 6.5 Projections and the Gram-Schmidt Process 163
- 6.6 Orthogonal Complements and Projections onto Subspaces 169
- 6.7 Least Squares Solutions 172
- 6.8 The Spaces ℓ_2 and $L^2[0,1]$ 179
- 6.9 Fourier Series on an Interval 184

7 Adjoints, Hermitian Operators, and Unitary Operators — 191
- 7.1 Adjoints and Transposes 192
- 7.2 Hermitian Operators 197
- 7.3 Quadratic Forms and Real Symmetric Matrices ... 202
- 7.4 Rotations, Orthogonal Operators, and Unitary Operators 210
- 7.5 How the Four Classes are Related 219

8 The Wave Equation — 225
- 8.1 Waves on the Line 226
- 8.2 Waves on the Half Line; Dirichlet and Neumann Boundary Conditions 229
- 8.3 The Vibrating String 235
- 8.4 Standing Waves and Fourier Series 238
- 8.5 Periodic Boundary Conditions 244
- 8.6 Equivalence of Traveling Waves and Standing Waves . 253
- 8.7 The Different Types of Fourier Series 257

9 Continuous Spectra and the Dirac Delta Function 263
9.1 The Spectrum of a Linear Operator 264
9.2 The Dirac δ Function 268
9.3 Distributions 274
9.4 Generalized Eigenfunction Expansions; The Spectral Theorem 277

10 Fourier Transforms 287
10.1 Existence of Fourier Transforms 287
10.2 Basic Properties of Fourier Transforms 293
10.3 Convolutions and Differential Equations 299
10.4 Partial Differential Equations 301
10.5 Bandwidth and Heisenberg's Uncertainty Principle . . 307
10.6 Fourier Transforms on the Half Line 313

11 Green's Functions 319
11.1 Delta Functions and the Superposition Principle . . . 319
11.2 Inverting Operators 322
11.3 The Method of Images 328
11.4 Initial Value Problems 334
11.5 Laplace's Equation on $\mathbb{R}^2$ 339

Index 346

Preface

The purpose of the book

This book was designed as a textbook for a junior-senior level second course in linear algebra at the University of Texas at Austin. In that course, and in this book, I try to show math, physics, and engineering majors the incredible power of linear algebra in the real world. The hope is that, when faced with a linear system (or a nonlinear system that can be reasonably linearized), future engineers will think to decompose the system into modes that they can understand. Usually this is done by diagonalization. Sometimes this is done by decomposing into a convenient orthonormal basis, such as Fourier series. Sometimes a continuous decomposition, into δ functions or by Fourier transforms, is called for. The underlying ideas of breaking a vector into modes (the Superposition Principle) and of decoupling a complicated system by a suitable choice of linear coordinates (the Decoupling Principle) appear throughout physics and engineering. My goal is to impress upon students the importance of these principles, while giving them enough tools to use them effectively.

There are many existing types of second linear algebra courses, and many books to match, but few if any make this goal a priority. Some courses are theoretical, going in the direction of functional analysis, Lie Groups or abstract algebra. "Applied" second courses tend to be heavily numerical, teaching efficient and robust algorithms for factorizing or diagonalizing matrices. Some courses split the difference, developing matrix theory in depth, proving classification theorems (e.g., Jordan form) and estimates (e.g., Gershegorin's Theorem). While each of these courses is well-suited for its chosen audience, none give a prospective physicist or engineer substantial insight into how or why to apply linear algebra at all.

Notes to the instructor

The readers of this book are assumed to have taken an introductory linear algebra class, and hence to be familiar with basic matrix operations such as row reduction, matrix multiplication and inversion, and taking determinants. The reader is also assumed to have had some exposure to vector spaces and linear transformations. This material is reviewed in Chapters 2 and 3, pretty much from the beginning, but a student who has never seen an abstract vector space will have trouble keeping up. The subject of Chapter 4, eigenvalues, is typically covered quite hastily at the end of a first course (if at all), so I work under the assumption that readers do not have any prior knowledge of eigenvalues.

The key concept of these introductory chapters is that a basis makes a vector space look like $\mathbb{R}^n$ (or sometimes $\mathbb{C}^n$) and makes linear transformations look like matrices. Some bases make the conversion process simple, while others make the end results simple. The standard basis in $\mathbb{R}^n$ makes coordinates easy to find, but may result in an operator being represented by an ugly matrix. A basis of eigenvectors, on the other hand, makes the operator appear simple but makes finding the coordinates of a vector difficult. To handle problems in linear algebra, one must be adept in coordinatization and in performing change-of-basis operations, both for vectors and for operators.

One premise of this book is that standard software packages (e.g., MATLAB, Maple or Mathematica) make it easy to diagonalize matrices without any knowledge of sophisticated numerical algorithms. This frees us to consider the *use* of diagonalization, and some general features of important classes of operators (e.g., Hermitian or unitary operators). Diagonalization, by computer or by hand, gives a set of coordinates in which a problem, even a problem with an infinite number of degrees of freedom, decouples into a collection of independent scalar equations. This is what I call the Decoupling Principle.

(Strictly speaking this is only true for diagonalizable operators. However, a matrix or operator coming from the real world is almost certainly diagonalizable, especially since Hermitian and unitary matrices are *always* diagonalizable. For completeness, I have included sections in the book about nondiagonalizable matrices, power vectors, and Jordan form, but these issues are not stressed, and these sections can be skipped with little loss of continuity.)

Preface

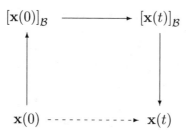

Figure 0.1: The decoupling strategy

The Decoupling Principle is first applied systematically in Chapter 5, where we consider a variety of coupled linear differential equations or difference equations. Students may have seen some of these problems in previous courses on differential equations, probability or classical mechanics, but typically have not understood that the right choice of coordinates (achieved by using a basis of eigenvectors) is independent of the type of problem. By presenting a sequence of problems solved via the Decoupling Principle, this point is driven home. It is also hoped that the examples are of interest in their own right, and provide an applied counterweight to the fairly theoretical introductory chapters.

In Chapter 5, students are also exposed to questions of linear and nonlinear stability. They learn to linearize nonlinear equations near fixed points, and to use their stability calculations to determine for how long linearized equations can adequately model an underlying nonlinear problem. These are questions of crucial importance to physicists and engineers.

Up through Chapter 5, our calculational model is as follows (see Figure 0.1). To solve a time evolution problem (say, $d\mathbf{x}/dt = L\mathbf{x}$) we find a basis $\mathcal{B}$ of eigenvectors of L. We then convert the initial vector $\mathbf{x}(0)$ into coordinates $[\mathbf{x}(0)]_\mathcal{B}$, compute the coordinates $[\mathbf{x}(t)]_\mathcal{B}$ of the vector at a later time, and from that reconstitute the vector $\mathbf{x}(t)$. The basis of eigenvectors makes the middle (horizontal) step easy, but the vertical steps, expecially finding the coordinates $[\mathbf{x}(0)]_\mathcal{B}$, can be difficult.

In Chapter 6 we introduce inner products and see how coordinatization becomes easy if our basis is orthogonal. Fourier series on L^2 of an interval is then a natural consequence. Chapter 6 also contains several subjects that, while interesting, may not fit into a course syllabus.

In Section 6.1, the subsection on nonstandard inner products may be skipped if desired. In Section 6.3, the general discussion of dual spaces is included mostly for reference, and may also be skipped, especially as many students find this material to be quite difficult. However, the beginning of the section should be covered thoroughly. It is important for students to understand that $|\mathbf{v}\rangle$ is a vector, while $\langle\mathbf{w}|$ is an operation, namely "take the inner product with $\mathbf{w}$". They should also understand the representation of bras and kets as rows and columns, respectively.

Section 6.7 on least squares also deserves comment. This topic is off the theme of the Decoupling Principle, but is far too useful to leave out. Instructors should feel free to spend as much or as little time on this digression as they see fit.

Chapter 5 demonstrates the utility of bases of eigenvectors, Chapter 6 demonstrates the utility of orthogonal bases, and Chapter 7 reconciles the two approaches, showing how several classes of important operators are diagonalizable with orthogonal eigenvectors. Fourier series, introduced in Chapter 6 as an expansion in an orthogonal basis, can then be reconsidered as an eigenfunction expansion for the Laplacian.

Another important premise of this book is that infinite dimensional systems are important, but that a full treatment of Banach spaces (or even just Hilbert spaces) would only distract students from the Decoupling Principle. The last third of the book is devoted to infinite dimensional problems (e.g., the wave equation in $1 + 1$ dimensions), with the idea of transfering intuition from finite to infinite dimensions. My attitude is summarized in the advice to the student at the end of Section 3.4, where infinite dimensional spaces are first introduced:

In short, infinite dimensional spaces and infinite dimensional operations are neither totally bizarre nor totally tame, but somewhere in between. If an argument or technique works in finite dimensions, it is probable, but by no means certain, that it will work in infinite dimensions. As a first approximation, applying your finite dimensional intuition to infinite dimensions is a very good idea. However, you should be prepared for an occasional surprise, almost always due to a lack of convergence of some sum.

Chapter 8 is the infinite dimensional sequel to Chapter 5, using the wave equation to demonstrate the Decoupling Principle for partial differential equations. The key idea is to think of a scalar-valued

Preface

partial differential equation as an ordinary differential equation on an infinite dimensional vector space. The results of Chapter 5 then carry over directly to give the general solution to the vibrating string problem in terms of standing waves. The wave equation can also be attacked in different ways, each demonstrating a different linear algebraic principle. Solving the wave equation on the whole line in terms of forward and backward traveling waves involves both the Superposition Principle and the properties of commuting operators. The wave equation on the half line gives us the method of images. Comparing the standing wave and traveling wave solutions to the vibrating string leads us naturally to consider two kinds of Fourier series on an interval $[0, L]$; the first in terms of $\sin(n\pi x/L)$, the second in terms of $\exp(2\pi i n x/L)$. Both are eigenfunction expansions for Hermitian operators, the first for the Laplacian with Dirichlet boundary conditions, the second for $-id/dx$ with periodic boundary conditions.

In Chapter 9 we make the transition from discrete to continuous spectra, introducing the Dirac δ function and expansions that involve integrating over generalized eigenfunctions. Fourier transforms (Chapter 10) then naturally appear as generalized eigenfunction expansions for the "momentum" operator $-id/dx$.

Finally, once we have a generalized basis of δ functions, we can decompose with respect to that basis to get integral kernels (a.k.a. Green's functions) for linear operators. Like the earlier discussion of least squares, this is a departure from the central theme of the book, but is much too useful to leave out.

Possible course outlines

There are three recommended courses that can be built from this book. The first 8 chapters, with an occasional section skipped, forms a coherent one semester course on diagonalization and on infinite dimensional problems with discrete spectra. This is essentially the course I have taught at the University of Texas at Austin. For such a course Chapters 2 and 3 should be presented quickly, as only the last section or two of each chapter is likely to be new material.

For universities on the quarter system, the entire book can be used for the second and third quarters of a year-long linear algebra course. In that case I recommend that the first quarter concentrate on matrix manipulations, solutions to linear equations, and vector space properties of $\mathbb{R}^n$ and its subspaces. (E.g., the first four chapters of David Lay's excellent text.) There is no need to discuss eigenvalues or inner products at all in the first quarter, as they are covered from scratch in Chapters 4 and 6, respectively. In such a sequence, Chapters 2 and 3 would be treated as new material and presented slowly, not as review material to be skimmed.

There are several sections (2.5, 3.4, 4.7, 4.9, 5.6, 5.7, 6.7, 6.8, 6.9) that can be skipped without too much loss of continuity. Some of these sections (especially 5.7: Linearization of nonlinear problems, 6.7: Least squares, and 6.9: Fourier series) are of tremendous importance in their own right and should be learned at some point, but it is certainly possible to construct a course without them. Which of these to include and which to skip is largely a matter of course pace and instructor taste.

As a third option, the first seven chapters of this book can make a substantial first course in linear algebra for strong students who have already learned about row reduction and matrix algebra in high school. For such a course I recommend emphasizing finite dimensional applications (e.g., Markov chains and least squares) and de-emphasizing infinite dimensional extensions.

Finally, this book can be used for self-study by advanced undergraduate or beginning graduate students who need more linear algebra than is typically taught in a first course. Chapters 6, 7, 9, and 10 are of particular interest to physics students struggling with the formalism of quantum mechanics, Chapter 11 to physics and engineering students studying electromagnetism, and Chapters 7–11 to students of applied math and functional analysis.

To serve the needs of such students, the last three chapters are written at a more sophisticated level than the earlier chapters. They are logically self-contained, treating each subject from scratch, but assume a significant background in general mathematics. For example, to appreciate Fourier transforms, it helps to be adept at computing them, and that often means doing contour integrals. These chapters are probably most useful to students who have been exposed to Fourier transforms and/or Green's functions in their physics and engineering coursework, but who lack a conceptual framework for these subjects.

Notes to the student

You will probably find the beginning of the book to be largely review. You may have seen much of Chapters 2 (Vector spaces) and 3

(Linear transformations) and some of Chapter 4 (Eigenvalues) in a first linear algebra course, but I do not assume that you have mastered these concepts. As befits review material, most of the concepts are presented quickly from the beginning. If you thoroughly understood your first course, you should be able to skim these chapters, concentrating on the last section or two of each. On the other hand, if your first course was not enough preparation, you should take the time to go through Chapters 2 and 3 carefully, and work out many of the problems. It's worth the extra effort, as the entire book depends strongly on the ideas of Chapters 2 and 3.

I typically present each major concept in three settings. The first setting is in $\mathbb{R}^n$, where the problem is essentially a (frequently familiar) matrix computation. The second setting is in a general n-dimensional vector space, where a choice of basis reduces the problem to one on $\mathbb{R}^n$. The key is to choose the right basis, and I put considerable emphasis on understanding what stays the same and what changes when you change basis. The third setting is in an infinite dimensional vector space. The goal is not to develop a general theory, but for you to see enough examples to start building up intuition. Infinite dimensional spaces appear more and more often later in the book.

While (almost) all results in finite dimensions are proven, most infinite dimensional theorems (such as the spectral theorem for bounded self-adjoint operators on a Hilbert space) are merely stated, and in some cases I just argue formally, by analogy to finite dimensions. Although such analogies can sometimes fail (I give examples), they are a very good intuitive starting point.

This book is aimed at a mixture of math, physics, computer science, engineering, and economics majors. The only absolute prerequisite is familiarity with matrix manipulation (Gaussian elimination, inverses, determinants, etc.). These topics are invariably covered in a first linear algebra course, but can also be picked up elsewhere. Basic vector space concepts are covered from the beginning in Chapters 2 and 3. However, if you have never seen a vector space (not even $\mathbb{R}^n$), you may have difficulty keeping up. These chapters are aimed primarily at students who once were exposed to vector spaces but may be rusty. In Chapters 4–8, no prior knowledge is expected or required. Chapters 9–11 are generally more sophisticated, but are self-contained, with no prior knowledge of the subject assumed.

Finally, software such as MATLAB, Mathematica and Maple can

make many linear algebra computations much easier, and I encourage you to use them. With technology, you can avoid the drudgery of, say, computing the eigenvalues and eigenvectors of a large matrix by hand. However, technology should enhance your thinking, not replace it! A computer can tell you what the eigenvalues of a matrix are, but it's up to you to figure out what to do with them.

Notation and terminology

Many objects in linear algebra are referred to by different names by mathematicians, physicists and engineers. What's worse, the same terms are often used by the different communities to mean different things. As Winston Churchill said of Americans and Englishmen, *We are divided by a common language.*

Here are some key differences. Others are noted in the text. In each case I have adopted the physics conventions and terminology, especially the terminology of quantum mechanics.

- Mathematicians usually denote their complex inner products $(\mathbf{x}, \mathbf{y})$, linear in $\mathbf{x}$ and conjugate-linear in $\mathbf{y}$. Physicists use Dirac's bra-ket notation $\langle \mathbf{x} | \mathbf{y} \rangle$, linear in $\mathbf{y}$ and conjugate-linear in $\mathbf{x}$, and refer to the individual factors $\langle \mathbf{x} |$ and $| \mathbf{y} \rangle$ as "bra"s and "ket"s, respectively.

- An eigenvalue whose geometric multiplicity is greater than one is called "repeated" or "multiple" by mathematicians and "degenerate" by physicists.

- An eigenvalue whose algebraic multiplicity is greater than its geometric multiplicity is called "degenerate" by mathematicians and "deficient" by physicists.

- If A is an operator, $\boldsymbol{\xi}$ is a vector and $(A - \lambda I)^p \boldsymbol{\xi} = 0$ for some exponent p, then $\boldsymbol{\xi}$ is often called a "generalized eigenvector" corresponding to the eigenvalue λ. I prefer Sheldon Axler's term "power vector", and use "generalized eigenvector" in the context of continuous spectrum to mean a non-normalizable eigenfunction. (E.g., on the real line, e^{ikx} is a generalized eigenfunction of $-id/dx$ with generalized eigenvalue k.)

Acknowledgments

This book grew out of lecture notes for a new class in Applied Linear Algebra at the University of Texas at Austin. I thank Karen Uhlenbeck and Jerry Bona for suggesting that I develop this class and to Martin Speight for helping design it. The students in this class, number M375 in Fall 1997 and M346 in Fall 1998, provided invaluable help, both as test subjects and as editors. In particular, their feedback led to a much needed overhaul of the early chapters.

Parts of this book were written at the Park City Math Institute in 1997 and 1998. In both summers, the participants in the Undergraduate Faculty Program were generous with their expertise and insight into what would or would not work in an actual classroom. I especially thank Jane Day, Daniel Goroff, Dan Kalman and John Polking. This book was completed at the Physics Department of the Israel Institute of Technology (Technion).

Finally, I am especially grateful to Margaret Combs, whose artistry, TEXnical expertise and many hours of work are responsible for most of the figures.

Lorenzo Sadun
sadun@math.utexas.edu

Chapter 1

The Decoupling Principle

Many phenomena in mathematics, physics, and engineering are described by linear evolution equations. Light and sound are linear waves. Population growth and radioactive decay are described well by first order linear differential equations. To understand the behavior of nonlinear systems near equilibrium, one always approximates them with linear systems, which can then be solved exactly. In short, to understand nature you have to understand linear equations.

In this book we cover a variety of linear evolution equations, beginning with the simplest equations in one variable, moving on to coupled equations in several variables, and culminating in problems such as wave propagation that involve an infinite number of degrees of freedom. Along the way we develop techniques, such as Fourier analysis, that allow us to decouple the equations into a set of scalar equations that we already know how to solve.

The general strategy is always the same. When faced with coupled equations involving variables $x_1, \ldots, x_m$, we define new variables $y_1, \ldots, y_m$. These variables can always be chosen so that the evolution of y_1 depends only on y_1 (and not on $y_2, \ldots, y_m$), the evolution of y_2 depends only on y_2, and so on. See Figure 1.1. To find $x_1(t), \ldots, x_m(t)$ in terms of the initial conditions $x_1(0), \ldots, x_m(0)$, we convert $\mathbf{x}(0)$ to $\mathbf{y}(0)$, then solve for $\mathbf{y}(t)$, then convert to $\mathbf{x}(t)$.

The first and third (vertical) steps are pure linear algebra and require no knowledge of evolution. The second step is done one variable at a time, and requires us to understand scalar evolution equations, but does not involve linear algebra. Instead of dealing with coupled equations all at once, we consider the coupling and the underlying scalar equations separately.

1

2   Chapter 1. The Decoupling Principle

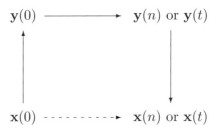

Figure 1.1: The decoupling strategy

Scalar linear evolution equations

The simplest linear evolution equations involve a single variable, which we will call x. Here are some examples of scalar linear evolution equations, most of which should already be familiar:

1. **Discrete time growth or decay.** Suppose $x(n)$ is the number of deer in a forest in year n, and that $x(n)$ is proportional to $x(n-1)$. That is,

$$x(n) = ax(n-1), \qquad (1.1)$$

for some constant a. Solving this equation we find that

$$x(n) = a^n x(0). \qquad (1.2)$$

Of course, equation (1.1) could describe a variety of phenomena. The variable x could denote the amount of money in the bank, or the size of a radioactive sample, or indeed anything that has the unifying property: *How much you have tomorrow is proportional to how much you have today.* I leave it to you to think up additional examples.

2. **Continuous time growth or decay.** If growth or decay is a continuous process (as is essentially the case with radioactive decay, or with population growth in humans, bacteria, and other species that never stop breeding), then equation (1.1) should be replaced by a linear differential equation:

$$\frac{dx}{dt} = ax. \qquad (1.3)$$

Chapter 1. The Decoupling Principle

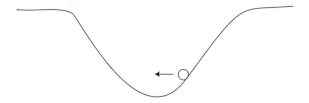

Figure 1.2: Acceleration is toward center

The solution to this equation is

$$x(t) = ce^{at}, \quad (1.4)$$

where the constant c is determined by the initial condition:

$$c = x(0). \quad (1.5)$$

3. **Oscillations near a stable equilibrium.** If x is the displacement of a mass on a spring, or the position of a ball rolling near the bottom of a hill (see Figure 1.2), or indeed *any* physical quantity near a stable equilibrium, then x is described by the second-order ordinary differential equation (ODE):

$$\frac{d^2x}{dt^2} = -ax, \quad (1.6)$$

where a is a positive constant. This equation says that x is being accelerated back towards the origin at a rate proportional to x. Let $\omega = \sqrt{a}$. The general solution to equation (1.6) is

$$x(t) = c_1 \cos(\omega t) + c_2 \sin(\omega t), \quad (1.7)$$

where c_1 and c_2 are determined by initial data. If our variable has initial value x_0 and initial velocity v_0, then

$$c_1 = x_0, \quad c_2 = v_0/\omega. \quad (1.8)$$

4. **Oscillations near an unstable equilibrium.** Let x be the position of a ball balanced near the *top* of a hill, as in Figure 1.3. The bigger x gets, the harder our ball gets pushed *away from* equilibrium. That is,

$$\frac{d^2x}{dt^2} = +ax, \quad (1.9)$$

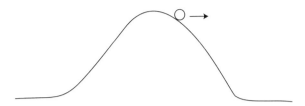

Figure 1.3: Acceleration is away from center

for some positive constant a. This phenomenon goes by many names, from the technical "positive feedback loop" to the commonplace "vicious cycle". We all know of many real-world examples where a delicate balance can be ruined by a small push either way. Let $\kappa = \sqrt{a}$. The general solution to equation (1.9) is

$$x(t) = c_1 e^{\kappa t} + c_2 e^{-\kappa t}, \qquad (1.10)$$

where once again c_1 and c_2 can be computed from the initial conditions.

Coupled linear evolution equations

Unfortunately, most real-world situations involve several quantities $x_1, x_2, \ldots, x_m$ coupled together. By "coupled" we mean that the evolution of each variable x_i depends on the other variables, not just on x_i itself.

For example, consider the population of owls and mice in a certain forest. Since owls eat mice, the more mice there are, the more the owls have to eat, and the more owls there will be next year. The more owls there are, the more mice get eaten, and the fewer mice there will be next year. If $x_1(n)$ and $x_2(n)$ describe the owl and mouse populations in year n, then the equations governing population growth might take the form

$$\begin{aligned} x_1(n) &= a_{11} x_1(n-1) + a_{12} x_2(n-1), \\ x_2(n) &= a_{21} x_1(n-1) + a_{22} x_2(n-1), \end{aligned} \qquad (1.11)$$

where a_{11}, a_{12} and a_{22} are positive constants and a_{21} is a negative constant.

In such situations it is convenient to combine the variables x_1 and x_2 into a vector $\mathbf{x} = \begin{pmatrix} x_1 \\ x_2 \end{pmatrix}$ and combine the coefficients $\{a_{ij}\}$

Chapter 1. The Decoupling Principle 5

into a matrix
$$A = \begin{pmatrix} a_{11} & a_{12} \\ a_{21} & a_{22} \end{pmatrix}. \qquad (1.12)$$
Equations (1.11) can then be rewritten as
$$\mathbf{x}(n) = A\mathbf{x}(n-1). \qquad (1.13)$$
Just as equation (1.13) is a multivariable extension of equation (1.1), there are multivariable extensions of equations (1.3), (1.6) and (1.9), namely
$$\frac{d\mathbf{x}}{dt} = A\mathbf{x}, \qquad (1.14)$$
$$\frac{d^2\mathbf{x}}{dt^2} = -A\mathbf{x}, \qquad (1.15)$$
and
$$\frac{d^2\mathbf{x}}{dt^2} = +A\mathbf{x}, \qquad (1.16)$$
respectively. Of course, in these equations $\mathbf{x}$ does not have to be a 2-component vector. $\mathbf{x}$ could be an m-component vector, in which case A would be an $m \times m$ matrix of coefficients. In fact, $\mathbf{x}$ could even be an element of an infinite dimensional vector space, with A an operator on that space.

Solving equations (1.13–1.16) certainly looks more difficult than solving the corresponding 1-variable equations, but there is a common occurrence where it is equally easy.

Decoupled equations

Suppose the matrix A is diagonal. For example, suppose that
$$A = \begin{pmatrix} 2 & 0 \\ 0 & 3 \end{pmatrix}. \qquad (1.17)$$
Then equation (1.13) becomes
$$\begin{aligned} x_1(n) &= 2x_1(n-1), \\ x_2(n) &= 3x_2(n-1). \end{aligned} \qquad (1.18)$$

Instead of having one matrix equation (or equivalently two coupled scalar equations), we have two *uncoupled* scalar equations, and we can immediately write down the solutions:
$$x_1(n) = 2^n x_1(0); \qquad x_2(n) = 3^n x_2(0). \qquad (1.19)$$
It is similarly easy to solve equations (1.14–1.16). What an improvement from the coupled situation!

Decoupling coupled equations

Next, we consider an example in which two equations *look* coupled, but with a judicious change of coordinates, can be changed into two uncoupled equations. Consider the equations

$$\begin{aligned} x_1(n) &= 2x_1(n-1) + x_2(n-1), \\ x_2(n) &= x_1(n-1) + 2x_2(n-1). \end{aligned} \quad (1.20)$$

These equations are of the form (1.13) with

$$A = \begin{pmatrix} 2 & 1 \\ 1 & 2 \end{pmatrix}. \quad (1.21)$$

We now define new variables

$$y_1 = (x_1 + x_2)/2; \quad y_2 = (x_1 - x_2)/2, \quad (1.22)$$

so that

$$x_1 = y_1 + y_2; \quad x_2 = y_1 - y_2. \quad (1.23)$$

Adding the two equations of (1.20) and dividing by 2 gives

$$y_1(n) = 3y_1(n-1), \quad (1.24)$$

while taking the difference of the two equations of (1.20) and dividing by 2 gives

$$y_2(n) = y_2(n-1). \quad (1.25)$$

The coupled equations $\mathbf{x}(n) = A\mathbf{x}(n-1)$ have been converted to decoupled equations

$$\mathbf{y}(n) = \begin{pmatrix} 3 & 0 \\ 0 & 1 \end{pmatrix} \mathbf{y}(n-1). \quad (1.26)$$

The decoupled equations can be solved by inspection, with solutions

$$y_1(n) = 3^n y_1(0); \quad y_2(n) = y_2(0). \quad (1.27)$$

To solve the original equations (1.20), we follow the 3-step procedure of Figure 1.1. We first use (1.22) to convert the initial data $\mathbf{x}(0)$ into initial data $\mathbf{y}(0)$. We then use equations (1.27) to find $\mathbf{y}(n)$. Finally, we use equations (1.23) to compute $\mathbf{x}(n)$. The end result is

$$\begin{aligned} x_1(n) &= \frac{1}{2}(3^n + 1)x_1(0) + \frac{1}{2}(3^n - 1)x_2(0), \\ x_2(n) &= \frac{1}{2}(3^n - 1)x_1(0) + \frac{1}{2}(3^n + 1)x_2(0). \end{aligned} \quad (1.28)$$

Using the y coordinates, one can similarly solve equations (1.14–1.16).

The moral of the story

The combinations $y_1 = (x_1 + x_2)/2$ and $y_2 = (x_1 - x_2)/2$, in terms of which everything simplified, were not chosen at random. The vectors $\begin{pmatrix} 1 \\ 1 \end{pmatrix}$ and $\begin{pmatrix} 1 \\ -1 \end{pmatrix}$ are the *eigenvectors* of the matrix $\begin{pmatrix} 2 & 1 \\ 1 & 2 \end{pmatrix}$, while 3 and 1 are the corresponding *eigenvalues*. In general, whenever we can find m eigenvalues (and corresponding eigenvectors) of an $m \times m$ matrix A, we can convert the coupled system of equations (1.13) (or 1.14 or 1.15 or 1.16) into m decoupled equations, which we can then solve, as above. (In Section 4.5 we determine when it is possible to find enough eigenvalues and eigenvectors.) The same ideas also apply (with a few modifications) to infinite dimensional problems, such as heat flow, wave propagation, and diffusion.

Note that the combinations $(x_1 + x_2)/2$ and $(x_1 - x_2)/2$ were the same for all four types of problems. In general, decoupling a system of equations does not require us to know anything about the class of equations; we just need to understand the eigenvalues and eigenvectors of the matrix A. This is the Decoupling Principle. Only after we have decoupled the variables do we need to examine the resulting (scalar!) equations.

This book is an extended exploration of this principle. After a review of elementary linear algebra in Chapters 2 and 3, we learn about eigenvalues and eigenvectors in Chapter 4. In Chapter 5 we use eigenvalues and eigenvectors to solve the kinds of problems we have just discussed. Thereafter we consider systems with additional structures, such as an inner product or a complex structure. We shall see that many powerful techniques, such as Fourier Analysis, are just special applications of the Decoupling Principle.

I have attempted to present each concept in three settings. The first setting is in $\mathbb{R}^n$, where the problem is essentially a (presumably familiar) matrix computation. The second setting is in a general n-dimensional vector space, where a choice of basis reduces the problem to one on $\mathbb{R}^n$ (or $\mathbb{C}^n$). The third setting is in an infinite dimensional vector space. The goal is not to understand the full theory of infinite dimensional vector spaces (something well beyond the scope of this book), but to see enough examples to start building up intuition. Infinite dimensional spaces appear with increasing frequency later in the book.

Exercises

1. With $A = \begin{pmatrix} 2 & 0 \\ 0 & 3 \end{pmatrix}$, write down the solution to equation (1.14). That is, express $x_1(t)$ and $x_2(t)$ in terms of initial conditions $x_1(0)$ and $x_2(0)$.

2. With $A = \begin{pmatrix} 2 & 0 \\ 0 & 3 \end{pmatrix}$, write down the solution to equation (1.15). That is, express $x_1(t)$ and $x_2(t)$ in terms of initial data (values and first derivatives at $t = 0$).

3. With $A = \begin{pmatrix} 2 & 0 \\ 0 & 3 \end{pmatrix}$, write down the most general solution to equation (1.16).

4. Let $x_1(0) = 5$ and let $x_2(0) = 3$. Solve equations (1.20) in the three steps indicated, first computing $y_1(0)$ and $y_2(0)$, then computing $y_1(n)$ and $y_2(n)$, and then computing $x_1(n)$ and $x_2(n)$. Check that your results agree with equation (1.28).

5. With $A = \begin{pmatrix} 2 & 1 \\ 1 & 2 \end{pmatrix}$, solve equation (1.14). That is, express $x_1(t)$ and $x_2(t)$ in terms of initial conditions $x_1(0)$ and $x_2(0)$.

6. With $A = \begin{pmatrix} 2 & 1 \\ 1 & 2 \end{pmatrix}$, solve equation (1.15). That is, express $x_1(t)$ and $x_2(t)$ in terms of initial data (values and first derivatives at $t = 0$).

7. With $A = \begin{pmatrix} 2 & 1 \\ 1 & 2 \end{pmatrix}$, find the most general solution to equation (1.16). You need not express $\mathbf{x}(t)$ in terms of initial data.

Chapter 2
Vector Spaces and Bases

This chapter and the next are an overview of elementary linear algebra. The reader is expected to be already familiar with basic matrix operations such as matrix multiplication, row reduction, taking determinants and finding inverses. However, no prior knowledge of vector spaces, bases, dimension, or linear transformations is assumed. That material is covered quickly from the beginning.

2.1 Vector Spaces

Linear algebra is the study of vector spaces. A vector space is a set whose elements (called *vectors*, naturally) can be added to each other and multiplied by scalars, such that the usual rules (such as "$\mathbf{x}+\mathbf{y} = \mathbf{y}+\mathbf{x}$") apply. By *scalars* we usually mean real numbers, in which case we say we are dealing with a *real vector space*. Sometimes by scalars we mean complex numbers, in which case we are dealing with a *complex vector space*.

In principle, we could consider scalars that belong to an arbitrary field, not just to $\mathbb{R}$ or $\mathbb{C}$. Algebraic geometers, number theorists and computer scientists often consider vector spaces over finite fields or over the rational numbers. In this book, however, we will only consider real and complex vector spaces.

For completeness, the axioms for a vector space are listed on page 10. Instead of dwelling on the axioms, however, consider some examples:

1. The set of all real n-tuples $\mathbf{x} = \begin{pmatrix} x_1 \\ \vdots \\ x_n \end{pmatrix}$, together with the

operations

$$\begin{pmatrix} x_1 \\ \vdots \\ x_n \end{pmatrix} + \begin{pmatrix} y_1 \\ \vdots \\ y_n \end{pmatrix} = \begin{pmatrix} x_1 + y_1 \\ \vdots \\ x_n + y_n \end{pmatrix}; \quad c \begin{pmatrix} x_1 \\ \vdots \\ x_n \end{pmatrix} = \begin{pmatrix} cx_1 \\ \vdots \\ cx_n \end{pmatrix}, \quad (2.1)$$

is a real vector space denoted $\mathbb{R}^n$. Since columns are difficult to type, we will usually write $\mathbf{x} = (x_1, \ldots, x_n)^T$. The superscript "T", which stands for "transpose", indicates that we are thinking of the n numbers forming a column rather than a row. (The reason for making $\mathbf{x}$ a column rather than a row is so we can take the product $A\mathbf{x}$, where A is an $m \times n$ matrix.)

2. If in example 1 we allow the entries x_i and the scalars c to be complex numbers, then we have a complex vector space denoted $\mathbb{C}^n$.

3. Let M_{nm} be the space of real $n \times m$ matrices. Matrices can be added and multiplied by real numbers, so M_{nm} is a real vector space.

4. Let $C^0[0,1]$ be the set of continuous real-valued functions on the closed interval $[0,1]$. Continuous functions may be added and multiplied by real numbers, yielding other continuous functions, so $C^0[0,1]$ is a real vector space.

In examples 3 and 4, we considered real-valued matrices and functions to obtain real vector spaces. If we instead consider complex-valued matrices or functions, we obtain a complex vector space.

With these examples under our belts, we now examine the formal definition:

Definition *Let V be a set on which addition and scalar multiplication are defined. (That is, if $\mathbf{x}$ and $\mathbf{y}$ are elements of V, and c is a scalar, then $\mathbf{x} + \mathbf{y}$ and $c\mathbf{x}$ are elements of V.) If the following eight axioms are satisfied by all elements $\mathbf{x}$, $\mathbf{y}$, and $\mathbf{z}$ of V and all scalars a and b, then V is called a* vector space *and the elements of V are called* vectors. *If these axioms apply to multiplication by real scalars, then V is called a* real vector space. *If the axioms apply to multiplication by complex scalars, then V is a* complex vector space.

1. *Commutativity of addition:* $\mathbf{x} + \mathbf{y} = \mathbf{y} + \mathbf{x}$.

2. *Associativity of addition:* $(\mathbf{x} + \mathbf{y}) + \mathbf{z} = \mathbf{x} + (\mathbf{y} + \mathbf{z})$.

2.1. Vector Spaces

3. *Additive identity:* There exists a vector, denoted **0**, such that, for every vector **x**, **0** + **x** = **x**.

4. *Additive inverses:* For every vector **x** there exists a vector (−**x**) such that **x** + (−**x**) = **0**.

5. *First distributive law:* $a(\mathbf{x} + \mathbf{y}) = a\mathbf{x} + a\mathbf{y}$.

6. *Second distributive law:* $(a + b)\mathbf{x} = a\mathbf{x} + b\mathbf{x}$.

7. *Multiplicative identity:* $1(\mathbf{x}) = \mathbf{x}$.

8. *Relation to ordinary multiplication:* $(ab)\mathbf{x} = a(b\mathbf{x})$.

In practice, checking these axioms is a tedious and often pointless task. Addition and scalar multiplication are usually defined in a straightforward way that makes these axioms obvious. What is far *less* obvious is that addition and scalar multiplication make sense as operations *on V*. One must check that the sum of two arbitrary elements of V is in V and that the product of an arbitrary scalar and an arbitrary element of V is in V.

Definition A set S is closed under addition *if the sum of any two elements of S is in S, and is* closed under scalar multiplication *if the product of an arbitrary scalar and an arbitrary element of S is in S.*

Frequently we consider a subset W of a vector space V. In this case, addition and scalar multiplication are already defined, and already satisfy the eight axioms. If W is closed under addition and scalar multiplication, then W is a vector space in its own right, and we call W a *subspace* of V.

With this in mind we consider a few more examples of vector spaces, as well as some sets that are not vector spaces. See Figure 2.1.

1. Let $W = \{\mathbf{x} \in \mathbb{R}^2 | x_1 + x_2 = 0\}$. The sum of any two vectors in W is in W, and any scalar multiple of a vector in W is in W. (Check this!) W is a subspace of the vector space $\mathbb{R}^2$.

2. More generally, let A be any fixed $m \times n$ matrix and let $W = \{\mathbf{x} \in \mathbb{R}^n | A\mathbf{x} = 0\}$. W is closed under addition and scalar multiplication (why?), so W is a subspace of $\mathbb{R}^n$.

Chapter 2. Vector Spaces and Bases

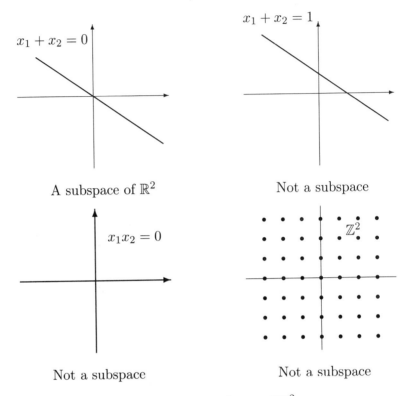

Figure 2.1: Four subsets of $\mathbb{R}^2$

3. Let $\mathbb{R}[t]$ be the set of all real-valued polynomials in a fixed variable t. Polynomials are continuous functions on $[0, 1]$, so $\mathbb{R}[t]$ is a subset of $C^0[0, 1]$. The sum of two polynomials is a polynomial, as is the product of scalar and a polynomial, so $\mathbb{R}[t]$ is a subspace of $C^0[0, 1]$.

4. Let $\mathbb{R}_n[t]$ be the set of real polynomials of degree n or less. For $n < m$, $\mathbb{R}_n[t]$ is a subspace of $\mathbb{R}_m[t]$, and $\mathbb{R}_n[t]$ is always a subspace of $\mathbb{R}[t]$.

5. Instead of considering polynomials with real coefficients, we could consider polynomials with complex coefficients to get examples of complex vector spaces. The space of polynomials with complex coefficients is usually denoted $\mathbb{C}[t]$.

6. Let $W' = \{\mathbf{x} \in \mathbb{R}^2 | x_1 + x_2 = 1\}$. W' is not a vector space, as the sum of two elements of W', or a scalar multiple of an element of W', is typically not in W'. (Again, check this!)

2.1. Vector Spaces

7. Let $\mathbb{Z}^2 = \{ \mathbf{x} \in \mathbb{R}^2 | x_1 \text{ and } x_2 \text{ are integers } \}$. $\mathbb{Z}^2$ is closed under addition, but not under scalar multiplication. $\mathbb{Z}^2$ is a sub*set* of $\mathbb{R}^2$, but not a sub*space*.

8. The set $\{ \mathbf{x} \in \mathbb{R}^2 | x_1 x_2 = 0 \}$ is the union of the coordinate axes in $\mathbb{R}^2$. This set is closed under scalar multiplication but not under addition, and so is not a subspace of $\mathbb{R}^2$.

Finally, notice that length is not an essential property of a vector. What is the length of a polynomial, or of a continuous function? It is true that many vector spaces come equipped with a notion of length, and this concept can be extremely useful. However, many vector spaces do not come so equipped. In Chapters 2–5 we concentrate on those operations that make sense in all vector spaces. In Chapters 6 and beyond we examine what more can be done when a vector space has an inner product and a notion of length.

Exercises

Explain why each of the following sets is or is not a vector space.

1. $\{ \mathbf{x} \in \mathbb{R}^2 | x_1 \leq x_2 \}$.
2. $\{ \mathbf{x} \in \mathbb{R}^2 | x_1 + x_2 = 0 \text{ and } x_1 \text{ is rational. } \}$
3. $\{ \mathbf{x} \in \mathbb{R}^n | x_1^2 + \cdots + x_n^2 = 1 \}$.
4. $\{ \mathbf{x} \in \mathbb{R}^n | x_1^2 \geq x_2^2 + \cdots + x_n^2 \}$.
5. All vectors $\mathbf{x} \in \mathbb{R}^4$ that satisfy $x_1 + x_2 = x_3 - x_4$ and $x_1 + 2x_2 + 3x_3 + 4x_4 = 0$.
6. All vectors in $\mathbb{R}^3$ of the form $c_1 \begin{pmatrix} 1 \\ 1 \\ -3 \end{pmatrix} + c_2 \begin{pmatrix} -2 \\ 5 \\ 2 \end{pmatrix}$, where c_1 and c_2 are arbitrary real numbers.
7. All vectors in a vector space V of the form $c_1 \mathbf{v} + c_2 \mathbf{w}$, where $\mathbf{v}$ and $\mathbf{w}$ are fixed vectors in V and c_1 and c_2 are arbitrary scalars.
8. All polynomials $\mathbf{p}(t) \in \mathbb{R}[t]$ such that $\mathbf{p}(3) = 0$.
9. All polynomials $\mathbf{p}(t) \in \mathbb{R}[t]$ such that $\mathbf{p}(3) = 1$.
10. All non-negative functions in $C^0[0,1]$.
11. All polynomials in $\mathbb{R}_5[t]$ with integer coefficients.
12. Let X be an arbitrary set. Is the set of all integer-valued functions on X a vector space? What about the set of all real-valued functions on X? Complex-valued functions on X?

13. Let **x** be an arbitrary vector. Show that $0\mathbf{x} = \mathbf{0}$.

14. Let V be a vector space and let W be a subset that is closed under scalar multiplication. Show that $\mathbf{0} \in W$. This provides us with a very useful rule: If a subset of a vector space does not contain the zero vector, it is not a subspace.

15. The axioms for a vector space demand the existence of an additive identity, but do not explicitly demand uniqueness. Prove that a vector space cannot have more than one additive identity.

16. Prove that an element of a vector space cannot have more than one additive inverse.

17. Prove that the additive inverse of **x** is the same as -1 times **x** (thereby justifying the notation $-\mathbf{x}$).

Some bizarre spaces satisfy the axioms of a vector space with operations that don't look like ordinary addition and scalar multiplication. These spaces are, in essence, ordinary vector spaces with the elements relabeled to disguise their nature. The following three exercises explore this frequently confusing possibility.

18. Let V be a vector space, let S be a set, and let $f : V \to S$ be a 1-1 and onto map, so that $f^{-1} : S \to V$ is well-defined. For every $\mathbf{x}, \mathbf{y} \in S$ and scalar c we define $\mathbf{x} \oplus \mathbf{y} = f(f^{-1}(\mathbf{x}) + f^{-1}(\mathbf{y}))$ and $c \otimes \mathbf{x} = f(cf^{-1}(\mathbf{x}))$. Show that S, with the operations $\oplus$ and $\otimes$ in place of ordinary addition and scalar multiplication, satisfies the axioms of a vector space.

19. Let S be the positive real numbers. Show that S is a vector space with the operations $x \oplus y = xy$ and $c \otimes x = x^c$. Which element of S is the additive identity? What does "$-x$" mean in this context?

20. Let S be the real numbers with the operations $x \oplus y = x+y+1$, $c \otimes x = cx + c$. Show that S is a vector space. Which element of S is the additive identity? What does "$-x$" mean in this context?

21. Show that the spaces of Exercises 19 and 20 can be obtained from $V = \mathbb{R}^1$ by the construction of Exercise 18 and the maps $f(x) = e^x$ and $f(x) = x+1$, respectively. In fact, *every* example of a bizarre vector space can be obtained from a simple vector space by this construction.

2.2. Linear Independence, Basis and Dimension

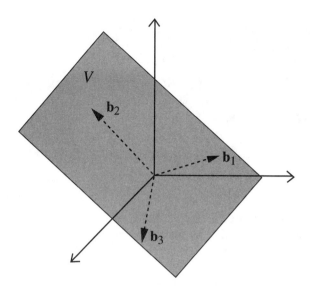

Figure 2.2: V is the span of b_1, b_2 and b_3

2.2 Linear Independence, Basis and Dimension

Definition *Let V be a vector space, and let $\mathcal{B} = \{b_1, b_2, \ldots, b_n\}$ be an ordered[1] set of vectors in V. A* linear combination *of the elements of $\mathcal{B}$ is any vector of the form*

$$\mathbf{v} = a_1 \mathbf{b}_1 + a_2 \mathbf{b}_2 + \cdots + a_n \mathbf{b}_n, \tag{2.2}$$

where the coefficients $a_1, \ldots, a_n$ are arbitrary scalars (some of which may be zero).

Definition *The* span *of $\mathcal{B}$, denoted $Span(\mathcal{B})$ is the (unordered) set of all linear combinations of the vectors in $\mathcal{B}$.*

Theorem 2.1 *Let $\mathcal{B} = \{b_1, \ldots, b_n\}$ be a finite collection of vectors in V. Then $Span(\mathcal{B})$ is a vector space (a subspace of V).*

---

[1] Technically one must distinguish between ordered and unordered sets. As unordered sets, $\{1, 2\}$ and $\{2, 1\}$ are the same, but as ordered sets they are different. As an unordered set, $\{1, 1, 2\}$ has two elements, while as an ordered set it has three elements — two of which happen to be equal. In this book, $\{b_1, \ldots, b_n\}$ will always denote an ordered set of vectors.

Proof: If **x** and **y** are in $Span(\mathcal{B})$, then we can write $\mathbf{x} = a_1\mathbf{b}_1 + \cdots + a_n\mathbf{b}_n$ and $\mathbf{y} = c_1\mathbf{b}_1 + \cdots + c_n\mathbf{b}_n$ for some coefficients $a_1, \ldots, a_n$ and $c_1, \ldots, c_n$. Let k be any scalar. Then $k\mathbf{x} = (ka_1)\mathbf{b}_1 + \cdots + (ka_n)\mathbf{b}_n$ is in $Span(\mathcal{B})$, as is $\mathbf{x} + \mathbf{y} = (a_1 + c_1)\mathbf{b}_1 + \cdots (a_n + c_n)\mathbf{b}_n$. Since $Span(\mathcal{B})$ is closed under addition and scalar multiplication, $Span(\mathcal{B})$ is a subspace of V. ∎

Definition The set $\mathcal{B}$ is said to be linearly independent if, whenever

$$a_1\mathbf{b}_1 + \cdots + a_n\mathbf{b}_n = 0, \tag{2.3}$$

we must also have $a_1 = a_2 = \cdots = a_n = 0$. If $\mathcal{B}$ is not linearly independent, it is linearly dependent. Note that linear independence is a property of the set $\mathcal{B}$, not of the individual vectors.

Example: In $\mathbb{R}^3$, the vectors $\mathbf{b}_1 = (-2, 1, 1)^T$, $\mathbf{b}_2 = (1, -2, 1)^T$ and $\mathbf{b}_3 = (1, 1, -2)^T$ are linearly dependent, because $1\mathbf{b}_1 + 1\mathbf{b}_2 + 1\mathbf{b}_3 = 0$. However, $\mathbf{b}_1$ and $\mathbf{b}_2$ are linearly independent, $\mathbf{b}_1$ and $\mathbf{b}_3$ are linearly independent, and $\mathbf{b}_2$ and $\mathbf{b}_3$ are linearly independent.

Example: In Figure 2.2, the three vectors shown are linearly dependent. In this example, each vector is in the span of the other two, namely the space V.

Theorem 2.2 A set $\mathcal{B} = \{\mathbf{b}_1, \ldots, \mathbf{b}_n\}$ with $n > 1$ is linearly dependent if, and only if, one of the vectors $\mathbf{b}_i$ can be written as a linear combination of the others.

Proof: If $\mathcal{B}$ is linearly dependent, then we have an expansion (2.3) with at least one of the a_i's nonzero. But if a_i is nonzero, then

$$\mathbf{b}_i = \frac{a_i \mathbf{b}_i}{a_i} = \frac{-a_1}{a_i}\mathbf{b}_1 + \cdots + \frac{-a_{i-1}}{a_i}\mathbf{b}_{i-1} + \frac{-a_{i+1}}{a_i}\mathbf{b}_{i+1} + \cdots + \frac{-a_n}{a_i}\mathbf{b}_n. \tag{2.4}$$

Hence $\mathbf{b}_i$ is a linear combination of the other elements of $\mathcal{B}$. Conversely, if

$$\mathbf{b}_i = c_1\mathbf{b}_1 + \cdots + c_{i-1}\mathbf{b}_{i-1} + c_{i+1}\mathbf{b}_{i+1} + \cdots + c_n\mathbf{b}_n, \tag{2.5}$$

then

$$0 = c_1\mathbf{b}_1 + \cdots + c_{i-1}\mathbf{b}_{i-1} + (-1)\mathbf{b}_i + c_{i+1}\mathbf{b}_{i+1} + \cdots + c_n\mathbf{b}_n, \tag{2.6}$$

so the set $\mathcal{B}$ is linearly dependent. ∎

2.2. Linear Independence, Basis and Dimension

Note that Theorem 2.2 does not say that a *particular* element of a linearly dependent set can be written as a linear combination of the others, only that *some* element can be so written. For example, in $\mathbb{R}^3$ the set $\{(1,0,0)^T, (0,1,0)^T, (0,0,1)^T, (0,1,1)^T\}$ is linearly dependent, and the last vector is the sum of the previous two. However, the first vector cannot be written as a linear combination of the other three.

Definition *If $Span(\mathcal{B}) = V$, the set $\mathcal{B}$ is said to* span *the vector space V.*

Definition *If $\mathcal{B}$ is linearly independent, and $\mathcal{B}$ spans V, then $\mathcal{B}$ is said to be a* basis *for the vector space V. (The plural of basis is bases.)*

We will see that finding a basis is the key to understanding a general vector space. Here are some examples of bases:

1. In $\mathbb{R}^n$, let $\mathbf{e}_1 = (1, 0, \ldots, 0)^T$, $\mathbf{e}_2 = (0, 1, \ldots, 0)^T$, and so on. Then $\mathcal{E} = \{\mathbf{e}_1, \mathbf{e}_2, \ldots, \mathbf{e}_n\}$ is a basis for $\mathbb{R}^n$. It is called the *standard basis* for $\mathbb{R}^n$. $\mathcal{E}$ is also the standard basis for the complex vector space $\mathbb{C}^n$.

2. Let $W = \{\mathbf{x} \in \mathbb{R}^3 : x_1 + x_2 + x_3 = 0\}$. The vectors $(1,1,-2)^T$ and $(1,-1,0)^T$ form a basis for W. This is not the only reasonable choice of basis. The basis $\{(0,1,-1)^T, (-2,1,1)^T\}$ is neither better nor worse, just different. In most spaces, which basis is best depends on which application you have in mind, and it is very important to learn how to change from one basis to another. We will revisit this question in Section 2.4.

3. In $\mathbb{R}_n[t]$ (the space of polynomials of degree n or less in a variable t), the vectors $\mathbf{p}_0(t) = 1, \mathbf{p}_1(t) = t, \mathbf{p}_2(t) = t^2, \ldots, \mathbf{p}_n(t) = t^n$ form a basis, which we call the standard basis for $\mathbb{R}_n[t]$.

4. In the space $\mathbb{R}[t]$, the set $\{\mathbf{p}_0, \mathbf{p}_1, \mathbf{p}_2, \ldots\}$ is a basis with an infinite number of elements. In such cases we say the vector space is infinite dimensional. There is nothing wrong with infinite dimensional spaces; they just take some getting used to.

5. In the space $M_{2,2}$ of 2×2 matrices, the matrices $\mathbf{e}_1 = \begin{pmatrix} 1 & 0 \\ 0 & 0 \end{pmatrix}$, $\mathbf{e}_2 = \begin{pmatrix} 0 & 1 \\ 0 & 0 \end{pmatrix}$, $\mathbf{e}_3 = \begin{pmatrix} 0 & 0 \\ 1 & 0 \end{pmatrix}$, $\mathbf{e}_4 = \begin{pmatrix} 0 & 0 \\ 0 & 1 \end{pmatrix}$ form a basis, which we call the standard basis. It is easy to write down a similar basis for the space M_{nm} of $n \times m$ matrices.

6. In physical 3-dimensional space, the unit vectors pointing forwards, left and up form a basis. This choice depends on which way you are facing, which again illustrates that a vector space may have more than one good choice of basis. This example also illustrates that physical space is not quite the same thing as $\mathbb{R}^3$, at least not until you've chosen your coordinate axes.

7. Since the real numbers are a subset of the complex numbers, every complex vector space may also be viewed as a real vector space. (If it makes sense to multiply a vector by an arbitrary complex number, it certainly makes sense to multiply it by a real number.) Viewed as a complex vector space, $\mathbb{C}^2$ has a basis $\{\mathbf{e}_1, \mathbf{e}_2\}$. Viewed as a real vector space, it has a basis $\{(1,0)^T, (i,0)^T, (0,1)^T, (0,i)^T\}$. The point is that $(1,0)^T$ and $(i,0)^T$ are linearly dependent over the complex numbers but linearly *independent* over the real numbers; the only way to have $a_1(1,0)^T + a_2(i,0)^T$ equal zero, with a_1 and a_2 both real, is to have $a_1 = a_2 = 0$.

If A is an $n \times m$ matrix and $\mathbf{x} \in \mathbb{R}^m$, then $A\mathbf{x}$ is a linear combination of the columns of A, namely x_1 times the first column plus x_2 times the second column plus ... plus x_m times the last column. Conversely, every linear combination of the columns of A is of the form $A\mathbf{x}$ for some $\mathbf{x}$. This observation allows us to turn questions about linear combinations of vectors in $\mathbb{R}^n$ into questions about solutions to linear equations $A\mathbf{x} = \mathbf{b}$, which are solved by row reduction. In $\mathbb{R}^n$, therefore, determining when a set of vectors is linearly independent, or spans, or is a basis, reduces to a matrix calculation.

Theorem 2.3 *Let A be an $n \times m$ matrix. Then the following statements are equivalent:*

1. *The columns of A are linearly independent.*

2. *The only solution to $A\mathbf{x} = 0$ is $\mathbf{x} = 0$.*

2.2. Linear Independence, Basis and Dimension

3. The reduced row echelon form of the matrix A has a pivot in each column.

Proof: The columns are linearly independent if (and only if) the only way to write 0 as a linear combination of the columns is as 0 times the first column plus ... plus 0 times the last column. That's exactly the same thing as saying that the only way to get $A\mathbf{x} = 0$ is to have $\mathbf{x} = 0$. But we know how to get all the solutions to $A\mathbf{x} = 0$ by row reduction. The solution $\mathbf{x} = 0$ always exists, and is unique if and only if the row-reduced form of the equations has a pivot in each column. ∎

For example, the reduced row echelon form of the 4×3 matrix
$$A = \begin{pmatrix} 1 & 1 & 1 \\ 1 & 0 & -1 \\ 1 & 2 & 1 \\ 3 & 2 & 4 \end{pmatrix} \text{ is } \begin{pmatrix} 1 & 0 & 0 \\ 0 & 1 & 0 \\ 0 & 0 & 1 \\ 0 & 0 & 0 \end{pmatrix},$$
which has three pivots, one in each column. Thus in $\mathbb{R}^4$ the vectors
$$\begin{pmatrix} 1 \\ 1 \\ 1 \\ 3 \end{pmatrix}, \begin{pmatrix} 1 \\ 0 \\ 2 \\ 2 \end{pmatrix} \text{ and } \begin{pmatrix} 1 \\ -1 \\ 1 \\ 4 \end{pmatrix} \text{ are}$$
linearly independent.

By contrast, the reduced row echelon form of the 3×4 matrix
$$B = \begin{pmatrix} 1 & 1 & 1 & 3 \\ 1 & 0 & 2 & 2 \\ 1 & -1 & 1 & 4 \end{pmatrix} \text{ is } \begin{pmatrix} 1 & 0 & 0 & 5 \\ 0 & 1 & 0 & -1/2 \\ 0 & 0 & 1 & -3/2 \end{pmatrix},$$
whose last column does not contain a pivot. Thus in $\mathbb{R}^3$, the vectors $\mathbf{b}_1 = \begin{pmatrix} 1 \\ 1 \\ 1 \end{pmatrix}$,
$\mathbf{b}_2 = \begin{pmatrix} 1 \\ 0 \\ -1 \end{pmatrix}, \mathbf{b}_3 = \begin{pmatrix} 1 \\ 2 \\ 1 \end{pmatrix}$ and $\mathbf{b}_4 = \begin{pmatrix} 3 \\ 2 \\ 4 \end{pmatrix}$ are linearly dependent.
In fact, $-10\mathbf{b}_1 + \mathbf{b}_2 + 3\mathbf{b}_3 + 2\mathbf{b}_4 = 0$.

Theorem 2.4 *Let A be an $n \times m$ matrix. Then the following statements are equivalent:*

1. *The columns of A span $\mathbb{R}^n$.*

2. *The equation $A\mathbf{x} = \mathbf{b}$ has a solution for every $\mathbf{b} \in \mathbb{R}^n$.*

3. *The reduced row echelon form of the matrix A has a pivot in each row.*

Proof: The columns span $\mathbb{R}^n$ if and only if every vector $\mathbf{b} \in \mathbb{R}^n$ can be written as a linear combination of the columns, namely as $A\mathbf{x}$ for some $\mathbf{x} \in \mathbb{R}^m$. That is, if $A\mathbf{x} = \mathbf{b}$ has a solution for every $\mathbf{b}$. We solve $A\mathbf{x} = \mathbf{b}$ by row reduction. If the row reduced form of A has a pivot in each row, then the row reduced form of the augmented matrix $[A|\mathbf{b}]$ has no contradictions, and there is at least one solution. However, if the row reduced form of A has a row of zeroes at the bottom, then, for some choices of $\mathbf{b}$, the reduction of the augmented matrix $[A|\mathbf{b}]$ will yield the contradiction $0 = 1$ in the bottom row. ∎

Revisiting the previous examples, the vectors $\begin{pmatrix} 1 \\ 1 \\ 1 \\ 3 \end{pmatrix}$, $\begin{pmatrix} 1 \\ 0 \\ 2 \\ 2 \end{pmatrix}$ and $\begin{pmatrix} 1 \\ -1 \\ 1 \\ 4 \end{pmatrix}$ do not span $\mathbb{R}^4$, since $A = \begin{pmatrix} 1 & 1 & 1 \\ 1 & 0 & -1 \\ 1 & 2 & 1 \\ 3 & 2 & 4 \end{pmatrix}$ row-reduces to a matrix with a row of zeroes at the bottom. However, the reduced row-echelon form of $B = \begin{pmatrix} 1 & 1 & 1 & 3 \\ 1 & 0 & 2 & 2 \\ 1 & -1 & 1 & 4 \end{pmatrix}$ does not have a row of zeroes at the bottom, so the columns of B do span $\mathbb{R}^3$.

Theorem 2.5 *Let A be an $n \times n$ matrix. Then the following statements are equivalent:*

1. *The columns of A form a basis for $\mathbb{R}^n$.*

2. *The equation $A\mathbf{x} = \mathbf{b}$ has a unique solution for every $\mathbf{b} \in \mathbb{R}^n$.*

3. *A is an invertible matrix.*

4. *The determinant of A is nonzero.*

5. *A is row equivalent to the identity matrix.*

Proof: The equivalence of the first, second, and last statements follows from Theorems 2.3 and 2.4. Now suppose that the equation $A\mathbf{x} = \mathbf{b}$ has a unique solution for every $\mathbf{b}$. Then the matrix whose i-th column is the solution to $A\mathbf{x} = \mathbf{e}_i$ is an inverse to A. (Check this!) Conversely, if A^{-1} exists, then $\mathbf{x} = A^{-1}\mathbf{b}$ is the unique solution

2.2. Linear Independence, Basis and Dimension

to $A\mathbf{x} = \mathbf{b}$. This shows the equivalence of the second and third statements. The equivalence of the third and fourth statements is standard. ∎

As an example, the vectors $\begin{pmatrix} 1 \\ 1 \\ 1 \end{pmatrix}$, $\begin{pmatrix} 1 \\ 0 \\ 2 \end{pmatrix}$ and $\begin{pmatrix} 1 \\ -1 \\ 1 \end{pmatrix}$ form a basis for $\mathbb{R}^3$ since the determinant of $\begin{pmatrix} 1 & 1 & 1 \\ 1 & 0 & -1 \\ 1 & 2 & 1 \end{pmatrix}$ equals 2.

Theorem 2.6 *Let $\mathcal{D} = \{\mathbf{d}_1, \mathbf{d}_2, \ldots, \mathbf{d}_m\}$ be a collection of vectors in $\mathbb{R}^n$.*

1. *If $m < n$, then $\mathcal{D}$ does not span $\mathbb{R}^n$.*

2. *If $m > n$, then $\mathcal{D}$ is linearly dependent.*

3. *If $\mathcal{D}$ is a basis for $\mathbb{R}^n$, then $m = n$.*

Proof: Let $\mathbf{d}_1, \ldots, \mathbf{d}_m$ be the columns of an $n \times m$ matrix A. The reduced row echelon form of A can have at most m pivots, one for each column. If $m < n$, it cannot have a pivot in each row, so, by Theorem 2.4, the vectors cannot span $\mathbb{R}^n$. Similarly, the reduced row echelon form can have at most n pivots, one for each row, so if $m > n$ there must be a column without a pivot, so by Theorem 2.3 the vectors must be linearly dependent. The third statement follows immediately from the first two. ∎

Exercises

Which of the following sets of vectors are linearly independent? Which span? Which are bases?

1. In an arbitrary space V, $\mathbf{b}_1 = 0$, $\mathbf{b}_2 \neq 0$.
2. In $\mathbb{R}^2$, $\mathbf{b}_1 = (r, 0)^T$, $\mathbf{b}_2 = (0, s)^T$, with r, s nonzero.
3. In $\mathbb{R}^2$, $\mathbf{b}_1 = (1, 2)^T$, $\mathbf{b}_2 = (2, 1)^T$.
4. In $\mathbb{R}^3$, $\mathbf{b}_1 = (1, 1, -2)^T$, $\mathbf{b}_2 = (1, -2, 1)^T$.
5. In $\mathbb{R}^3$, $\mathbf{b}_1 = (1, 1, -2)^T$, $\mathbf{b}_2 = (1, -2, 1)^T$, and $\mathbf{b}_3 = (-2, 1, 1)^T$.
6. In $\mathbb{R}^2$, $\mathbf{b}_1 = (1, 1)^T$, $\mathbf{b}_2 = (1, 2)^T$, $\mathbf{b}_3 = (2, 1)^T$.
7. In $\mathbb{R}^3$, $\mathbf{b}_1 = (1, 1, 1)^T$, $\mathbf{b}_2 = (1, 2, 3)^T$, $\mathbf{b}_3 = (1, 4, 9)^T$.
8. In $\mathbb{R}^3$, $\mathbf{b}_1 = (1, 2, 3)^T$, $\mathbf{b}_2 = (2, 3, 4)^T$, $\mathbf{b}_3 = (3, 4, 5)^T$.

9. In $\mathbb{R}^3$, the columns of $\begin{pmatrix} 1 & 2 & 3 \\ 4 & 5 & 6 \\ 7 & 8 & 9 \end{pmatrix}$.

10. In $\mathbb{R}^3$, the columns of $\begin{pmatrix} 1 & 4 & 7 \\ 2 & 5 & 8 \\ 3 & 6 & 9 \end{pmatrix}$.

11. In $\mathbb{R}^3$, the columns of $\begin{pmatrix} 1 & 2 & 3 & 1 \\ 4 & 5 & 6 & 0 \\ 7 & 8 & 9 & 0 \end{pmatrix}$.

12. In $\mathbb{R}^3$, the columns of $\begin{pmatrix} 1 & 2 & 2 & 3 \\ 2 & 1 & 2 & 5 \\ 3 & 3 & 4 & 8 \end{pmatrix}$.

2.3 Properties and Uses of a Basis

A basis is a tool for making an abstract real vector space look just like $\mathbb{R}^n$ (or making a complex vector space look like $\mathbb{C}^n$). Let $\mathcal{B}$ be a basis for a vector space V. Since $\mathcal{B}$ spans V, every vector $\mathbf{v} \in V$ can be written in the form (2.2). I claim this form is unique, so that $\mathbf{v}$ determines the coefficients $a_1, \ldots, a_n$. We then associate the vector $\mathbf{v} \in V$ with the vector $(a_1, \ldots, a_n)^T \in \mathbb{R}^n$.

To see this uniqueness, suppose that

$$\mathbf{v} = a_1 \mathbf{b}_1 + \cdots + a_n \mathbf{b}_n \tag{2.7}$$

and that

$$\mathbf{v} = a_1' \mathbf{b}_1 + \cdots + a_n' \mathbf{b}_n. \tag{2.8}$$

Then

$$0 = \mathbf{v} - \mathbf{v} = (a_1 - a_1') \mathbf{b}_1 + \cdots (a_n - a_n') \mathbf{b}_n. \tag{2.9}$$

But $\mathcal{B}$ is linearly independent, so we must have $a_1 - a_1' = \cdots = a_n - a_n' = 0$, and thus $(a_1, \ldots, a_n)^T = (a_1', \ldots, a_n')^T$.

We will be associating $(a_1, \ldots, a_n)^T$ with $\mathbf{v}$ so often that we set up a special notation for it:

Definition *If* $\mathbf{v} = a_1 \mathbf{b}_1 + \cdots + a_n \mathbf{b}_n$, *we write* $[\mathbf{v}]_\mathcal{B} = (a_1, \ldots, a_n)^T$. *The numbers* $a_1, \ldots, a_n$ *are called the* coordinates *of the vector* $\mathbf{v}$ *in the $\mathcal{B}$ basis.*

Theorem 2.7 *1. If* $\mathbf{x}, \mathbf{y} \in V$ *and c is a scalar, then* $[\mathbf{x} + \mathbf{y}]_\mathcal{B} = [\mathbf{x}]_\mathcal{B} + [\mathbf{y}]_\mathcal{B}$ *and* $[c\mathbf{x}]_\mathcal{B} = c[\mathbf{x}]_\mathcal{B}$.

2.3. Properties and Uses of a Basis

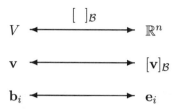

Figure 2.3: A basis makes a vector space look like $\mathbb{R}^n$

2. Let $\mathbf{v}_1, \ldots, \mathbf{v}_m$ and $\mathbf{w}$ be elements of V. Then $\mathbf{w}$ is a linear combination of the $\mathbf{v}_i$'s if, and only if, $[\mathbf{w}]_\mathcal{B}$ is a linear combination of the $[\mathbf{v}_i]_\mathcal{B}$'s.

3. The $\mathbf{v}_i$'s are linearly independent in V if, and only if, the $[\mathbf{v}_i]_\mathcal{B}$'s are linearly independent in $\mathbb{R}^n$.

Proof: Exercises 1–3.

The upshot of Theorem 2.7 is that a basis makes a vector space look like $\mathbb{R}^n$, with the basis $\mathcal{B}$ of V corresponding to the standard basis of $\mathbb{R}^n$. See Figure 2.3. Item 1 tells us that addition and scalar multiplication in V correspond to addition and scalar multiplication in $\mathbb{R}^n$. Items 2 and 3 say that questions about linear combinations in V (which might seem abstract and difficult) can be converted to questions about linear combinations in $\mathbb{R}^n$ (which are concrete and easier). In short, coordinates provide us with a concrete way to manipulate abstract vectors.

Example: In $\mathbb{R}_2[t]$, consider the vectors $\mathbf{x}(t) = 1 + t$, $\mathbf{y}(t) = 1 + t^2$ and $\mathbf{z}(t) = 1 + t + t^2$. To see if these are linearly independent in $\mathbb{R}_2[t]$, we pick a basis $\mathcal{B} = \{1, t, t^2\}$ and convert to $\mathbb{R}^3$. Now the three vectors $[\mathbf{x}]_\mathcal{B} = (1,1,0)^T$, $[\mathbf{y}]_\mathcal{B} = (1,0,1)^T$ and $[\mathbf{z}]_\mathcal{B} = (1,1,1)^T$ are linearly independent in $\mathbb{R}^3$ by Theorem 2.3. Therefore $\mathbf{x}$, $\mathbf{y}$ and $\mathbf{z}$ are linearly independent in $\mathbb{R}_2[t]$.

We have said that a basis makes a vector space "look like" $\mathbb{R}^n$. The mathematical term for "looks like" is *isomorphic*. Two vector spaces V and W are isomorphic if to every vector in V there corresponds a vector in W, and vice-versa, with addition and scalar multiplication in one vector space corresponding exactly to addition and scalar multiplication in the other. More precisely:

Definition *Let V and W be vector spaces. An* isomorphism *between V and W is a map $L : V \to W$ that is 1-1 and onto, and* such that, for any vectors $\mathbf{x}, \mathbf{y} \in V$ and any scalar c, $L(\mathbf{x} + \mathbf{y}) = L(\mathbf{x}) + L(\mathbf{y})$ and $L(c\mathbf{x}) = cL(\mathbf{x})$. *If an isomorphism from V to W exists, then V and W are said to be* isomorphic.

If V is a vector space with a basis $\mathcal{B} = \{\mathbf{b}_1, \ldots, \mathbf{b}_n\}$, then the assignment $\mathbf{x} \to [\mathbf{x}]_\mathcal{B}$ is an isomorphism from V to $\mathbb{R}^n$. From this, we can infer results about a general vector space V from known results about $\mathbb{R}^n$. For example, the following result follows from Theorem 2.6:

Corollary 2.8 *Suppose $\mathcal{B} = (\mathbf{b}_1, \ldots, \mathbf{b}_n)$ is a basis for a vector space V, and suppose $\mathcal{D} = (\mathbf{d}_1, \mathbf{d}_2, \ldots, \mathbf{d}_m)$ is another set of vectors in V. Then*

1. *If $m < n$, then $\mathcal{D}$ does not span V.*

2. *If $m > n$, then $\mathcal{D}$ is linearly dependent.*

3. *If $\mathcal{D}$ is a basis for V, then $m = n$.*

Proof: Exercise 5.

Definition *If a vector space V has a basis with n elements, we say that V is n* dimensional, *or that the* dimension *of V is n. The third statement of Corollary 2.8 shows that the dimension does not depend on the choice of basis, but is an intrinsic property of the vector space itself. If V has a basis with an infinite number of elements, we say that V is* infinite dimensional.

Exercises

1. Prove statement 1 of Theorem 2.7.
2. Prove statement 2 of Theorem 2.7.
3. Prove statement 3 of Theorem 2.7.
4. If $V = \mathbb{R}^n$ and $\mathcal{B} = \mathcal{E}$, show that $[\mathbf{v}]_\mathcal{B} = \mathbf{v}$ for every vector $\mathbf{v} \in \mathbb{R}^n$. This should not be surprising, since $[\cdot]_\mathcal{B}$ associates V with $\mathbb{R}^n$ and $\mathcal{B}$ with $\mathcal{E}$, and since V and $\mathcal{B}$ are already the same as $\mathbb{R}^n$ and $\mathcal{E}$.
5. Show how Corollary 2.8 follows from Theorem 2.6 and Theorem 2.7.

In Exercises 6–9, use Theorem 2.7 to determine which of the following sets of vectors are linearly independent, which span, and which are bases.

6. In $\mathbb{R}_2[t]$, $\mathbf{b}_1 = 1 + t + t^2$, $\mathbf{b}_2 = 1 + 2t + 3t^2$, $\mathbf{b}_3 = 1 + 4t + 9t^2$. How does this compare to Exercise 7 of Section 2.2?

7. Let V be the subspace of $R_3[t]$ consisting of polynomials $\mathbf{p}$ with $\mathbf{p}(0) = 0$. Let $\mathbf{b}_1 = t^2 - t$, $\mathbf{b}_2 = t^3 + t^2 + t$, $\mathbf{b}_3 = 2t^3 - 5t^2 - 7t$.

8. Let V be the subspace of $R_3[t]$ consisting of polynomials $\mathbf{p}$ with $\mathbf{p}(1) = 0$. Let $\mathbf{b}_1 = t^2 - t$, $\mathbf{b}_2 = t^3 + t^2 + t - 3$, $\mathbf{b}_3 = 2t^3 - 5t^2 - 7t + 10$.

9. In $M_{2,2}$, $\mathbf{b}_1 = \begin{pmatrix} 1 & 0 \\ 0 & -1 \end{pmatrix}$, $\mathbf{b}_2 = \begin{pmatrix} 1 & 2 \\ 2 & -1 \end{pmatrix}$, $\mathbf{b}_3 = \begin{pmatrix} 0 & 1 \\ -1 & 0 \end{pmatrix}$.

10. Let V be the subspace of $M_{2,2}$ consisting of symmetric matrices ($A = A^T$). What is the dimension of V? Find a basis for V.

11. If L is an isomorphism from V to W, show that L^{-1} is well-defined and is an isomorphism from W to V.

12. If V_1 and V_2 are isomorphic, and V_2 and V_3 are isomorphic, show that V_1 and V_3 are isomorphic.

From Exercises 11 and 12 it follows that all n-dimensional real vector spaces are isomorphic to $\mathbb{R}^n$, and hence to each other.

2.4 Change of Basis

Decoupling coupled equations is a matter of picking the right basis for a vector space and computing coordinates of vectors relative to that basis. This usually means computing coordinates relative to a standard basis and then converting from one basis to the other. This section is about how to do that conversion.

Suppose we have two different bases, $\mathcal{B} = \{\mathbf{b}_1, \ldots, \mathbf{b}_n\}$ and $\mathcal{D} = \{\mathbf{d}_1, \ldots, \mathbf{d}_n\}$ for an n-dimensional vector space V. These bases allow us to associate a vector $\mathbf{v} \in V$ with vectors $[\mathbf{v}]_\mathcal{B}$ and $[\mathbf{v}]_\mathcal{D}$ in $\mathbb{R}^n$. The two vectors $[\mathbf{v}]_\mathcal{B}$ and $[\mathbf{v}]_\mathcal{D}$ are related by the following theorem.

Theorem 2.9 *There exists a unique $n \times n$ matrix $P_{\mathcal{DB}}$ such that, for any vector $\mathbf{v} \in V$,*

$$[\mathbf{v}]_\mathcal{D} = P_{\mathcal{DB}} [\mathbf{v}]_\mathcal{B}. \qquad (2.10)$$

Furthermore, $P_{\mathcal{DB}}$ is given by the explicit formula

$$P_{\mathcal{DB}} = \begin{pmatrix} [\mathbf{b}_1]_{\mathcal{D}} & [\mathbf{b}_2]_{\mathcal{D}} & \cdots & [\mathbf{b}_n]_{\mathcal{D}} \end{pmatrix}. \qquad (2.11)$$

Proof: First we show that, if $P_{\mathcal{DB}}$ exists, then it must be given by the formula (2.11), thereby proving uniqueness. Then we check that the formula (2.11) does indeed work, establishing existence.

Note that $[\mathbf{b}_i]_{\mathcal{B}} = \mathbf{e}_i$. Taking $\mathbf{v} = \mathbf{b}_i$, we see that $[\mathbf{b}_i]_{\mathcal{D}} = P_{\mathcal{DB}}[\mathbf{b}_i]_{\mathcal{B}} = P_{\mathcal{DB}}\mathbf{e}_i$, which is the i-th column of $P_{\mathcal{DB}}$. Thus $P_{\mathcal{DB}}$ must take the form (2.11).

To see that formula (2.11) actually works, let $\mathbf{v} = a_1\mathbf{b}_1 + \cdots + a_n\mathbf{b}_n$, so that $[\mathbf{v}]_{\mathcal{B}} = (a_1, \ldots, a_n)^T$. Since $\mathcal{B}$ is a basis, every vector in V takes this form. By Theorem 2.7 (and induction) we know that

$$\begin{aligned}
[\mathbf{v}]_{\mathcal{D}} &= a_1[\mathbf{b}_1]_{\mathcal{D}} + \cdots + a_n[\mathbf{b}_n]_{\mathcal{D}} \\
&= \begin{pmatrix} [\mathbf{b}_1]_{\mathcal{D}} & [\mathbf{b}_2]_{\mathcal{D}} & \cdots & [\mathbf{b}_n]_{\mathcal{D}} \end{pmatrix} \begin{pmatrix} a_1 \\ \vdots \\ a_n \end{pmatrix} \\
&= P_{\mathcal{DB}}[\mathbf{v}]_{\mathcal{B}}. \blacksquare
\end{aligned} \qquad (2.12)$$

The matrix $P_{\mathcal{DB}}$ is called the *change of basis matrix*. The situation of Theorem 2.9 is summarized in the following diagram:

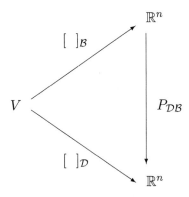

Note that $P_{\mathcal{DB}}$ changes $[\mathbf{v}]_{\mathcal{B}}$ into $[\mathbf{v}]_{\mathcal{D}}$, not the other way around. That is, we read the subscripts on P right to left. The reason for this peculiar notation is to make the composition law (Theorem 2.11, below) work out neatly.

For example, in $\mathbb{R}^2$ let $\mathcal{E}$ be the standard basis and let $\mathcal{B} = \{\mathbf{b}_1, \mathbf{b}_2\}$ be another basis, with $\mathbf{b}_1 = (1,1)^T$, $\mathbf{b}_2 = (1,-1)^T$. See

2.4. Change of Basis

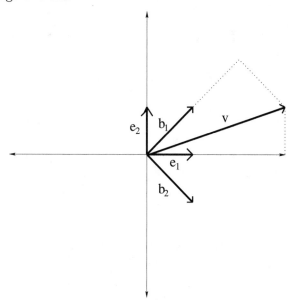

Figure 2.4: Two bases for $\mathbb{R}^2$

Figure 2.4. Since $\mathbf{b}_1 = \mathbf{e}_1 + \mathbf{e}_2$, $[\mathbf{b}_1]_{\mathcal{E}} = \begin{pmatrix} 1 \\ 1 \end{pmatrix}$. Similarly, $[\mathbf{b}_2]_{\mathcal{E}} = \begin{pmatrix} 1 \\ -1 \end{pmatrix}$, so

$$P_{\mathcal{E}\mathcal{B}} = ([\mathbf{b}_1]_{\mathcal{E}} \ [\mathbf{b}_2]_{\mathcal{E}}) = \begin{pmatrix} 1 & 1 \\ 1 & -1 \end{pmatrix}. \tag{2.13}$$

Similarly, $\mathbf{e}_1 = (\mathbf{b}_1 + \mathbf{b}_2)/2$ and $\mathbf{e}_2 = (\mathbf{b}_1 - \mathbf{e}_2)/2$, so

$$P_{\mathcal{B}\mathcal{E}} = ([\mathbf{e}_1]_{\mathcal{B}} \ [\mathbf{e}_2]_{\mathcal{B}}) = \begin{pmatrix} 1/2 & 1/2 \\ 1/2 & -1/2 \end{pmatrix}. \tag{2.14}$$

In Figure 2.4, $\mathbf{v} = (3,1)^T = 3\mathbf{e}_1 + \mathbf{e}_2 = 2\mathbf{b}_1 + \mathbf{b}_2$, so $[\mathbf{v}]_{\mathcal{E}} = \begin{pmatrix} 3 \\ 1 \end{pmatrix}$ and $[\mathbf{v}]_{\mathcal{B}} = \begin{pmatrix} 2 \\ 1 \end{pmatrix}$. You should check that $[\mathbf{v}]_{\mathcal{B}}$ is indeed equal to $P_{\mathcal{B}\mathcal{E}}[\mathbf{v}]_{\mathcal{E}}$ and that $[\mathbf{v}]_{\mathcal{E}}$ is equal to $P_{\mathcal{E}\mathcal{B}}[\mathbf{v}]_{\mathcal{B}}$.

The operation of converting from $\mathcal{B}$ coordinates to $\mathcal{E}$ coordinates is the inverse of converting from $\mathcal{E}$ coordinates to $\mathcal{B}$ coordinates. This is reflected in the fact that the matrices $P_{\mathcal{B}\mathcal{E}}$ and $P_{\mathcal{E}\mathcal{B}}$ are inverses of one another. This is true in general.

Theorem 2.10 *If $\mathcal{B}$ and $\mathcal{D}$ are bases for a vector space V, then $P_{\mathcal{B}\mathcal{D}} = P_{\mathcal{D}\mathcal{B}}^{-1}$.*

Proof: We must show that $P_{\mathcal{DB}}P_{\mathcal{BD}}$ is the identity. Every vector $\mathbf{y} \in \mathbb{R}^n$ equals $[\mathbf{x}]_{\mathcal{D}}$ for some vector $\mathbf{x}$ in V, so

$$P_{\mathcal{DB}}P_{\mathcal{BD}}\mathbf{y} = (P_{\mathcal{DB}}P_{\mathcal{BD}})[\mathbf{x}]_{\mathcal{D}} = P_{\mathcal{DB}}(P_{\mathcal{BD}}[\mathbf{x}]_{\mathcal{D}}) = P_{\mathcal{DB}}[\mathbf{x}]_{\mathcal{B}} = [\mathbf{x}]_{\mathcal{D}} = \mathbf{y}. \tag{2.15}$$

Since $P_{\mathcal{DB}}P_{\mathcal{BD}}\mathbf{y} = \mathbf{y}$ for every $\mathbf{y}$, $P_{\mathcal{DB}}P_{\mathcal{BD}}$ is the identity. ∎

Theorem 2.11 *If $\mathcal{B}$, $\mathcal{C}$ and $\mathcal{D}$ are bases for the same vector space, then $P_{\mathcal{BD}} = P_{\mathcal{BC}}P_{\mathcal{CD}}$.*

Proof: Exercise 1.

Worked exercise: In $\mathbb{R}^3$, let $\mathbf{b}_1 = (1, 2, 3)^T$, $\mathbf{b}_2 = (4, 5, 6)^T$, and $\mathbf{b}_3 = (7, 8, 8)^T$ form the basis $\mathcal{B}$, and let $\mathcal{E}$ be the standard basis. Let $\mathbf{v} = (7, 2, 4)^T$. Find $[\mathbf{v}]_{\mathcal{B}}$.

$[\mathbf{v}]_{\mathcal{E}} = \mathbf{v} = (7, 2, 4)^T$. Using the formula (2.11) to directly find $P_{\mathcal{BE}}$ is tedious, but using it to find $P_{\mathcal{EB}}$ is downright easy:

$$P_{\mathcal{EB}} = \begin{pmatrix} [\mathbf{b}_1]_{\mathcal{E}} & [\mathbf{b}_2]_{\mathcal{E}} & [\mathbf{b}_3]_{\mathcal{E}} \end{pmatrix} = \begin{pmatrix} 1 & 4 & 7 \\ 2 & 5 & 8 \\ 3 & 6 & 8 \end{pmatrix}. \tag{2.16}$$

But $P_{\mathcal{BE}} = P_{\mathcal{EB}}^{-1}$ so, using any algorithm for finding matrix inverses (e.g., giving the problem to MATLAB), we see that

$$P_{\mathcal{BE}} = P_{\mathcal{EB}}^{-1} = \begin{pmatrix} -8/3 & 10/3 & -1 \\ 8/3 & -13/3 & 2 \\ -1 & 2 & -1 \end{pmatrix}, \tag{2.17}$$

and hence that $[\mathbf{v}]_{\mathcal{B}} = P_{\mathcal{BE}}[\mathbf{v}]_{\mathcal{E}} = (-16, 18, -7)^T$. You should check that $\mathbf{v}$ is indeed equal to $-16\mathbf{b}_1 + 18\mathbf{b}_2 - 7\mathbf{b}_3$.

Exercises

1. Prove Theorem 2.11.

In Exercises 2–8 you are given a basis $\mathcal{B}$ for $\mathbb{R}^n$ ($n = 2$, 3 or 4) and a vector $\mathbf{v}$. Find $P_{\mathcal{EB}}$, $P_{\mathcal{BE}}$ and $[\mathbf{v}]_{\mathcal{B}}$.

2. $\mathbf{b}_1 = \begin{pmatrix} 3 \\ 2 \end{pmatrix}$, $\mathbf{b}_2 = \begin{pmatrix} 4 \\ 3 \end{pmatrix}$, and $\mathbf{v} = \begin{pmatrix} 2 \\ 5 \end{pmatrix}$.

3. $\mathbf{b}_1 = \begin{pmatrix} 1 \\ 3 \end{pmatrix}$, $\mathbf{b}_2 = \begin{pmatrix} 3 \\ 1 \end{pmatrix}$, and $\mathbf{v} = \begin{pmatrix} 3 \\ 7 \end{pmatrix}$.

4. $\mathbf{b}_1 = \begin{pmatrix} 1.03 \\ 3.29 \end{pmatrix}$, $\mathbf{b}_2 = \begin{pmatrix} 1.41 \\ 2.73 \end{pmatrix}$, and $\mathbf{v} = \begin{pmatrix} 3.14 \\ -2.72 \end{pmatrix}$. Express your answers to 4 decimal places.

2.4. Change of Basis

5. $\mathbf{b}_1 = \begin{pmatrix} 1 \\ 1 \\ 1 \end{pmatrix}$, $\mathbf{b}_2 = \begin{pmatrix} 2 \\ 3 \\ 1 \end{pmatrix}$, $\mathbf{b}_3 = \begin{pmatrix} 1 \\ 2 \\ 1 \end{pmatrix}$ and $\mathbf{v} = \begin{pmatrix} 5 \\ -2 \\ 3 \end{pmatrix}$.

6. $\mathbf{b}_1 = \begin{pmatrix} 2 \\ 1 \\ 1 \end{pmatrix}$, $\mathbf{b}_2 = \begin{pmatrix} 1 \\ 2 \\ 1 \end{pmatrix}$, $\mathbf{b}_3 = \begin{pmatrix} 1 \\ 1 \\ 2 \end{pmatrix}$ and $\mathbf{v} = \begin{pmatrix} 1 \\ 0 \\ -1 \end{pmatrix}$.

7. $\mathbf{b}_1 = \begin{pmatrix} 1 \\ 0 \\ 0 \\ 1 \end{pmatrix}$, $\mathbf{b}_2 = \begin{pmatrix} 1 \\ 0 \\ 0 \\ -1 \end{pmatrix}$, $\mathbf{b}_3 = \begin{pmatrix} 0 \\ 1 \\ 1 \\ 0 \end{pmatrix}$, $\mathbf{b}_4 = \begin{pmatrix} 0 \\ 1 \\ -1 \\ 0 \end{pmatrix}$, and

$\mathbf{v} = \begin{pmatrix} 1 \\ 2 \\ 3 \\ 4 \end{pmatrix}$.

8. $\mathbf{b}_1 = \begin{pmatrix} 2 \\ 1 \\ 1 \\ 1 \end{pmatrix}$, $\mathbf{b}_2 = \begin{pmatrix} 1 \\ 2 \\ 1 \\ 1 \end{pmatrix}$, $\mathbf{b}_3 = \begin{pmatrix} 1 \\ 1 \\ 2 \\ 1 \end{pmatrix}$, $\mathbf{b}_4 = \begin{pmatrix} 1 \\ 1 \\ 1 \\ 2 \end{pmatrix}$, and

$\mathbf{v} = \begin{pmatrix} 5 \\ 6 \\ 5 \\ 4 \end{pmatrix}$.

Exercises 9–12 are similar, only in abstract vector spaces.

9. In $\mathbb{R}_2[t]$, let $\mathbf{b}_1(t) = t^2 + t + 1$, $\mathbf{b}_2(t) = t^2 + 3t + 2$, $\mathbf{b}_3(t) = t^2 + 2t + 1$, and $\mathbf{v}(t) = 3t^2 - 2t + 5$. Let $\mathcal{E} = \{1, t, t^2\}$ be the standard basis. Find $P_{\mathcal{E}B}$, $P_{B\mathcal{E}}$ and $[\mathbf{v}]_B$. Compare to Exercise 5.

10. In $\mathbb{R}_2[t]$, let $\mathbf{b}_1(t) = t^2$, $\mathbf{b}_2(t) = (t-1)^2$, $\mathbf{b}_3(t) = (t+1)^2$, and $\mathbf{v}(t) = t^2 - 4t + 1$. Let $\mathcal{E} = \{1, t, t^2\}$ be the standard basis. Find $P_{\mathcal{E}B}$, $P_{B\mathcal{E}}$ and $[\mathbf{v}]_B$.

11. In $M_{2,2}$, let $\mathcal{E}$ be the standard basis, and let $\mathcal{D} = $ be the basis $\left\{\begin{pmatrix} 1 & 0 \\ 0 & 1 \end{pmatrix}, \begin{pmatrix} 1 & 0 \\ 0 & -1 \end{pmatrix}, \begin{pmatrix} 0 & 1 \\ 1 & 0 \end{pmatrix}, \begin{pmatrix} 0 & 1 \\ -1 & 0 \end{pmatrix}\right\}$. Let $\mathbf{v} = \begin{pmatrix} 1 & 2 \\ 4 & 3 \end{pmatrix}$. Find $P_{\mathcal{D}\mathcal{E}}$, $P_{\mathcal{E}\mathcal{D}}$, $[\mathbf{v}]_\mathcal{D}$ and $[\mathbf{v}]_\mathcal{E}$. Compare to Exercise 7.

12. In $M_{2,2}$, let $\mathcal{E}$ be the standard basis, and let $\mathcal{D} = $ be the basis $\left\{\begin{pmatrix} 2 & 1 \\ 1 & 1 \end{pmatrix}, \begin{pmatrix} 1 & 2 \\ 1 & -1 \end{pmatrix}, \begin{pmatrix} 1 & 1 \\ 2 & 1 \end{pmatrix}, \begin{pmatrix} 1 & 1 \\ 1 & 2 \end{pmatrix}\right\}$. Let $\mathbf{v} = \begin{pmatrix} 5 & 6 \\ 5 & 4 \end{pmatrix}$. Find $P_{\mathcal{D}\mathcal{E}}$, $P_{\mathcal{E}\mathcal{D}}$, $[\mathbf{v}]_\mathcal{D}$ and $[\mathbf{v}]_\mathcal{E}$. Compare to Exercise 8.

2.5 Building New Vector Spaces from Old Ones[2]

In this section we consider two operations, the direct sum and the quotient, that allow us to build new vector spaces out of old ones.

The external direct sum

Definition *Suppose that V and W are two vector spaces, either both real or both complex. The external direct sum of V and W, denoted $V \oplus W$, is the set of pairs $\begin{pmatrix} \mathbf{v} \\ \mathbf{w} \end{pmatrix}$, with $\mathbf{v} \in V$ and $\mathbf{w} \in W$, with the following operations:*

$$\begin{pmatrix} \mathbf{v} \\ \mathbf{w} \end{pmatrix} + \begin{pmatrix} \mathbf{v}' \\ \mathbf{w}' \end{pmatrix} = \begin{pmatrix} \mathbf{v} + \mathbf{v}' \\ \mathbf{w} + \mathbf{w}' \end{pmatrix}; \qquad c \begin{pmatrix} \mathbf{v} \\ \mathbf{w} \end{pmatrix} = \begin{pmatrix} c\mathbf{v} \\ c\mathbf{w} \end{pmatrix}. \qquad (2.18)$$

The zero vector in $V \oplus W$ is of course $\begin{pmatrix} 0 \\ 0 \end{pmatrix}$.

Similarly, if we have three vector spaces, V_1, V_2 and V_3, we can define the direct sum $V_1 \oplus V_2 \oplus V_3$ as the set of all triples $\begin{pmatrix} \mathbf{v}_1 \\ \mathbf{v}_2 \\ \mathbf{v}_3 \end{pmatrix}$, where $\mathbf{v}_i \in V_i$. You should check that $(V_1 \oplus V_2) \oplus V_3$ and $V_1 \oplus (V_2 \oplus V_3)$ are both the same thing as $V_1 \oplus V_2 \oplus V_3$. Here are some examples of direct sums:

1. $\mathbb{R} \oplus \mathbb{R}$ is just $\mathbb{R}^2$. Similarly $\mathbb{C} \oplus \mathbb{C} = \mathbb{C}^2$.

2. $\mathbb{R}^n \oplus \mathbb{R}^m = \mathbb{R}^{n+m}$. If $\mathbf{v} \in \mathbb{R}^{n+m}$, you can think of the first n entries of $\mathbf{v}$ as defining a vector in $\mathbb{R}^n$, and the last m entries as defining a vector in $\mathbb{R}^m$. Similarly, $\mathbb{C}^n \oplus \mathbb{C}^m = \mathbb{C}^{n+m}$.

3. From examples 1 and 2 we see that $\mathbb{R}^n$ is the direct sum of $\mathbb{R}$ with itself n times, while $\mathbb{C}^n$ is the direct sum of $\mathbb{C}$ with itself n times.

4. Suppose V is a vector space with basis $\{\mathbf{b}_1, \ldots, \mathbf{b}_n\}$, and that W is a vector space with basis $\{\mathbf{d}_1, \ldots \mathbf{d}_m\}$. Then $V \oplus W$

---

[2]The material in this section is considerably more abstract than the rest of the chapter and may be postponed until after Section 3.4. The direct sum operation appears repeatedly throughout the book, beginning in Section 3.5, while the quotient operation is mostly needed for Section 3.5.

2.5. Building New Vector Spaces from Old Ones

has a basis $\left\{ \begin{pmatrix} \mathbf{b}_1 \\ 0 \end{pmatrix}, \ldots, \begin{pmatrix} \mathbf{b}_n \\ 0 \end{pmatrix}, \begin{pmatrix} 0 \\ \mathbf{d}_1 \end{pmatrix}, \ldots, \begin{pmatrix} 0 \\ \mathbf{d}_m \end{pmatrix} \right\}$. Using these bases makes taking the direct sum of V and W look exactly like example 2. In particular, the direct sum of an n dimensional space and an m dimensional space is always an $n+m$ dimensional space.

5. Let V be the space of continuous functions on a domain $U \subset \mathbb{R}^3$. Then $V \oplus V \oplus V$ is the space of continuous $\mathbb{R}^3$-valued functions on U.

There are natural projections P_1 from $V_1 \oplus V_2$ to V_1 and P_2 from $V_1 \oplus V_2$ to V_2. Namely,

$$P_1\begin{pmatrix} \mathbf{v}_1 \\ \mathbf{v}_2 \end{pmatrix} = \mathbf{v}_1; \qquad P_2\begin{pmatrix} \mathbf{v}_1 \\ \mathbf{v}_2 \end{pmatrix} = \mathbf{v}_2. \tag{2.19}$$

The internal direct sum

Definition *Suppose V is a vector space, and that W_1 and W_2 are subspaces of V with the property that any vector $\mathbf{v} \in V$ can be uniquely decomposed as*

$$\mathbf{v} = \mathbf{w}_1 + \mathbf{w}_2, \tag{2.20}$$

where $\mathbf{w}_i \in W_i$. Then V is said to be the internal direct sum *of W_1 and W_2.*

The existence of the decomposition (2.20) means that the two subspaces W_1 and W_2 are large enough. Uniqueness is equivalent to $W_1 \cap W_2 = \{0\}$. (Why?) The connection between internal and external direct sums is based on the following theorem:

Theorem 2.12 *Suppose V is the internal direct sum of W_1 and W_2. Then V is isomorphic to the external direct sum of W_1 and W_2.*

Proof: We construct a map from the external direct sum $W_1 \oplus W_2$ to V and then show that this map is an isomorphism. Let

$$L\begin{pmatrix} \mathbf{w}_1 \\ \mathbf{w}_2 \end{pmatrix} = \mathbf{w}_1 + \mathbf{w}_2. \tag{2.21}$$

It is clear that L respects addition and scalar multiplication. Since every vector in V can be decomposed as in (2.20), L is onto. Since this decomposition is unique, L is 1-1. ∎

Motivated by Theorem 2.12, we write $W_1 \oplus W_2$ for the internal direct sum of W_1 and W_2 as well as the external direct sum. We also define projection maps, as with external direct sums. If $\mathbf{v} = \mathbf{w}_1 + \mathbf{w}_2$, we let
$$P_1 \mathbf{v} = \mathbf{w}_1; \qquad P_2 \mathbf{v} = \mathbf{w}_2. \tag{2.22}$$

We use the same notation as in (2.19) because the projections of (2.22) play exactly the same role for internal direct sums as the projections of (2.19) play for external direct sums. The projection P_1 is called *the projection onto W_1 along W_2*. The following examples demonstrate that this projection depends on the subspace W_2 as well as on the subspace W_1.

1. In $V = \mathbb{R}^2$, let W_1 be all vectors of the form $(x_1, 0)^T$, and let W_2 be all vectors of the form $(0, x_2)^T$. Then $\mathbb{R}^2$ is the internal direct sum of W_1 and W_2, and the decomposition (2.20) amounts to $(x_1, x_2)^T = (x_1, 0)^T + (0, x_2)^T$. The projections are $P_1(x_1, x_2)^T = (x_1, 0)$ and $P_2(x_1, x_2)^T = (0, x_2)$.

2. In $V = \mathbb{R}^2$, let W_1 be all multiples of $(1, 0)^T$, as in example 1. Let W_2 be all multiples of $(1, 1)^T$. Then the decomposition (2.20) is
$$\begin{pmatrix} x_1 \\ x_2 \end{pmatrix} = \begin{pmatrix} x_1 - x_2 \\ 0 \end{pmatrix} + \begin{pmatrix} x_2 \\ x_2 \end{pmatrix}, \tag{2.23}$$
and the projections are $P_1 \mathbf{x} = (x_1 - x_2, 0)^T$ and $P_2 \mathbf{x} = (x_2, x_2)^T$. See Figure 2.5. Notice that, while the subspace W_1 is the same as in example 1, the projection P_1 is different, because the subspace W_2 is different.

3. Let V be an $n + m$ dimensional space. We can write V as an internal direct sum in many different ways, as follows: Let W_1 be any n-dimensional subspace of V and let W_2 be any m dimensional subspace of V such that $W_1 \cap W_2 = \{0\}$. Then V is the internal direct sum of W_1 and W_2. Unfortunately we do not yet have the tools to prove this. We will revisit this construction in Section 3.5.

4. Let V be the space of continuous functions on $\mathbb{R}$. Let W_1 be the space of even functions, that is those with $f(-t) = f(t)$ for all t. Let W_2 be the space of odd functions, that is those with $f(-t) = -f(t)$. Then V is the internal direct sum of W_1 and

2.5. Building New Vector Spaces from Old Ones

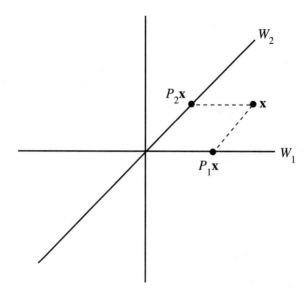

Figure 2.5: $\mathbb{R}^2$ is the internal direct sum of W_1 and W_2

W_2, since any function can be uniquely decomposed into even and odd parts as

$$f(t) = \frac{f(t) + f(-t)}{2} + \frac{f(t) - f(-t)}{2}. \tag{2.24}$$

Notice that examples 1 and 2 are special cases of example 3. In example 1, the subspaces are the x_1 and x_2 axes, which intersect only at the origin. In example 2, the subspaces are the x_1 axis and the line $x_1 = x_2$, which again meet only at the origin.

Quotient spaces

Now let V be a vector space and let W be a subspace of V. We will construct a vector space called the *quotient of V by W* and denoted V/W. This construction is not quite as intuitive as direct sums, so it is best to keep a simple example in mind. The example is $V = \mathbb{R}^2$, with W being all multiples of $(1,0)^T$, that is the x_1 axis.

We define an equivalence relation $\sim$ on V:

$$\mathbf{x} \sim \mathbf{y} \text{ if } \mathbf{x} - \mathbf{y} \in W. \tag{2.25}$$

You should check that $\sim$ satisfies all the requirements of an equivalence relation, namely: 1) $\mathbf{x} \sim \mathbf{x}$ (reflexivity), 2) if $\mathbf{x} \sim \mathbf{y}$ then $\mathbf{y} \sim \mathbf{x}$

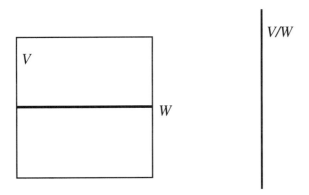

Figure 2.6: The quotient of $\mathbb{R}^2$ by the x_1-axis

(symmetry), and 3) if $\mathbf{x} \sim \mathbf{y}$ and $\mathbf{y} \sim \mathbf{z}$ then $\mathbf{x} \sim \mathbf{z}$ (transitivity). In our example, two elements of V are equivalent if they differ by a multiple of $(1, 0)^T$. In other words, $\mathbf{x} \sim \mathbf{y}$ if and only if $x_2 = y_2$. We denote by $[\mathbf{x}]$ the equivalence class of $\mathbf{x}$. That is,

$$[\mathbf{x}] = \{\mathbf{y} \in V | \mathbf{y} - \mathbf{x} \in W\}. \tag{2.26}$$

(Do not confuse this with the representation of a vector in a basis, which we denote $[\mathbf{x}]_\mathcal{B}$. These very different concepts are both traditionally denoted by square brackets.) In our example, $[(0, 2)^T]$ is the set of all vectors with second component 2. $[(5, 2)^T]$ is the same set, since $(1, 2)^T \sim (5, 2)^T$. However, $[(1, 3)^T]$ is a different set.

Definition *The quotient space V/W is the set of all equivalence classes of $\sim$, equipped with the following operations of addition and scalar multiplication:*

$$[\mathbf{x}] + [\mathbf{y}] = [\mathbf{x} + \mathbf{y}]; \qquad c[\mathbf{x}] = [c\mathbf{x}]. \tag{2.27}$$

In our example, every equivalence set is of the form $[(0, x_2)^T]$ for some x_2, and our rules are just $[(0, x_2)^T] + [(0, y_2)^T] = [(0, x_2 + y_2)^T]$ and $c[(0, x_2)^T] = [(0, cx_2)^T]$. In other words, V/W is essentially $\mathbb{R}$, where the single variable is x_2. See Figure 2.6.

It is not immediately clear that the rules (2.27) make sense in all cases. To add two equivalence classes we are told to pick a vector $\mathbf{x}$ in the first class, pick a vector $\mathbf{y}$ in the second class, add them, and then take the class containing that sum. We must show that this answer does not depend on which vectors we pick. That is Exercise 6.

2.5. Building New Vector Spaces from Old Ones

A second example: If $V = W_1 \oplus W_2$, let $W = W_1$. (If V is an internal direct sum this makes sense directly. If V is an external direct sum we identify W_1 with the subspace of vectors of the form $\begin{pmatrix} \mathbf{w}_1 \\ 0 \end{pmatrix}$.) Then the equivalence class $[\begin{pmatrix} \mathbf{w}_1 \\ \mathbf{w}_2 \end{pmatrix}]$ is the same as $[\begin{pmatrix} 0 \\ \mathbf{w}_2 \end{pmatrix}]$, and addition and scalar multiplication of equivalence classes is essentially the same thing as addition and scalar multiplication of elements of W_2. In other words, $(W_1 \oplus W_2)/W_1$ is essentially W_2. Since the dimension of a direct sum $W_1 \oplus W_2$ is the sum of the dimensions of W_1 and W_2, this shows that the dimension of a quotient V/W is the dimension of V *minus* the dimension of W.

Exercises

1. Exhibit $\mathbb{R}^2$ as the internal direct sum of the lines $x_2 = x_1$ and $x_2 = 2x_1$. That is, show explicitly that, for any $\mathbf{x} \in \mathbb{R}^2$, the decomposition (2.20) exists and is unique. What is $P_1(1, -1)^T$? What is $P_2(1, -1)^T$?

2. Show that $\mathbb{R}_2[t]$ is the internal direct sum of the polynomials that vanish at $t = 0$ and the span of $t^2 + 2t + 1$.

3. Show that $M_{3,3}$, the space of 3×3 real matrices, is the direct sum of the subspace of symmetric matrices ($A^T = A$), and the subspace of antisymmetric matrices $A^T = -A$. What are the dimensions of these subspaces? What is the dimension of $M_{3,3}$?

4. Let $A = \begin{pmatrix} 1 & 2 & 3 \\ 4 & 5 & 6 \\ 7 & 8 & 8 \end{pmatrix}$. In the decomposition of Exercise 3, what is $P_1 A$? $P_2 A$?

5. Show that, for general V and W, the equivalence relation $\sim$ of (2.25) satisfies the linearity properties: 1) If $\mathbf{x}_1 \sim \mathbf{y}_1$ and $\mathbf{x}_2 \sim \mathbf{y}_2$, then $\mathbf{x}_1 + \mathbf{x}_2 \sim \mathbf{y}_1 + \mathbf{y}_2$, and 2) If $\mathbf{x} \sim \mathbf{y}$, then $c\mathbf{x} \sim c\mathbf{y}$.

6. Show that, for general V and W, the addition operation of (2.27) are well defined. That is, show that, if $\mathbf{x}$ and $\mathbf{x}'$ are both in $[\mathbf{x}]$, and if $\mathbf{y}$ and $\mathbf{y}'$ are both in $[\mathbf{y}]$, then $[\mathbf{x} + \mathbf{y}] = [\mathbf{x}' + \mathbf{y}']$.

7. Let $V = \mathbb{R}[t]$, and let W be the subspace consisting of all polynomials divisible by $(x - 1)^2$. Show that W is a subspace of V and that V/W is 2-dimensional, and exhibit a basis for V/W. [A point

in V/W is the set of all polynomials that have a specified value and derivative at $x = 1$.]

8. Let $V = \mathbb{R}[t]$, let **p** be a fixed polynomial of degree n, and let W be all polynomials divisible by **p**. Show that V/W is an n-dimensional vector space and exhibit a basis for V/W.

Chapter 3

Linear Transformations and Operators

3.1 Definitions and Examples

Definition *Let V and W be vector spaces, and let $L : V \to W$ be a map. That is, for every vector $\mathbf{v} \in V$ we have a vector $L(\mathbf{v}) \in W$. If, for every $\mathbf{v}_1, \mathbf{v}_2 \in V$ and every scalar c we have*

$$L(\mathbf{v}_1 + \mathbf{v}_2) = L(\mathbf{v}_1) + L(\mathbf{v}_2); \qquad L(c\mathbf{v}_1) = cL(\mathbf{v}_1), \qquad (3.1)$$

then we say that L is a linear transformation *from V to W, or equivalently that L is a* linear map. *An* operator *on a vector space V is a linear transformation from V to itself.*

The situation is depicted in Figure 3.1. The vectors $L(\mathbf{v})$ and $L(\mathbf{w})$ are unrelated; we can pick the lengths of $L(\mathbf{v})$ and $L(\mathbf{w})$, and the angle between $L(\mathbf{v})$ and $L(\mathbf{w})$, to be anything we wish. However, once $L(\mathbf{v})$ and $L(\mathbf{w})$ are fixed, then L of any linear combination of $\mathbf{v}$ and $\mathbf{w}$ is determined. In particular, $L(-2\mathbf{v})$ must be -2 times $L(\mathbf{v})$ and $L(\mathbf{v} + \mathbf{w})$ must be the sum of $L(\mathbf{v})$ and $L(\mathbf{w})$.

Here are some examples of linear transformations:

1. If A is an $m \times n$ matrix, then the map $L(\mathbf{x}) = A\mathbf{x}$ is a linear transformation from $\mathbb{R}^n$ to $\mathbb{R}^m$. (Why?) We shall soon see that every linear transformation from $\mathbb{R}^n$ to $\mathbb{R}^m$ is of this form.

2. Various geometrical operations in physical 3-dimensional space (with a choice of an origin) are described by linear operators. Examples include rotation by a fixed angle θ about a fixed axis

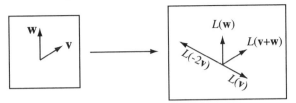

Figure 3.1: A linear transformation

through the origin, reflection about a fixed plane through the origin, and stretching in a fixed direction.

3. Although the word "transformation" suggests change and the word "isomorphism" implies sameness, every isomorphism is a linear transformation. If V is an n-dimensional vector space with basis $\mathcal{B}$, the map $\mathbf{x} \to [\mathbf{x}]_\mathcal{B}$ is a linear transformation from V to $\mathbb{R}^n$.

4. Let $V = C^0[0,1]$. The *evaluation* map $L(f) = f(1/2)$ is a linear transformation from V to $\mathbb{R}$.

5. Let $V = \mathbb{R}^3$, and let $L(\mathbf{x}) = x_1$. L is a linear transformation from $\mathbb{R}^3$ to $\mathbb{R}^1$.

6. Let $V = \mathbb{R}[t]$. The derivative map $L(\mathbf{p}) = d\mathbf{p}/dt$ is a linear operator on V.

7. If $m \geq n - 1$, then d/dt is a linear transformation from $\mathbb{R}_n[t]$ to $\mathbb{R}_m[t]$.

8. The transpose map, sending a matrix A to A^T, is a linear transformation from M_{nm} to M_{mn}.

At first glance, matrix multiplication, rotations in physical space, finding the coordinates of a vector, evaluating a function at a point, taking the derivative of a function and taking the transpose of a matrix seem like very different operations. Remarkably, the machinery we are about to develop handles all five cases on the same footing.

9. On $\mathbb{R}^1$, the map $L(x) = |x|$ is not a linear transformation, since $|-x| \neq -|x|$, and $|x+y|$ is often not equal to $|x| + |y|$.

A key property of linear transformations is that they are determined by their action on a basis. If $\mathcal{B}$ is a basis for V,

3.1. Definitions and Examples

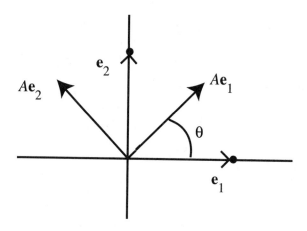

Figure 3.2: Rotation is a linear transformation

then knowing $L(\mathbf{b}_i)$ for each $\mathbf{b}_i$ tells us what L does to an arbitrary vector in V. If $\mathbf{v} = a_1\mathbf{b}_1 + \cdots a_n\mathbf{b}_n$, then

$$L(\mathbf{v}) = L(a_1\mathbf{b}_1 + \cdots a_n\mathbf{b}_n) = a_1 L(\mathbf{b}_1) + \cdots + a_n L(\mathbf{b}_n). \quad (3.2)$$

In particular, suppose that $V = \mathbb{R}^n$ and $W = \mathbb{R}^m$. Let $\mathbf{v} = (a_1, \ldots, a_n)^T = a_1\mathbf{e}_1 + \cdots + a_n\mathbf{e}_n$, where $\{\mathbf{e}_i\}$ is the standard basis. Then

$$\begin{aligned} L(\mathbf{v}) &= a_1 L(\mathbf{e}_1) + \cdots + a_n L(\mathbf{e}_n) \\ &= \begin{pmatrix} L(\mathbf{e}_1) & \cdots & L(\mathbf{e}_n) \end{pmatrix} \begin{pmatrix} a_1 \\ \vdots \\ a_n \end{pmatrix} \\ &= A\mathbf{x}, \end{aligned} \quad (3.3)$$

where

$$A = \begin{pmatrix} L(\mathbf{e}_1) & \cdots & L(\mathbf{e}_n) \end{pmatrix}. \quad (3.4)$$

A is called the matrix of the linear transformation L. In other words, every linear transformation from $\mathbb{R}^n$ to $\mathbb{R}^m$ is multiplication by an $m \times n$ matrix. **The i-th column of the matrix tells what the transformation does to $\mathbf{e}_i$.** In particular, the geometric operations of example 2 can all be described by 3×3 matrices.

Euclidean $\mathbb{R}^n$, sometimes denoted E^n, is $\mathbb{R}^n$ together with our usual notions of length and angle. On Euclidean $\mathbb{R}^2$, let L_θ be a counterclockwise rotation by an angle θ (see Figure 3.2). Since $L_\theta \begin{pmatrix} 1 \\ 0 \end{pmatrix} = \begin{pmatrix} \cos(\theta) \\ \sin(\theta) \end{pmatrix}$ and since $L_\theta \begin{pmatrix} 0 \\ 1 \end{pmatrix} = \begin{pmatrix} -\sin(\theta) \\ \cos(\theta) \end{pmatrix}$, the matrix of the linear transformation L_θ is

$$A = \begin{pmatrix} \cos(\theta) & -\sin(\theta) \\ \sin(\theta) & \cos(\theta) \end{pmatrix}. \tag{3.5}$$

Computer Graphics[1]

To move a picture on a computer screen, you must transform the coordinates of every point in the picture in an appropriate way. We have already seen how to rotate about the origin. A simple trick allows us to also implement side-to-side motion, up-and-down motion, and rotations about points other than the origin.

We represent points in the plane, not as vectors $(x, y)^T$ in $\mathbb{R}^2$, but as vectors $(x, y, 1)^T$ in $\mathbb{R}^3$. The following matrices, multiplying $(x, y, 1)^T$, have the following effects:

$$R_\theta = \begin{pmatrix} \cos(\theta) & -\sin(\theta) & 0 \\ \sin(\theta) & \cos(\theta) & 0 \\ 0 & 0 & 1 \end{pmatrix} \tag{3.6}$$

rotates by an angle θ counterclockwise around the origin. Notice that in this operation the third coordinate (which equals 1) just comes along for the ride.

$$P_x = \begin{pmatrix} -1 & 0 & 0 \\ 0 & 1 & 0 \\ 0 & 0 & 1 \end{pmatrix} \tag{3.7}$$

flips an image side-to-side, since $P_x \begin{pmatrix} x \\ y \\ 1 \end{pmatrix} = \begin{pmatrix} -x \\ y \\ 1 \end{pmatrix}$. Similarly,

$$P_y = \begin{pmatrix} 1 & 0 & 0 \\ 0 & -1 & 0 \\ 0 & 0 & 1 \end{pmatrix} \tag{3.8}$$

turns things upside-down.

$$T_{(a,b)} = \begin{pmatrix} 1 & 0 & a \\ 0 & 1 & b \\ 0 & 0 & 1 \end{pmatrix} \tag{3.9}$$

---

[1] This material is not used later in the book, and may be skipped on first reading.

3.1. Definitions and Examples

slides things a distance a to the right and a distance b upwards, since

$$T_{(a,b)} \begin{pmatrix} x \\ y \\ 1 \end{pmatrix} = \begin{pmatrix} x+a \\ y+b \\ 1 \end{pmatrix}, \text{ and}$$

$$E_{(c,d)} = \begin{pmatrix} c & 0 & 0 \\ 0 & d & 0 \\ 0 & 0 & 1 \end{pmatrix} \tag{3.10}$$

expands things by a factor of c in the horizontal direction and d in the vertical direction.

To rotate or reflect about a point other than the origin, you need to combine these basic operations. To rotate about the point $(a,b)^T$, first move $(a,b)^T$ to the origin, then rotate about the origin, and then move the origin back to $(a,b)^T$. In other words, the product $T_{(a,b)} R_\theta T_{(-a,-b)}$ (in that order!) describes a counterclockwise rotation by an angle θ about the point $(a,b)^T$. Similarly, $T_{(a,b)} P_x T_{(-a,-b)}$ and $T_{(a,b)} P_y T_{(-a,-b)}$ flip horizontally and vertically, respectively, about the point $(a,b)^T$.

To summarize, while simple rotations in Euclidean $\mathbb{R}^2$ can be described by 2×2 matrices such as (3.5), which multiply the vector $(x,y)^T$, more complicated geometrical manipulations require 3×3 matrices, multiplying the vector $(x,y,1)^T$. Similarly, rotations in Euclidean $\mathbb{R}^3$ can be described by 3×3 matrices acting on $(x,y,z)^T$, but more complicated manipulations require 4×4 matrices, which act on $(x,y,z,1)^T$. Video cards on computers often contain special processors dedicated to doing 2×2, 3×3 and 4×4 matrix multiplications, in order to present 2 and 3-dimensional motions realistically on the screen.

Exercises

1. Find 3 qualitatively different examples of maps between vector spaces that are not linear transformations.

2. Show that multiplication by a fixed $m \times n$ matrix is a linear transformation from $\mathbb{R}^n$ to $\mathbb{R}^m$, as claimed in Example 1.

3. Let I be indefinite integration on $C[0,1]$: $(If)(x) = \int_0^x f(t)dt$. Show the I is a linear operator.

4. Show that, on $\mathbb{R}[x]$, the operation $d^2/dx^2 + 17 d/dx$ is a linear operator.

5. Suppose that $L : \mathbb{R}^2 \to \mathbb{R}^2$ is a linear transformation, that $L\begin{pmatrix}1\\0\end{pmatrix} = \begin{pmatrix}1\\1\end{pmatrix}$ and that $L\begin{pmatrix}0\\1\end{pmatrix} = \begin{pmatrix}1\\2\end{pmatrix}$. What is $L\begin{pmatrix}3\\5\end{pmatrix}$?

6. Let $L : \mathbb{R}^2 \to \mathbb{R}^3$ be defined by $L(\mathbf{x}) = (x_1+x_2, -2x_1, -x_1+4x_2)^T$, where $\mathbf{x} = (x_1, x_2)^T$. Find the matrix of L.

7. Let $P(x, y, z)^T = (x, y)^T$ be the natural projection from $\mathbb{R}^3$ to $\mathbb{R}^2$. Find the matrix of P.

Representing points in the plane as 3-vectors $(x, y, 1)^T$, find 3×3 matrices that represent the following motions:

8. A counterclockwise rotation by 45 degrees ($\pi/4$ radians) about the origin followed by a rightwards shift of one unit.

9. Horizontal reflection about the line $x = 1$.

10. Counterclockwise rotation by 90 degrees ($\pi/2$ radians) about the point $x = y = 1$.

11. Reflection about the x-axis, followed by counterclockwise rotation by an angle θ, followed by reflection about the y-axis.

12. Reflection about the x-axis followed by reflection about the y-axis.

Representing points in Euclidean $\mathbb{R}^3$ as 3-vectors $(x, y, z)^T$, find 3×3 matrices that represent the following motions:

13. Rotation by an angle θ about the x-axis.

14. Rotation by an angle θ about the y-axis.

15. Rotation by an angle θ about the z-axis.

16. Rotation by an angle $\pi/3$ about the x-axis followed by rotation by an angle $\pi/2$ about the z-axis.

17. Rotation by an angle $\pi/2$ about the z-axis followed by rotation by an angle $\pi/3$ about the x-axis. Compare this result to the answer to Exercise 16.

18. Rotation by $\pi/2$ about the x axis followed by rotation by θ about the z axis followed by rotation by $-\pi/2$ about the x axis.

3.2 The Matrix of a Linear Transformation

In the last section we saw that every linear transformation from $\mathbb{R}^n$ to $\mathbb{R}^m$ is described by an $n \times m$ matrix. In this section we extend this idea to linear transformations between two general vector spaces, call them V and W. We will see that the matrix form of the linear

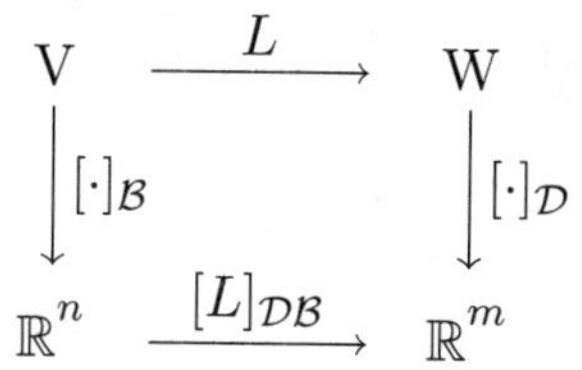

3.2. The Matrix of a Linear Transformation

$$\begin{array}{ccc} V & \xrightarrow{L} & W \\ \downarrow [\cdot]_\mathcal{B} & & \downarrow [\cdot]_\mathcal{D} \\ \mathbb{R}^n & \xrightarrow{[L]_{\mathcal{DB}}} & \mathbb{R}^m \end{array}$$

Figure 3.3: The matrix of a linear transformation

transformation depends on the basis. Much of this book is about picking the right bases to make linear transformations look as simple as possible.

Suppose that V is an n-dimensional vector space with basis $\mathcal{B} = \{\mathbf{b}_1, \ldots, \mathbf{b}_n\}$, that W is an m-dimensional vector space with basis $\mathcal{D} = \{\mathbf{d}_1, \ldots, \mathbf{d}_m\}$, and that L is a linear transformation from V to W. We have already seen that $\mathcal{B}$ makes V look like $\mathbb{R}^n$, and that $\mathcal{D}$ makes W look like $\mathbb{R}^m$. Taken together, they make L look like a linear transformation from $\mathbb{R}^n$ to $\mathbb{R}^m$, in other words an $m \times n$ matrix.

This idea is depicted in Figure 3.3. While L takes a vector in V (call if $\mathbf{v}$) to a vector $\mathbf{w} = L(\mathbf{v})$ in W, the matrix $[L]_{\mathcal{DB}}$ takes the corresponding vector in $\mathbb{R}^n$ (namely $[\mathbf{v}]_\mathcal{B}$) to the vector $[\mathbf{w}]_\mathcal{D}$ in $\mathbb{R}^m$.

Theorem 3.1 *Let $\mathcal{B} = \{\mathbf{b}_1, \ldots, \mathbf{b}_n\}$ and $\mathcal{D} = \{\mathbf{d}_1, \ldots, \mathbf{d}_m\}$ be bases for V and W, respectively, and let L be a linear transformation from V to W. Then there is a unique matrix $[L]_{\mathcal{DB}}$ such that, for any $\mathbf{v} \in V$,*

$$[L\mathbf{v}]_\mathcal{D} = [L]_{\mathcal{DB}}[\mathbf{v}]_\mathcal{B}. \tag{3.11}$$

Furthermore, this matrix is given by the formula

$$[L]_{\mathcal{DB}} = \Big([L\mathbf{b}_1]_\mathcal{D} \quad \cdots \quad [L\mathbf{b}_n]_\mathcal{D} \Big). \tag{3.12}$$

Proof: The proof is almost identical to the proof of Theorem 2.9. First we show that, if such a matrix exists, it must take the form (3.12). Then we show that the formula (3.12) actually works.

Assuming the matrix exists, we must have

$$[L\mathbf{b}_i]_\mathcal{D} = [L]_{\mathcal{DB}}[\mathbf{b}_i]_\mathcal{B} = [L]_{\mathcal{DB}}\mathbf{e}_i, \tag{3.13}$$

which is the i-th column of $[L]_{\mathcal{DB}}$. Thus $[L]_{\mathcal{DB}}$ has to take the form (3.12). To see that this formula actually works, let $\mathbf{v} = a_1\mathbf{b}_1 + \cdots +$

$a_n\mathbf{b}_n$, so that $[\mathbf{v}]_\mathcal{B} = (a_1, \ldots, a_n)^T$. Then

$$\begin{aligned}[L\mathbf{v}]_\mathcal{D} &= [a_1 L\mathbf{b}_1 + \cdots a_n L\mathbf{b}_n]_\mathcal{D} \\ &= a_1[L\mathbf{b}_1]_\mathcal{D} + \cdots + a_n[L\mathbf{b}_n]_\mathcal{D} \\ &= \begin{pmatrix} [L\mathbf{b}_1]_\mathcal{D} & [L\mathbf{b}_2]_\mathcal{D} & \cdots & [L\mathbf{b}_n]_\mathcal{D} \end{pmatrix} \begin{pmatrix} a_1 \\ \vdots \\ a_n \end{pmatrix} \\ &= [L]_{\mathcal{DB}}[\mathbf{v}]_\mathcal{B}. \blacksquare \end{aligned} \quad (3.14)$$

Example 1: Let $V = W = \mathbb{R}_3[t]$, and let $L = d/dt$. We use the basis $\mathcal{B} = \{1, t, t^2, t^3\}$ for both V and W. We compute $L(\mathbf{b}_1) = 0$, $L(\mathbf{b}_2) = 1 = \mathbf{b}_1$, $L(\mathbf{b}_3) = 2t = 2\mathbf{b}_2$ and $L(\mathbf{b}_4) = 3t^2 = 3\mathbf{b}_3$. The matrix of d/dt is then

$$[d/dt]_{\mathcal{BB}} = \begin{pmatrix} 0 & 1 & 0 & 0 \\ 0 & 0 & 2 & 0 \\ 0 & 0 & 0 & 3 \\ 0 & 0 & 0 & 0 \end{pmatrix}. \quad (3.15)$$

If we look at a vector $\mathbf{p} \in \mathbb{R}_3[t]$, we can compute $[L\mathbf{p}]_\mathcal{B}$ in two different ways, corresponding to the two different paths from V to $\mathbb{R}^m$ in Figure 3.3. We can first apply L (i.e. take a derivative) and then compute the coordinates of the resulting function $\mathbf{p}'(t)$, or we can first find the coordinates of $\mathbf{p}$ and then multiply by the matrix $[L]_{\mathcal{BB}}$. For example, if $\mathbf{p}(t) = 7 + 5t - 2t^2 + t^3$, then $L(\mathbf{p})(t) = d\mathbf{p}(t)/dt = 5 - 4t + 3t^2$ and $[L(\mathbf{p})]_\mathcal{B} = (5, -4, 3, 0)^T$. On the other hand, $[\mathbf{p}]_\mathcal{B} = (7, 5, -2, 1)^T$, so

$$[L]_{\mathcal{BB}}[\mathbf{p}]_\mathcal{B} = \begin{pmatrix} 0 & 1 & 0 & 0 \\ 0 & 0 & 2 & 0 \\ 0 & 0 & 0 & 3 \\ 0 & 0 & 0 & 0 \end{pmatrix} \begin{pmatrix} 7 \\ 5 \\ -2 \\ 1 \end{pmatrix} = \begin{pmatrix} 5 \\ -4 \\ 3 \\ 0 \end{pmatrix}. \quad (3.16)$$

If L is a linear operator on a single space V, as in the last example, then we have only one basis to worry about instead of two. We usually denote the matrix of the operator by $[L]_\mathcal{B}$ instead of $[L]_{\mathcal{BB}}$, so that, for any $\mathbf{v} \in V$,

$$[L\mathbf{v}]_\mathcal{B} = [L]_\mathcal{B}[\mathbf{v}]_\mathcal{B}. \quad (3.17)$$

Example 2: Let $V = M_{2,2}$, the space of real 2×2 matrices, with the standard basis $\mathcal{E} = \left\{ \begin{pmatrix} 1 & 0 \\ 0 & 0 \end{pmatrix}, \begin{pmatrix} 0 & 1 \\ 0 & 0 \end{pmatrix}, \begin{pmatrix} 0 & 0 \\ 1 & 0 \end{pmatrix}, \begin{pmatrix} 0 & 0 \\ 0 & 1 \end{pmatrix} \right\}$ and let

3.2. The Matrix of a Linear Transformation

L be a linear operator on V given by $L(A) = A + A^T$. We compute $[L]_{\mathcal{E}}$ as follows. $L(\mathbf{e}_1) = 2\mathbf{e}_1$, $L(\mathbf{e}_2) = \mathbf{e}_2 + \mathbf{e}_3$, $L(\mathbf{e}_3) = \mathbf{e}_2 + \mathbf{e}_3$, $L(\mathbf{e}_4) = 2\mathbf{e}_4$, so

$$[L]_{\mathcal{E}} = \begin{pmatrix} 2 & 0 & 0 & 0 \\ 0 & 1 & 1 & 0 \\ 0 & 1 & 1 & 0 \\ 0 & 0 & 0 & 2 \end{pmatrix}. \quad (3.18)$$

If $A = \begin{pmatrix} 1 & 2 \\ 3 & 4 \end{pmatrix}$, then $L(A) = A + A^T = \begin{pmatrix} 2 & 5 \\ 5 & 8 \end{pmatrix}$ and $[L(A)]_{\mathcal{E}} = (2, 5, 5, 8)^T$. Alternatively we can use the fact that $[A]_{\mathcal{E}} = (1, 2, 3, 4)^T$ to compute

$$[L(A)]_{\mathcal{E}} = [L]_{\mathcal{E}}[A]_{\mathcal{E}} = \begin{pmatrix} 2 & 0 & 0 & 0 \\ 0 & 1 & 1 & 0 \\ 0 & 1 & 1 & 0 \\ 0 & 0 & 0 & 2 \end{pmatrix} \begin{pmatrix} 1 \\ 2 \\ 3 \\ 4 \end{pmatrix} = \begin{pmatrix} 2 \\ 5 \\ 5 \\ 8 \end{pmatrix}. \quad (3.19)$$

Exercises

1. Let V be a vector space with a basis $\mathcal{B}$, and let L be an operator on V. L^2 means the operator applied twice: $L^2(\mathbf{v}) = L(L(\mathbf{v}))$. Show that the matrix of L^2 is the square of the matrix of L. That is, $[L^2]_{\mathcal{B}} = [L]_{\mathcal{B}}^2$.

2. On $\mathbb{R}^2$ (with the standard basis), let L be a rotation by $45°$. Compute the matrices of L, L^2, L^4 and L^8.

3. On $\mathbb{R}_3[t]$ with basis $\{1, t, t^2, t^3\}$, compute the matrix of d^2/dt^2 and compare to the square of the matrix in (3.15).

4. Let V be a vector space with basis $\mathcal{B}$ and let L and M be operators on V. Show that the composition $L \circ M$ defined by $L \circ M(\mathbf{v}) = L(M(\mathbf{v}))$ is also an operator on V. Prove that $[L \circ M]_{\mathcal{B}} = [L]_{\mathcal{B}}[M]_{\mathcal{B}}$. In other words, composition of operators corresponds to multiplication of square matrices.

5. On $\mathbb{R}^2$, let L be rotation by an angle θ and let M be rotation by an angle ϕ. From the results of Exercise 4, derive the formulas for the sine and cosine of a sum of two angles.

6. Exercise 4 concerned operators on a single vector space. A similar result applies to general linear transformations. Let U, V and W be vector spaces, with bases $\mathcal{B}$, $\mathcal{C}$ and $\mathcal{D}$, respectively. Let $L_1 : U \to V$ and $L_2 : V \to W$ be linear transformations. We can define the composition $L_2 \circ L_1 : U \to W$ by $L_2 \circ L_1(\mathbf{u}) = L_2(L_1(\mathbf{u}))$. Show that $L_2 \circ L_1$ is a linear transformation from U to W. Show that the matrix of $L_2 \circ L_1$ is given by $[L_2 \circ L_1]_{\mathcal{DB}} = [L_2]_{\mathcal{DC}}[L_1]_{\mathcal{CB}}$.

7. On $\mathbb{R}_2[t]$, let $(T_1\mathbf{p})(t) = \mathbf{p}(t-1)$. Find the matrix of T_1.

8. More generally, for each value of a, let $(T_a\mathbf{p})(t) = \mathbf{p}(t-a)$. Compute the matrix of T_a acting on $\mathbb{R}_2[t]$. What is the matrix of T_a^{-1}?

9. Compute the matrix of T_a acting on $\mathbb{R}_4[t]$. Where have you seen this pattern before? Can you guess what the matrix of T_a acting on $\mathbb{R}_n[t]$ looks like?

10. On $\mathbb{R}_2[t]$, let $L\mathbf{p} = p(0)t^2 + p(1)t + p(2)$. Find the matrix of L relative to the standard basis $\{1, t, t^2\}$.

11. Let $L : \mathbb{R}_3[t] \to \mathbb{R}_2[t]$ be defined by $L(\mathbf{p}(t)) = \mathbf{p}'(t) + p(1) + p(2)t + p(3)t^2$. Find the matrix of L relative to the standard bases of $\mathbb{R}_2[t]$ and $\mathbb{R}_3[t]$.

12. Let $L : \mathbb{R}_2[t] \to \mathbb{R}^4$ be given by $L(\mathbf{p}) = (\mathbf{p}(0), \mathbf{p}(1), \mathbf{p}(2), \mathbf{p}(3))^T$. Find the matrix of L relative to the standard bases of $\mathbb{R}_2[t]$ and $\mathbb{R}^4$.

13. Let W be the external direct sum of $V_1 = \mathbb{R}^3$ and $V_2 = \mathbb{R}^2$. Write down the matrices for the projections P_1 and P_2 (relative to the standard bases).

14. Suppose $A \in M_{m,n}$ and $B \in M_{m',n'}$. Define a linear map $L : \mathbb{R}^n \oplus \mathbb{R}^{n'} \to \mathbb{R}^m \oplus \mathbb{R}^{m'}$ by $L\begin{pmatrix}\mathbf{x}\\\mathbf{y}\end{pmatrix} = \begin{pmatrix}A\mathbf{x}\\B\mathbf{y}\end{pmatrix}$. Find the matrix of L (relative to the standard bases) in terms of the matrices A and B.

3.3 The Effect of a Change of Basis

The matrix of a linear operator on a vector space V depends both on the operator itself and on the basis we choose for V. Some bases make it easy to compute coordinates of vectors, while others make the operators we are interested in look simple. Unfortunately, these bases are usually not the same, so we need to understand how to convert from one basis to another. The conversion for coordinates of vectors was discussed in Section 2.4. Here we consider how to convert matrices of operators, or more generally matrices of linear transformations, from one basis to another.

We begin with operators. Suppose V is a vector space with bases $\mathcal{B}$ and $\mathcal{D}$, that L is an operator on V, and that we have computed

3.3. The Effect of a Change of Basis

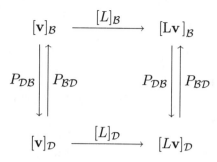

Figure 3.4: A change of basis

the matrix $[L]_\mathcal{B}$. Now, for whatever reason, we wish to change our basis from $\mathcal{B}$ to $\mathcal{D}$. How does that change our matrix?

The situation is described in Figure 3.4. We are looking for a matrix $[L]_\mathcal{D}$ that converts the coordinates $[\mathbf{v}]_\mathcal{D}$ of $\mathbf{v}$ in the $\mathcal{D}$ basis into the coordinates $[L\mathbf{v}]_\mathcal{D}$. We already know that multiplication by the change of basis matrix $P_{\mathcal{BD}}$ converts $[\mathbf{v}]_\mathcal{D}$ to $[\mathbf{v}]_\mathcal{B}$, that multiplication by $[L]_\mathcal{B}$ converts $[\mathbf{v}]_\mathcal{B}$ to $[L\mathbf{v}]_\mathcal{B}$, and that multiplication by the matrix $P_{\mathcal{DB}}$ converts $[L\mathbf{v}]_\mathcal{B}$ to $[L\mathbf{v}]_\mathcal{D}$. The product of the three matrices therefore converts $[\mathbf{v}]_\mathcal{D}$ to $[L\mathbf{v}]_\mathcal{D}$. This product therefore equals $[L]_\mathcal{D}$. In other words, we have proven:

Theorem 3.2 *Let L be an operator on a vector space with alternate bases $\mathcal{B}$ and $\mathcal{D}$. The matrix of L relative to $\mathcal{D}$ is*

$$[L]_\mathcal{D} = P_{\mathcal{DB}}[L]_\mathcal{B} P_{\mathcal{BD}}. \tag{3.20}$$

Since $P_{\mathcal{DB}}$ and $P_{\mathcal{BD}}$ are inverses of one another, we can also write this result as

$$[L]_\mathcal{D} = P_{\mathcal{DB}}[L]_\mathcal{B} P_{\mathcal{DB}}^{-1} \tag{3.21}$$

or as

$$[L]_\mathcal{D} = P_{\mathcal{BD}}^{-1}[L]_\mathcal{B} P_{\mathcal{BD}}. \tag{3.22}$$

Example: Let V be a 2-dimensional vector space with basis $\mathcal{B}$, (e.g., $\mathbb{R}^2$ with the standard basis) and let $\mathcal{D}$ be the alternate basis $\{\mathbf{b}_1 + \mathbf{b}_2, \mathbf{b}_1 - \mathbf{b}_2\}$. The change-of-basis matrices are

$$P_{\mathcal{BD}} = \begin{pmatrix} 1 & 1 \\ 1 & -1 \end{pmatrix}, \quad P_{\mathcal{DB}} = P_{\mathcal{BD}}^{-1} = \begin{pmatrix} 1/2 & 1/2 \\ 1/2 & -1/2 \end{pmatrix}. \tag{3.23}$$

If L an operator that takes $\mathbf{b}_1$ to $2\mathbf{b}_1 + \mathbf{b}_2$ and takes $\mathbf{b}_2$ to $\mathbf{b}_1 + 2\mathbf{b}_2$, then the matrix of L in the $\mathcal{B}$ basis is

$$[L]_\mathcal{B} = \begin{pmatrix} 2 & 1 \\ 1 & 2 \end{pmatrix}, \tag{3.24}$$

while the matrix of L in the $\mathcal{D}$ basis is

$$[L]_{\mathcal{D}} = \begin{pmatrix} 1/2 & 1/2 \\ 1/2 & -1/2 \end{pmatrix} \begin{pmatrix} 2 & 1 \\ 1 & 2 \end{pmatrix} \begin{pmatrix} 1 & 1 \\ 1 & -1 \end{pmatrix} = \begin{pmatrix} 3 & 0 \\ 0 & 1 \end{pmatrix}. \quad (3.25)$$

You should check for yourself that $L\mathbf{d}_1 = 3\mathbf{d}_1$ and that $L\mathbf{d}_2 = \mathbf{d}_2$, as implied by (3.25).

In this example, the basis $\mathcal{D}$ makes our operator look very simple. In fact, we will soon see that every operator can be made to look simple through the right choice of basis. Finding these bases and expressing everything in terms of them is the key to decoupling.

Although most of our applications will involve operators on a single vector space, sometimes we also need to express linear transformations between different vector spaces as matrices, and to see how these matrices change when we change our bases. The following is a generalization of Theorem 3.2.

Theorem 3.3 *Let V be a vector space with bases $\mathcal{B}$ and $\mathcal{D}$, let W be a vector space with bases $\mathcal{B}'$ and $\mathcal{D}'$, and let L be an linear transformation from V to W. Let $[L]_{\mathcal{B}'\mathcal{B}}$ be the matrix of L relative to the $\mathcal{B}$ and $\mathcal{B}'$ bases. The matrix of L relative to the $\mathcal{D}$ and $\mathcal{D}'$ bases is*

$$[L]_{\mathcal{D}'\mathcal{D}} = P_{\mathcal{D}'\mathcal{B}'}[L]_{\mathcal{B}'\mathcal{B}}P_{\mathcal{B}\mathcal{D}}. \quad (3.26)$$

***Proof*:** The argument is almost identical to the proof of Theorem 3.2. Let $\mathbf{v}$ be an arbitrary vector in V. Then

$$\begin{aligned} P_{\mathcal{B}\mathcal{D}}[\mathbf{v}]_{\mathcal{D}} &= [\mathbf{v}]_{\mathcal{B}} \\ [L]_{\mathcal{B}'\mathcal{B}}P_{\mathcal{B}\mathcal{D}}[\mathbf{v}]_{\mathcal{D}} &= [L]_{\mathcal{B}'\mathcal{B}}[\mathbf{v}]_{\mathcal{B}} = [L\mathbf{v}]_{\mathcal{B}'}, \\ P_{\mathcal{D}'\mathcal{B}'}[L]_{\mathcal{B}'\mathcal{B}}P_{\mathcal{B}\mathcal{D}}[\mathbf{v}]_{\mathcal{D}} &= P_{\mathcal{D}'\mathcal{B}'}[L\mathbf{v}]_{\mathcal{B}'} = [L\mathbf{v}]_{\mathcal{D}'}. \end{aligned} \quad (3.27)$$

Since multiplication by $P_{\mathcal{D}'\mathcal{B}'}[L]_{\mathcal{B}'\mathcal{B}}P_{\mathcal{B}\mathcal{D}}$ converts $[\mathbf{v}]_{\mathcal{D}}$ to $[L\mathbf{v}]_{\mathcal{D}'}$ for every vector $\mathbf{v}$, $P_{\mathcal{D}'\mathcal{B}'}[L]_{\mathcal{B}'\mathcal{B}}P_{\mathcal{B}\mathcal{D}}$ must equal $[L]_{\mathcal{D}'\mathcal{D}}$. ∎

Exercises

1. On $\mathbb{R}^2$, let L be multiplication by the matrix $\begin{pmatrix} 1 & 4 \\ 1 & 1 \end{pmatrix}$. That is, $[L]_{\mathcal{E}} = \begin{pmatrix} 1 & 4 \\ 1 & 1 \end{pmatrix}$. Let $\mathcal{B} = \{\begin{pmatrix} 2 \\ 1 \end{pmatrix}, \begin{pmatrix} 2 \\ -1 \end{pmatrix}\}$. Compute $[L]_{\mathcal{B}}$.

2. In the setup of Exercise 1, let $\mathbf{v} = (3,3)^T$. Compute $[L\mathbf{v}]_{\mathcal{B}}$ in two ways. One way is to compute $L\mathbf{v}$ and then convert to the $\mathcal{B}$ basis. The other way is to compute $[\mathbf{v}]_{\mathcal{B}}$ and then multiply by $[L]_{\mathcal{B}}$.

3.4. Infinite Dimensional Vector Spaces

3. In the vector space $\mathbb{R}_2[t]$, let $\mathbf{b}_1(t) = t^2 + t + 1$, let $\mathbf{b}_2(t) = t^2 + 3t + 2$, and let $\mathbf{b}_3(t) = t^2 + 2t + 1$, as in Exercise 9 of Section 2.4. Let $d/dt : \mathbb{R}_2[t] \to \mathbb{R}_2[t]$ be the derivative map. Find the matrix of d/dt relative to the basis $\mathcal{B}$. [Hint: First find the matrix of d/dt relative to the standard basis and then convert.]

4. In $\mathbb{R}_2[t]$, let $\mathcal{B}$ be as in Exercise 3 and let $\mathcal{E}$ be the standard basis. Consider the map $L : \mathbb{R}_2[t] \to \mathbb{R}^3$ given by $L(\mathbf{p}) = (p(1), p(2), p(3))^T$. Find $[L]_{\mathcal{E}\mathcal{E}}$ and $[L]_{\mathcal{E}\mathcal{B}}$ and show explicitly that $[L]_{\mathcal{E}\mathcal{B}} = [L]_{\mathcal{E}\mathcal{E}} P_{\mathcal{E}\mathcal{B}}$.

5. On $\mathbb{R}_2[t]$, let L be the operator of Exercise 7 of Section 3.2. Find the matrix of L relative to the basis $\mathcal{B}$ of Exercise 3.

6. On $M_{2,2}$, let $\mathcal{B} = \{\begin{pmatrix} 1 & 0 \\ 0 & 1 \end{pmatrix}, \begin{pmatrix} 1 & 0 \\ 0 & -1 \end{pmatrix}, \begin{pmatrix} 0 & 1 \\ 1 & 0 \end{pmatrix}, \begin{pmatrix} 0 & 1 \\ -1 & 0 \end{pmatrix}\}$.
Let $L(A) = (A + A^T)/2$. Find $[L]_{\mathcal{B}\mathcal{B}}$.

7. On $\mathbb{R}_2[t]$, consider the bases $\mathcal{B} = \{1, t, t^2\}$ and $\mathcal{D} = \{1, t-1, t^2 - 2t + 1\}$. Let $L\mathbf{p}(t) = \mathbf{p}(1-t)$. Compute $P_{\mathcal{B}\mathcal{D}}$, $P_{\mathcal{D}\mathcal{B}}$, and $[L]_{\mathcal{B}}$ and $[L]_{\mathcal{D}}$ and verify equation 3.20 in this instance.

8. Draw a figure, similar to Figure 3.4, that summarizes the proof of Theorem 3.3.

3.4 Infinite Dimensional Vector Spaces

In describing vectors and linear transformations by columns of numbers and matrices, respectively, we have always assumed that our vector spaces were finite dimensional. What happens if the spaces are infinite dimensional, as with the space $\mathbb{R}[t]$ of polynomials of arbitrary degree? In most common examples, the preceding discussion goes through unchanged, only with matrices and column vectors of infinite extent.

If we have an infinite basis $\mathbf{b}_1, \mathbf{b}_2, \ldots$, then we can associate the vector $\mathbf{v} = a_1\mathbf{b}_1 + a_2\mathbf{b}_2 + \cdots$ with the infinite column vector $(a_1, a_2, \ldots)^T$. In writing $\mathbf{v} = a_1\mathbf{b}_1 + a_2\mathbf{b}_2 + \cdots$ we either assume that all but a finite number of coefficients are zero, or that we know how to take infinite sums as appropriate limits of finite sums. In the latter case, the expression only makes sense if the infinite sum converges.

Multiplying an infinite matrix by an infinite column vector is easy. Instead of having $(A\mathbf{x})_i = \sum_{j=1}^{n} A_{ij}x_j$, as in the n-dimensional case, we just have

$$(A\mathbf{x})_i = \sum_{j=1}^{\infty} A_{ij} x_j. \tag{3.28}$$

We multiply two infinite matrices by the formula:

$$(AB)_{ik} = \sum_{j=1}^{\infty} A_{ij} B_{jk}. \tag{3.29}$$

In both cases the operation makes sense only if the infinite sums all converge.

If we have a linear transformation from an infinite dimensional space V, with basis $\mathcal{B}$, to an infinite dimensional space W, with basis $\mathcal{D}$, then it can be described by a matrix of infinite size. As usual, the i-th column of this matrix is just $[L\mathbf{b}_i]_\mathcal{D}$, and multiplication by this matrix sends $[\mathbf{v}]_\mathcal{B}$ to $[L\mathbf{v}]_\mathcal{D}$. For instance, if we take $\mathbb{R}[t]$ with standard basis $\mathcal{B} = \mathcal{D} = \{1, t, t^2, \ldots\}$, then the operator d/dt is described by the matrix:

$$\begin{pmatrix} 0 & 1 & 0 & 0 & \cdots \\ 0 & 0 & 2 & 0 & \cdots \\ 0 & 0 & 0 & 3 & \cdots \\ \vdots & \vdots & \vdots & \vdots & \ddots \end{pmatrix}. \tag{3.30}$$

In dealing with infinite-size matrices, one does have to be careful about convergence questions. As a result, some standard results about finite dimensional matrices no longer hold. For example, whenever two square matrices A, B of finite size satisfy $AB = I$, then they also satisfy $BA = I$. However, consider the infinite matrix

$$A = \begin{pmatrix} 0 & 1 & 0 & 0 & \cdots \\ 0 & 0 & 1 & 0 & \cdots \\ 0 & 0 & 0 & 1 & \cdots \\ \vdots & \vdots & \vdots & \vdots & \ddots \end{pmatrix} \tag{3.31}$$

and its transpose

$$B = \begin{pmatrix} 0 & 0 & 0 & 0 & \cdots \\ 1 & 0 & 0 & 0 & \cdots \\ 0 & 1 & 0 & 0 & \cdots \\ \vdots & \vdots & \vdots & \vdots & \ddots \end{pmatrix}. \tag{3.32}$$

3.4. Infinite Dimensional Vector Spaces

It is easy to check that AB is the identity, but that

$$BA = \begin{pmatrix} 0 & 0 & 0 & 0 & \cdots \\ 0 & 1 & 0 & 0 & \cdots \\ 0 & 0 & 1 & 0 & \cdots \\ 0 & 0 & 0 & 1 & \cdots \\ \vdots & \vdots & \vdots & \vdots & \ddots \end{pmatrix} \ne I. \tag{3.33}$$

Another standard property that can fail concerns traces (see Section 4.6). If A and B are $n \times n$ matrices, then the trace of $AB - BA$ is always zero. But in our infinite example the trace of $AB - BA$ is one.

In short, infinite dimensional spaces and infinite dimensional operations are neither totally bizarre nor totally tame, but somewhere in between. If an argument or technique works in finite dimensions, it is likely, but by no means certain, that it will work in infinite dimensions. As a first approximation, applying your finite dimensional intuition to infinite dimensions is a very good idea. However, you should be prepared for an occasional surprise, almost always due to a lack of convergence of some sum.

Exercises

Exercises 1–6 are all about $\mathbb{R}[t]$ with the standard basis $\mathcal{E} = \{1, t, t^2, \ldots\}$, which is our simplest example of an infinite dimensional vector space.

1. Let $L_1 f(t) = t f(t)$. Find the matrix of L_1.
2. Let $L_2 f(t) = (f(t) - f(0))/t$. Find the matrix of L_2.
3. Let A_1 be the matrix of L_1, and let A_2 be the matrix of L_2. Compute $A_1 A_2$ and $A_2 A_1$. How do these compare to the matrices of $L_1 \circ L_2$ and $L_2 \circ L_1$?
4. Let $L_3 f(t) = \int_0^t f(t') dt'$. Find the matrix of L_3. What is the matrix of d/dt times the matrix of L_3? What is the matrix of L_3 times the matrix of d/dt? How does this jibe with the idea that integration and differentiation are inverse operations?
5. Show that the traces of $L_3^T L_3$ and $L_3 L_3^T$ exist and are equal.
6. Let $L_4 f(t) = f(1) + t f(t)$. Find the matrix of L_4. Calling this matrix A_4, can you make sense of $A_4^T A_4$?
7. An infinite matrix A is said to be Hilbert-Schmidt if the double sum $\sum_{i,j=1}^{\infty} |A_{ij}|^2$ converges. Show that if A is Hilbert-Schmidt,

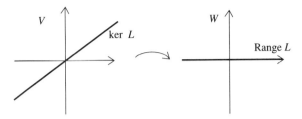

Figure 3.5: The kernel and range of $L : (x, y)^T \to (x - y, 0)^T$.

then the traces of $A^T A$ and AA^T are well-defined, and $Tr(A^T A) = Tr(AA^T)$.

3.5 Kernels, Ranges, and Quotient Maps

Definition *Let V and W be vector spaces, and let L be a linear transformation from V to W. The* kernel *of L, denoted $Ker(L)$, is the set of all vectors $\mathbf{v} \in V$ for which $L(\mathbf{v}) = 0$. The* range *of L, denoted $Range(L)$, is the set of all vectors of the form $L(\mathbf{v})$, where $\mathbf{v} \in V$.*

Theorem 3.4 *$Ker(L)$ is a subspace of V, and $Range(L)$ is a subspace of W.*

***Proof*:** Suppose $\mathbf{x}, \mathbf{y} \in Ker(L)$. Then $L(\mathbf{x}) = L(\mathbf{y}) = 0$, so $L(\mathbf{x} + \mathbf{y}) = L(\mathbf{x}) + L(\mathbf{y}) = 0 + 0 = 0$, so $\mathbf{x} + \mathbf{y} \in Ker(L)$, so $Ker(L)$ is closed under addition. Now let c be a scalar. $L(c\mathbf{x}) = cL(\mathbf{x}) = c0 = 0$, so $c\mathbf{x} \in Ker(L)$, and $Ker(L)$ is closed under scalar multiplication. Thus $Ker(L)$ is a subspace of V.

Now suppose that $\mathbf{x}$ and $\mathbf{y}$ are in $Range(L)$, so there exist vectors $\mathbf{v}$ and $\mathbf{w}$ in V with $L(\mathbf{v}) = \mathbf{x}$ and $L(\mathbf{w}) = \mathbf{y}$. Then $L(\mathbf{v}+\mathbf{w}) = L(\mathbf{v}) + L(\mathbf{w}) = \mathbf{x} + \mathbf{y}$, so $\mathbf{x} + \mathbf{y} \in Range(L)$. Similarly, $L(c\mathbf{v}) = cL(\mathbf{v}) = c\mathbf{x}$, so $c\mathbf{x} \in Range(L)$. Since $Range(L)$ is closed under addition and scalar multiplication, it is a subspace of W. ∎

Example: Suppose $V = \mathbb{R}^4$, $W = \mathbb{R}^2$ and L is multiplication by the matrix
$$A = \begin{pmatrix} 1 & 2 & 3 & 4 \\ 1 & 4 & 7 & 8 \end{pmatrix}. \tag{3.34}$$
The kernel of L is the set of all vectors $\mathbf{x}$ for which $A\mathbf{x} = 0$. But those equations are easily solved by row reduction. The matrix A row-reduces to $\begin{pmatrix} 1 & 2 & 0 & 4 \\ 0 & 0 & 1 & 0 \end{pmatrix}$. The first row says that $x_1 = -2x_2 - 4x_4$,

3.5. Kernels, Ranges, and Quotient Maps

the second row say $x_3 = 0$, and there are no constraints on x_2 or x_4. The kernel is therefore a 2-dimensional subspace of $\mathbb{R}^4$, spanned by $\mathbf{b}_1 = (2, -1, 0, 0)^T$ and $\mathbf{b}_2 = (4, 0, 0, -1)^T$. You should check explicitly that any linear combination of $\mathbf{b}_1$ and $\mathbf{b}_2$, say $2\mathbf{b}_1 + 5\mathbf{b}_2$, solves $A\mathbf{x} = 0$.

When $V = \mathbb{R}^n$ and $W = \mathbb{R}^m$ and L is multiplication by an $m \times n$ matrix A, as in the example, then there are special names for the kernel and range of L. The range of L is called the *column space of A*, and the kernel of L is called the *null space of A*. The column space of A can also be viewed as the span of the columns of A (hence the name), a fact you are asked to prove in the exercises.

Many of our simplest examples of subspaces are kernels and ranges of linear transformations. For instance, the subspace of $\mathbb{R}^3$ defined by $\mathbf{x}_1 + \mathbf{x}_2 + \mathbf{x}_3 = 0$ is the kernel of the linear transformation $L : \mathbb{R}^3 \to \mathbb{R}; L(\mathbf{x}) = \mathbf{x}_1 + \mathbf{x}_2 + \mathbf{x}_3$. The set of all polynomials $\mathbf{p}$ such that $\mathbf{p}(3) = 0$ is the kernel of the evaluation map $L : \mathbb{R}[t] \to \mathbb{R}$; $L(\mathbf{p}) = \mathbf{p}(3)$. The span of the vectors $\begin{pmatrix} 1 \\ 1 \\ -3 \end{pmatrix}$ and $\begin{pmatrix} -2 \\ 5 \\ 2 \end{pmatrix}$ in $\mathbb{R}^3$ is the column space of the matrix $A = \begin{pmatrix} 1 & -2 \\ 1 & 5 \\ -3 & 2 \end{pmatrix}$, hence the range of the linear transformation $L : \mathbb{R}^2 \to \mathbb{R}^3; L(\mathbf{x}) = A\mathbf{x}$.

The kernel of a linear transformation measures the extent to which that linear transformation fails to be 1-1:

Theorem 3.5 *A linear transformation is 1-1 if and only if its kernel is $\{0\}$.*

Proof: Let L be a linear transformation. $L(0) = L(0\mathbf{x}) = 0L(\mathbf{x}) = 0$, so $0 \in \text{Ker}(L)$. If L is 1-1, then $0 \in V$ is the only vector to be mapped to $0 \in W$, so $\text{Ker}(L) = \{0\}$. Conversely, suppose $\text{Ker}(L) = \{0\}$. Then, if $L(\mathbf{x}) = L(\mathbf{y})$, we must have $L(\mathbf{x}-\mathbf{y}) = L(\mathbf{x})-L(\mathbf{y}) = 0$, so $\mathbf{x} - \mathbf{y} \in \text{Ker}(L)$. But $\text{Ker}(L) = \{0\}$, so this implies $\mathbf{x} = \mathbf{y}$, so L is 1-1. ∎

Kernels and ranges are further related by the quotient operation of Section 2.5.

Theorem 3.6 *The range of a linear transformation $L : V \to W$ is isomorphic to the quotient space $V/\text{Ker}(L)$.*

Proof: Let $\hat{V} = V/\text{Ker}(L)$. We construct a map $\hat{L}$ from $\hat{V}$ to W, and show that this map is 1-1. This map is then an isomorphism from $\hat{V}$ to $Range(\hat{L})$. Finally we show that the range of $\hat{L}$ is the same as the range of L. Let $\hat{L}$ be defined by

$$\hat{L}([\mathbf{x}]) = L(\mathbf{x}). \tag{3.35}$$

This formula says that, to compute $\hat{L}$ of an equivalence class, pick a vector in that class and apply L to it. To see that this is well-defined, we must show that the answer does not depend on which vector we pick. If $[\mathbf{y}] = [\mathbf{x}]$, then $\mathbf{y} - \mathbf{x} \in \text{Ker}(L)$, so $L(\mathbf{y} - \mathbf{x}) = 0$. But then $L(\mathbf{y}) = L(\mathbf{x} + (\mathbf{y} - \mathbf{x})) = L(\mathbf{x}) + L(\mathbf{y} - \mathbf{x}) = L(\mathbf{x})$, so $\hat{L}$ is well-defined. To see that $\hat{L}$ is 1-1, suppose that $\hat{L}([\mathbf{x}]) = 0$. Then $L(\mathbf{x}) = 0$, so $\mathbf{x} \in \text{Ker}(L)$, so $[\mathbf{x}] = [0]$. Since the kernel of $\hat{L}$ is trivial, $\hat{L}$ is 1-1. The range of $\hat{L}$ is contained in the range of L, since every vector of the form $\hat{L}([\mathbf{x}])$ is also of the form $L(\mathbf{x})$. But every vector of the form $L(\mathbf{x})$ is also equal to $\hat{L}([\mathbf{x}])$, so the range of L is contained in the range of $\hat{L}$. Thus the two ranges are the same. ∎

In Section 2.5 we saw that the dimension of a quotient space V/W is the dimension of V minus the dimension of W. Applying that to $V/Ker(L)$ we obtain

Theorem 3.7 (The Dimension Theorem) *The dimension of V equals the dimension of* $\text{Ker}(L)$ *plus the dimension of* $Range(L)$.

In the example above, V was $\mathbb{R}^4$, the range of L was all of $\mathbb{R}^2$, and the kernel of L was 2 dimensional.

Definition *The rank of a linear transformation L is the dimension of $Range(L)$.*

If $V = \mathbb{R}^n$, $W = \mathbb{R}^m$ and L is multiplication by an $m \times n$ matrix A, then the rank has an additional interpretation. In that case the rank is the number of pivots in the reduced row-echelon form of A. If the rank of a matrix A is k, null space of A is then $n - k$ dimensional. This is seen by looking at the row reduced form of the equations $A\mathbf{x} = 0$. Each pivot constrains one of the variables, leaving the other $n - k$ variables free. The number of pivots is also equal to the number of linearly independent columns of A, hence to the dimension of the column space of A, hence to the rank of L. Since there is at most one pivot per row, and at most one pivot per column, the rank can never be bigger than the smaller of m and n.

3.5. Kernels, Ranges, and Quotient Maps

In general, if V is n-dimensional and W is m-dimensional, then V and W are isomorphic to $\mathbb{R}^n$ and $\mathbb{R}^m$, and the isomorphism reduces L to multiplication by a matrix, so the exact same arguments apply. If the rank of L is k, then $Range(L)$ has dimension k and $Ker(L)$ has dimension $n - k$. L is 1-1 if (and only if) $k = n$, and L is onto if (and only if) $k = m$. In short, we have proven

Theorem 3.8 *Suppose L is a linear transformation from an n-dimensional space V to an m-dimensional space W.*

- *If $n < m$, then L is not onto.*

- *If $n > m$, then L is not 1-1.*

- *If $n = m$, then L is either an isomorphism (if the rank of L equals n) or is neither 1-1 nor onto (if the rank is less than n).*

Exercises

1. Show that the column space of a matrix A is the same as the span of the columns of A.

2. Let V be an n-dimensional vector space with basis $\mathcal{B}$, let W be an m-dimensional vector space with basis $\mathcal{D}$, and let $L : V \to W$ be a linear transformation. Show that the rank of the linear transformation L is the same as the number of pivots in the row-reduced form of the matrix $[L]_{\mathcal{DB}}$.

3. Consider the operator $L : \mathbb{R}^2 \to \mathbb{R}^2$, $L(\mathbf{x}) = \begin{pmatrix} 1 & 1 \\ 1 & 1 \end{pmatrix} \mathbf{x}$. Find the kernel and the range of L, and construct an explicit isomorphism between $\mathbb{R}^2/(\text{Ker}L)$ and $Range(L)$.

4. Consider the operator $L : \mathbb{R}^4 \to \mathbb{R}^3$, $L(\mathbf{x}) = \begin{pmatrix} 1 & 2 & 3 & 4 \\ 2 & 3 & 4 & 5 \\ 3 & 4 & 5 & 6 \end{pmatrix} \mathbf{x}$. Find a basis for $Ker(L)$ and a basis for $Range(L)$.

5. Let L be the operator on $M_{2,2}$ given by $L(A) = A + A^T$. Find bases for $Ker(L)$ and $Range(L)$. What is the rank of L?

6. Repeat Exercise 5 for 3×3 matrices. Can you generalize your results to $n \times n$ matrices?

7. Consider the operator $L : \mathbb{R}_3[t] \to \mathbb{R}^2$, $L(\mathbf{p}) = (p(0), p(1))^T$. Show that L is onto. What is the dimension of $Ker(L)$?

8. Consider the derivative operator $D : \mathbb{R}_4[t] \to \mathbb{R}_3[t]$, $D\mathbf{p}(t) = d\mathbf{p}(t)/dt$. What is the rank of D? What are the dimensions of the kernel and range of D? Is D 1-1? Onto?

9. Repeat Exercise 8, only with D now mapping $\mathbb{R}_4[t]$ to $R_4[t]$.

10. Let $C^\infty(\mathbb{R})$ denote the space of infinitely differentiable functions on the real line. Let $L = d^2/dt^2 + 3d/dt + 2$ be an operator on $C^\infty(\mathbb{R})$. Find a basis for Ker(L).

11. Let W_1 and W_2 be subspaces of V such that the dimensions of W_1 and W_2 add up to the dimension of V, and such that $W_1 \cap W_2 = \{0\}$. Show that V is the internal direct sum of W_1 and W_2. [Hint: Construct an isomorphism from the external direct sum of W_1 and W_2 to V.]

Chapter 4

An Introduction to Eigenvalues

In this chapter we define eigenvalues and eigenvectors, learn how to find them in simple cases, and learn some of their properties. This chapter is not about what eigenvalues are good for. That question is huge; most of the book is devoted to the answer.

4.1 Definitions and Examples

Definition *Let V be a vector space, L a linear operator on V, $\boldsymbol{\xi}$ a nonzero vector in V, and λ a scalar. If*

$$L(\boldsymbol{\xi}) = \lambda \boldsymbol{\xi}, \qquad (4.1)$$

then we say that λ is an eigenvalue *of L and that $\boldsymbol{\xi}$ is an* eigenvector. *We also refer to λ as the eigenvalue corresponding to $\boldsymbol{\xi}$, and to $\boldsymbol{\xi}$ as an eigenvector corresponding to λ.*

Definition *If A is an $n \times n$ matrix, then the eigenvalues and eigenvectors of A are defined to be the eigenvalues and eigenvectors of the linear operator $L(\mathbf{x}) = A\mathbf{x}$.*

Eigenvalues are usually denoted λ, in the same way that unknowns in high school algebra are usually called x.[1] The notation for eigenvectors is less standard; $\boldsymbol{\xi}$ is common but not universal.

---
[1] Eigenvalues were first discovered in connection with maximizing a function on a circle. In that setting, the letter λ actually referred to a Lagrange multiplier.

Chapter 4. An Introduction to Eigenvalues

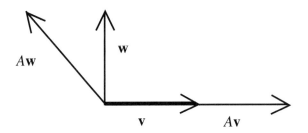

Figure 4.1: **v** is an eigenvector of A; **w** is not

The word *eigen* is German for "proper" or "characteristic", and some older texts refer to "proper values" and "proper vectors" or to "characteristic" values and vectors. These archaic usages seem to be dying out.

Here are some examples of eigenvalues and eigenvectors:

1. Let $A = \begin{pmatrix} 2 & 0 \\ 0 & 3 \end{pmatrix}$. Then $\mathbf{e}_1$ is an eigenvector of A with eigenvalue 2, and $\mathbf{e}_2$ is an eigenvector with eigenvalue 3, insofar as $A\mathbf{e}_1 = 2\mathbf{e}_1$ and $A\mathbf{e}_2 = 3\mathbf{e}_2$. The vector $\mathbf{v} = (1,1)^T$ is not an eigenvector, since $A\mathbf{v} = (2,3)^T$ is not a scalar multiple of $\mathbf{v}$.

2. Let $V = \mathbb{R}^2$, and let $L(\mathbf{x}) = 2\mathbf{x}$. Then every nonzero vector in V is an eigenvector of L with eigenvalue 2.

3. Let $A = \begin{pmatrix} 2 & 1 \\ 1 & 2 \end{pmatrix}$. Then $(1,1)^T$ is an eigenvector of A with eigenvalue 3, while $(1,-1)^T$ is an eigenvector of A with eigenvalue 1. (Check this!)

4. Let $A = \begin{pmatrix} 2 & -1 \\ 0 & 1 \end{pmatrix}$. Then $\mathbf{v} = (1,0)^T$ is an eigenvector with eigenvalue 2, while $\mathbf{w} = (0,1)^T$ is not an eigenvector. See Figure 4.1.

5. Let $V = \mathbb{R}[t]$, and let $L(\mathbf{p}) = t d\mathbf{p}/dt$. Since $L(t^k) = kt^k$, t^k is an eigenvector of L with eigenvalue k.

6. Let $V = C^\infty(\mathbb{R})$, the space of infinitely differentiable functions of a variable x, and let $L(f) = d^2f/dx^2$. Then $\sin(kx)$ is an eigenvector of L with eigenvalue $-k^2$, since $d^2(\sin(kx))/dx^2 = -k^2 \sin(kx)$. Similarly, $\cos(kx)$ is an eigenvector with eigenvalue $-k^2$, while $e^{\pm kx}$ are eigenvectors with eigenvalue $+k^2$.

4.2. Bases of Eigenvectors

7. Let V be physical 3-dimensional space, and let L be a rotation by 30° about the vertical axis. Since a vector pointing straight up is unchanged by this rotation, that vector is an eigenvector with eigenvalue 1.

Definition *If λ is an eigenvalue, the set of $\boldsymbol{\xi}$'s such that $L(\boldsymbol{\xi}) = \lambda \boldsymbol{\xi}$ is called the* eigenspace *corresponding to λ, and is denoted E_λ.*

Notice that the zero vector always satisfies $L(\boldsymbol{\xi}) = \lambda \boldsymbol{\xi}$, regardless of λ. This is why we required $\boldsymbol{\xi}$ to be nonzero in the definition of eigenvalues and eigenvectors. As a result, the zero vector is not an eigen*vector* for any eigenvalue, but is in the eigen*space* of every eigenvalue. Thus E_λ and the set of eigenvectors corresponding to λ are slightly different sets; the first contains the zero vector, but the second does not. In practice, most people ignore this distinction and use "set of eigenvectors" and "eigenspace" interchangably.

Exercises

1. Show that E_λ is a subspace (not just a subset) of V.
2. Let $A = \begin{pmatrix} 1 & 2 \\ 5 & -2 \end{pmatrix}$. $\lambda = 3$ is an eigenvalue of A. Find a corresponding eigenvector.
3. Let $A = \begin{pmatrix} 0 & 1 & 1 \\ 1 & 0 & 1 \\ 1 & 1 & 0 \end{pmatrix}$. Find a basis for E_2 and a basis for E_{-1}.
4. On $C^\infty(\mathbb{R})$, consider the operator $L = d^2/dx^2$. Find a basis for E_0.
5. On $C^\infty(\mathbb{R})$, let $L = d^2/dx^2 + d/dx$. Find a basis for E_0 and a basis for E_2.
6. On $M_{2,2}$, let $L(A) = A^T$. Find a basis for E_1 and a basis for E_{-1}.
7. On $\mathbb{R}_2[t]$, let $L(\mathbf{p})(t) = \mathbf{p}(t) + \mathbf{p}(-t)$. Find a basis for E_2.

4.2 Bases of Eigenvectors

Suppose that the vector space V has a basis $\mathcal{B} = \{\mathbf{b}_1, \ldots, \mathbf{b}_n\}$ consisting entirely of eigenvectors of the linear operator L. That is, $L(\mathbf{b}_i) = \lambda_i \mathbf{b}_i$, where $\{\lambda_1, \ldots, \lambda_n\}$ are eigenvalues of L. As usual, we expand a typical element $\mathbf{v}$ of V as

$$\mathbf{v} = a_1 \mathbf{b}_1 + \cdots + a_n \mathbf{b}_n, \tag{4.2}$$

or equivalently write $[\mathbf{v}]_\mathcal{B} = (a_1, \ldots, a_n)^T$. Then
$$\begin{aligned} L(\mathbf{v}) &= a_1 L(\mathbf{b}_1) + \cdots + a_n L(\mathbf{b}_n) \\ &= a_1 \lambda_1 \mathbf{b}_1 + \cdots a_n \lambda_n \mathbf{b}_n. \end{aligned} \quad (4.3)$$

That is, $[L(\mathbf{v})]_\mathcal{B}$ is simply $(\lambda_1 a_1, \ldots, \lambda_n a_n)^T$. No matter how complicated V and L are, once we express things in the right basis the action of L reduces to simply multiplying the i-th coefficient a_i by λ_i. In other words, we have proven

Theorem 4.1 *If the vector space V has a basis $\mathcal{B} = \{\mathbf{b}_1, \ldots, \mathbf{b}_n\}$ consisting of eigenvectors of the linear operator L, with eigenvalues $(\lambda_1, \ldots, \lambda_n)$, then*

$$[L]_\mathcal{B} = \begin{pmatrix} \lambda_1 & 0 & \cdots & 0 \\ 0 & \lambda_2 & \cdots & 0 \\ \vdots & \vdots & \ddots & 0 \\ 0 & 0 & \cdots & \lambda_n \end{pmatrix}. \quad (4.4)$$

Because of Theorem 4.1, finding all the eigenvalues and eigenvectors of a linear operator is sometimes called *diagonalizing* the operator.

Now would be a good time to review Chapter 1 with eigenvalues and eigenvectors in mind. We decoupled the equations (1.20) by finding the eigenvectors of the matrix A of (1.21). The eigenvectors were $\mathbf{b}_1 = (1,1)^T$ and $\mathbf{b}_2 = (1,-1)^T$, with eigenvalues $\lambda_1 = 3$ and $\lambda_2 = 1$. The numbers y_1 and y_2 were just the coefficients of $\mathbf{x}$ with respect to the $\mathcal{B}$ basis. That is, $\mathbf{y} = [\mathbf{x}]_\mathcal{B}$. Multiplying $\mathbf{x}$ by A was equivalent to multiplying y_i by λ_i, and the equations expressed in terms of $\mathbf{y}$ were easy to solve.

Had A been a different $m \times m$ matrix, with (1.20) a correspondingly different set of coupled equations, the same method would have worked, as long as A had enough eigenvectors to form a basis $\mathcal{B}$ of $\mathbb{R}^m$. We still would have let $\mathbf{y} = [\mathbf{x}]_\mathcal{B}$, and the coupled equations for $\mathbf{x}$ would still have turned into decoupled equations

$$\mathbf{y}(n) = \begin{pmatrix} \lambda_1 & 0 & \cdots & 0 \\ 0 & \lambda_2 & \cdots & 0 \\ \vdots & \vdots & \ddots & 0 \\ 0 & 0 & \cdots & \lambda_m \end{pmatrix} \mathbf{y}(n-1), \quad (4.5)$$

or equivalently
$$y_i(n) = \lambda_i y_i(n-1), \quad (4.6)$$

with solutions
$$y_i(n) = \lambda_i^n y_i(0). \tag{4.7}$$
The only difference would be that converting from $\mathbf{x}(0)$ to $\mathbf{y}(0)$, and from $\mathbf{y}(n)$ to $\mathbf{x}(n)$, could be messier than in the simple example (1.21).

We will return to applications of eigenvalues in Chapter 5. First we must learn how to actually find the eigenvalues and eigenvectors of a linear operator.

Exercises

1. Find a 2×2 matrix A whose eigenvalues are 3 and 7 and whose eigenvectors are $(1,1)^T$ and $(1,-1)^T$. [Hint: first find $[A]_\mathcal{B}$, where $\mathcal{B}$ is a basis of eigenvectors, and then convert to the standard basis for $\mathbb{R}^2$.]

2. Find a 3×3 matrix A whose eigenvalues are 1, 0 and -1, with eigenvectors $(1,1,1)^T$, $(1,1,0)^T$, and $(1,0,1)^T$, respectively.

3. Find the eigenvalues and corresponding eigenvectors of the matrix PDP^{-1}, where $P = \begin{pmatrix} 1 & 2 & 3 \\ 4 & 5 & 6 \\ 7 & 8 & 10 \end{pmatrix}$ and $D = \begin{pmatrix} 3 & 0 & 0 \\ 0 & -14 & 0 \\ 0 & 0 & 137 \end{pmatrix}$.

4. Suppose A is a 3×3 matrix, with eigenvalues 1, 2, and 3, and corresponding eigenvectors $\mathbf{b}_1$, $\mathbf{b}_2$, and $\mathbf{b}_3$. Suppose that $\mathbf{v} = \mathbf{b}_1 - 4\mathbf{b}_2 + 3\mathbf{b}_3$. Compute $A^5 \mathbf{v}$.

5. The matrix $A = \begin{pmatrix} 2 & -3 \\ 0 & -1 \end{pmatrix}$ has eigenvalues 2 and -1 and corresponding eigenvectors $(1,0)^T$ and $(1,1)^T$. Compute $A^{10}\mathbf{e}_2$.

6. Suppose we have a basis $\mathbf{b}_1, \ldots, \mathbf{b}_n$ of eigenvectors of A, with corresponding eigenvalues λ_i. Show that the matrix product $(A - \lambda_1 I) \cdots (A - \lambda_n I)$ equals the zero matrix.

4.3 Eigenvalues and the Characteristic Polynomial

As a warmup to working with a general vector space V and a general linear operator L, we first work with $V = \mathbb{R}^n$, with $L(\mathbf{x}) = A\mathbf{x}$ for some matrix A. As we have seen, this isn't really a restriction, because picking a basis (*any* basis) for a finite dimensional real vector space makes that space look like $\mathbb{R}^n$, and makes a linear operator look like matrix multiplication. Once we understand eigenvalues and eigenvectors of matrices, we can convert our results to understanding eigenvalues and eigenvectors of linear operators on general vector spaces.

We are trying to find solutions to

$$A\mathbf{x} = \lambda\mathbf{x}, \tag{4.8}$$

with $\mathbf{x}$ nonzero. However,

$$\lambda\mathbf{x} = \lambda I\mathbf{x}, \tag{4.9}$$

where I is the identity matrix, so we can rewrite (4.8) as

$$(\lambda I - A)\mathbf{x} = 0. \tag{4.10}$$

That is, we are looking for the values of λ for which $(\lambda I - A)$ is not 1-1. However, we know (Theorem 2.5) that a square matrix fails to be 1-1 if and only if its determinant is zero.

Definition *The* characteristic polynomial *of the matrix A is* $p_A(\lambda) = \det(\lambda I - A)$.

If A is an $n \times n$ matrix, then p_A is a polynomial in λ with leading term λ^n. We have shown that λ_0 is an eigenvalue if and only if $p_A(\lambda_0) = 0$. That is,

Theorem 4.2 *The eigenvalues of the matrix A are precisely the roots of the characteristic polynomial $p_A(\lambda)$.*

A 2×2 example: If

$$A = \begin{pmatrix} 2 & 1 \\ 1 & 2 \end{pmatrix}, \tag{4.11}$$

then

$$p_A(\lambda) = \det(\lambda I - A) = \begin{vmatrix} \lambda - 2 & -1 \\ -1 & \lambda - 2 \end{vmatrix} = \lambda^2 - 4\lambda + 3 = (\lambda - 3)(\lambda - 1). \tag{4.12}$$

Thus the eigenvalues of A are 3 and 1. To find the eigenvectors, we need to solve $A\mathbf{x} = \lambda\mathbf{x}$, or equivalently $(\lambda I - A)\mathbf{x} = 0$. For $\lambda = 3$, this is

$$\begin{pmatrix} 1 & -1 \\ -1 & 1 \end{pmatrix} \begin{pmatrix} x_1 \\ x_2 \end{pmatrix} = 0, \tag{4.13}$$

4.3. Eigenvalues and the Characteristic Polynomial

or equivalently $x_1 = x_2$. The eigenspace E_3 is all multiples of the basic eigenvector $(1,1)^T$. Similarly, for $\lambda = 1$ we get the equations $x_1 = -x_2$, and the eigenspace E_1 is all multiples of the basic eigenvector $(1,-1)^T$. (We could just as well have said that E_3 was all multiples of $(-1,-1)^T$, or of $(1/2, 1/2)^T$. In making a basis of eigenvectors we have a free choice as to scale. We found it convenient to take $\mathbf{b}_1 = (1,1)^T$ and $\mathbf{b}_2 = (1,-1)^T$, but rescaling these choices would not have any significant effect.)

A 3×3 example: If

$$A = \begin{pmatrix} 0 & 1 & 0 \\ 1 & 0 & 1 \\ 0 & 1 & 0 \end{pmatrix}, \tag{4.14}$$

then

$$p_A(\lambda) = \begin{vmatrix} \lambda & -1 & 0 \\ -1 & \lambda & -1 \\ 0 & -1 & \lambda \end{vmatrix} = \lambda^3 - 2\lambda = \lambda(\lambda^2 - 2). \tag{4.15}$$

The eigenvalues of A are the roots of $p_A(\lambda)$, namely 0, $\sqrt{2}$ and $-\sqrt{2}$. To find E_0, we solve $(0I - A)\mathbf{x} = 0$, or equivalently $A\mathbf{x} = 0$. The first row and the third row say $x_2 = 0$, while the second says $x_1 + x_3 = 0$, so E_0 is all multiples of $\mathbf{b}_1 = (1, 0, -1)^T$. To find $E_{\sqrt{2}}$ we solve

$$\begin{pmatrix} \sqrt{2} & -1 & 0 \\ -1 & \sqrt{2} & -1 \\ 0 & -1 & \sqrt{2} \end{pmatrix} \begin{pmatrix} x_1 \\ x_2 \\ x_3 \end{pmatrix} = 0. \tag{4.16}$$

The first row says that $x_2 = \sqrt{2}x_1$, and the third says that $x_2 = \sqrt{2}x_3$, so our solutions are all multiples of $\mathbf{b}_2 = (1, \sqrt{2}, 1)^T$. By a similar calculation, the eigenspace $E_{-\sqrt{2}}$ is all multiples of $\mathbf{b}_3 = (1, -\sqrt{2}, 1)^T$.

To sum up, we have developed a 3-step method for finding eigenvalues and eigenvectors of any square matrix A:

1. Write down $\lambda I - A$ and take its determinant to get the characteristic polynomial $p_A(\lambda)$.

2. Find the roots of $p_\lambda(A)$ to get the eigenvalues of A.

3. For each eigenvalue λ_i, the eigenspace E_{λ_i} is the set of solutions to $(\lambda_i I - A)\mathbf{x} = 0$. Solve these linear equations by Gaussian elimination or by inspection.

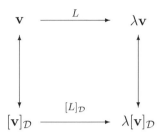

Figure 4.2: $L\mathbf{v} = \lambda\mathbf{v}$ if and only if $[L]_{\mathcal{D}}[\mathbf{v}]_{\mathcal{D}} = \lambda[\mathbf{v}]_{\mathcal{D}}$

This is an extremely effective method for diagonalizing small matrices by hand. 2×2 matrices are easily handled this way and 3×3 matrices are usually not too hard. Even 4×4 matrices can often be diagonalized this way. However, this method is not very good for handling anything 5×5 or larger. The first step, taking the determinant of a large matrix, is very time consuming. The second step, finding the roots of a large polynomial, can be even harder.

Fortunately, there are many sophisticated and efficient numerical algorithms for diagonalizing larger matrices. You may not know about these algorithms, but MATLAB and Maple (and other software packages) do, and you can take advantage of these built-in routines. It is important to understand the properties of eigenvalues and eigenvectors yourself, but you can often turn the task of finding the eigenvalues and eigenvectors over to a computer.

Now that we understand how to diagonalize operators on $\mathbb{R}^n$ (i.e., square matrices), we can turn our attention to linear operators on general vector spaces. As usual, we exploit the isomorphism between a general vector space and $\mathbb{R}^n$. See Figure 4.2.

Theorem 4.3 *Let V be a vector space with basis $\mathcal{D}$, and let L be a linear operator on V. A vector $\mathbf{v}$ is an eigenvector of L if and only if $[\mathbf{v}]_{\mathcal{D}}$ is an eigenvector of $[L]_{\mathcal{D}}$, and the corresponding eigenvalues are the same. In particular, the eigenvalues of the operator L are the same as the eigenvalues of the matrix $[L]_{\mathcal{D}}$.*

In other words, to diagonalize L, pick a basis $\mathcal{D}$ for V (any basis will do) and diagonalize the matrix $[L]_{\mathcal{D}}$. Then convert the eigenvectors of $[L]_{\mathcal{D}}$ to eigenvectors of L as follows: if $(a_1, \ldots, a_n)^T$ is an eigenvector of $[L]_{\mathcal{D}}$, then $a_1 \mathbf{d}_1 + \cdots + a_n \mathbf{d}_n$ is an eigenvector of L.

4.3. Eigenvalues and the Characteristic Polynomial

Proof: If $\mathbf{v}$ is an eigenvector of L with eigenvalue λ, then

$$[L]_\mathcal{D}[\mathbf{v}]_\mathcal{D} = [L(\mathbf{v})]_\mathcal{D} = [\lambda \mathbf{v}]_\mathcal{D} = \lambda[\mathbf{v}]_\mathcal{D}, \qquad (4.17)$$

so $[\mathbf{v}]_\mathcal{D}$ is an eigenvector of $[L]_\mathcal{D}$ with eigenvalue λ. Conversely, if $[\mathbf{v}]_\mathcal{D}$ is an eigenvector of $[L]_\mathcal{D}$ with eigenvalue λ, then

$$[L(\mathbf{v}) - \lambda \mathbf{v}]_\mathcal{D} = [L(\mathbf{v})]_\mathcal{D} - \lambda[\mathbf{v}]_\mathcal{D} = [L]_\mathcal{D}[\mathbf{v}]_\mathcal{D} - \lambda[\mathbf{v}]_\mathcal{D} = 0, \qquad (4.18)$$

so $L(\mathbf{v}) = \lambda \mathbf{v}$ and $\mathbf{v}$ is an eigenvector of L with eigenvalue λ. ∎

Example: On $\mathbb{R}_1[t]$, let $L(\mathbf{p}) = \mathbf{p}(0)(1-t) + \mathbf{p}(1)(2t+1)$. We will find the eigenvalues and eigenvectors of L. First we pick the standard basis $\mathcal{D} = \{1, t\}$ for V. We compute $L(\mathbf{d}_1) = L(1) = 1(1-t) + 1(2t+1) = 2+t$. $L(\mathbf{d}_2) = L(t) = 0(1-t) + 1(2t+1) = 2t+1$. Thus the matrix of L is

$$[L]_\mathcal{D} = \begin{pmatrix} [L\mathbf{d}_1]_\mathcal{D} & [L\mathbf{d}_2]_\mathcal{D} \end{pmatrix} = \begin{pmatrix} 2 & 1 \\ 1 & 2 \end{pmatrix}. \qquad (4.19)$$

But we already know that the eigenvalues of this matrix are 3 and 1, and that the corresponding eigenvectors are $(1,1)^T$ and $(1,-1)^T$. Thus the eigenvalues of L are 3 and 1, and the corresponding eigenvectors are $\mathbf{d}_1 + \mathbf{d}_2 = 1+t$ and $\mathbf{d}_1 - \mathbf{d}_2 = 1-t$. You should compute $L(1+t)$ and $L(1-t)$ yourself, to confirm that $1+t$ and $1-t$ are indeed eigenvectors with eigenvalues 3 and 1.

Conjugates of matrices

If $\mathcal{B}$ and $\mathcal{D}$ are two bases for a vector space V, then $[L]_\mathcal{B}$ and $[L]_\mathcal{D} = P_{\mathcal{DB}}[L]_\mathcal{B} P_{\mathcal{DB}}^{-1}$ must have the same eigenvalues, since both have the same eigenvalues as L. Furthermore, their eigenvectors are related to each other, since both are related to the eigenvectors of L. This is a special case of the following theorem, depicted in Figure 4.3.

Theorem 4.4 Let A be an $n \times n$ matrix, let P be an invertible $n \times n$ matrix, and let $B = PAP^{-1}$. Then $\mathbf{v}$ is an eigenvector of A with eigenvalue λ if, and only if, $P\mathbf{v}$ is an eigenvector of B with eigenvalue λ. In particular, A and B have the same eigenvalues.

Proof: If $A\mathbf{v} = \lambda \mathbf{v}$, then

$$B(P\mathbf{v}) = PAP^{-1}P\mathbf{v} = PA\mathbf{v} = P\lambda\mathbf{v} = \lambda(P\mathbf{v}). \qquad (4.20)$$

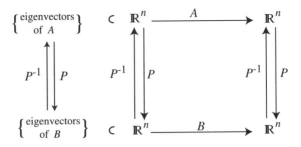

Figure 4.3: The eigenvectors of A and of $B = PAP^{-1}$

Conversely, if $B(P\mathbf{v}) = \lambda P\mathbf{v}$, then

$$A\mathbf{v} = P^{-1}PAP^{-1}P\mathbf{v} = P^{-1}BP\mathbf{v} = P^{-1}\lambda P\mathbf{v} = \lambda\mathbf{v}. \blacksquare \quad (4.21)$$

Since A and B have the same eigenvalues, it should be no surprise that they have the same characteristic polynomial. This can be seen directly:

$$\begin{aligned} p_B(\lambda) &= \det(\lambda I - B) = \det(P(\lambda I - A)P^{-1}) \\ &= \det P \det(\lambda I - A) \det P^{-1} = p_A(\lambda). \end{aligned} \quad (4.22)$$

Definition *If $B = PAP^{-1}$, we say that B is the* conjugate *of A by P, or that A and B are* conjugates.

Definition *If L is a linear operator on a finite dimensional space V, let $p_L(\lambda) = p_{[L]_\mathcal{B}}(\lambda)$ for any basis $\mathcal{B}$. By equation (4.22), this polynomial is the same for every basis. By Theorem 4.3, the eigenvalues of L are the roots of $p_L(\lambda)$.*

Theorem 4.5 *An $n \times n$ matrix A has n linearly independent eigenvectors if and only if it can be written as $A = PDP^{-1}$, where D is a diagonal matrix. In that case, the eigenvalues of A are the diagonal entries of D and the eigenvectors of A are the columns of P.*

Proof: By Theorem 4.1, if a matrix has n linearly independent eigenvectors, then it is conjugate to a diagonal matrix. For the converse, note that a diagonal matrix D has eigenvectors $\{\mathbf{e}_1, \ldots, \mathbf{e}_n\}$. If $A = PDP^{-1}$, then Theorem 4.4 implies that the i-th column of P, which equals $P\mathbf{e}_i$, is an eigenvector of A. Since P is invertible, the n columns of P are linearly independent. $\blacksquare$

4.3. Eigenvalues and the Characteristic Polynomial

Example: The eigenvectors of $\begin{pmatrix} 2 & 1 \\ 1 & 2 \end{pmatrix}$ are $(1,1)^T$ and $(1,-1)^T$, with eigenvalues 3 and 1. Therefore

$$\begin{pmatrix} 2 & 1 \\ 1 & 2 \end{pmatrix} = \begin{pmatrix} 1 & 1 \\ 1 & -1 \end{pmatrix} \begin{pmatrix} 3 & 0 \\ 0 & 1 \end{pmatrix} \begin{pmatrix} 1 & 1 \\ 1 & -1 \end{pmatrix}^{-1}. \qquad (4.23)$$

Exercises

1. Find, by hand, the eigenvalues of $A = \begin{pmatrix} 4 & 2 \\ -1 & 1 \end{pmatrix}$. For each eigenvalue, find a corresponding eigenvector. When you are done, check your results using the MATLAB eig command (or the corresponding command of another software package). You eigenvectors may look a little different from MATLAB's. Why? (For information on the "eig" command, type "help eig" at the MATLAB prompt.)

2. Check equation (4.23) explicitly.

For each of the following matrices, find the characteristic polynomial, find the eigenvalues, and find a basis for each eigenspace.

3. $\begin{pmatrix} 1 & 2 \\ 1 & 2 \end{pmatrix}$. **4.** $\begin{pmatrix} 3 & 4 \\ 4 & -3 \end{pmatrix}$. **5.** $\begin{pmatrix} 0 & 1 \\ 1 & 1 \end{pmatrix}$.

6. $\begin{pmatrix} 0 & 0 & 1 \\ 0 & 0 & 0 \\ 1 & 0 & 0 \end{pmatrix}$. **7.** $\begin{pmatrix} 1 & 2 & 3 \\ 2 & 4 & 6 \\ 3 & 6 & 9 \end{pmatrix}$. **8.** $\begin{pmatrix} 2 & 1 & 0 \\ 1 & 1 & 1 \\ 0 & 1 & 2 \end{pmatrix}$.

9. Show that the matrix $A = \begin{pmatrix} 0 & 1 \\ 0 & 0 \end{pmatrix}$ does not have two linearly independent eigenvectors. A matrix such as A is called "nondiagonalizable".

In Exercises 10–12, diagonalize the given operators on $\mathbb{R}_2[t]$.

10. $L(\mathbf{p})(t) = \mathbf{p}(3t)$.

11. $L(\mathbf{p})(t) = \mathbf{p}(1-t)$.

12. $L(\mathbf{p})(t) = \mathbf{p}(0) + \mathbf{p}(1)t + \mathbf{p}(2)t^2$.

13. Show that the matrices $\begin{pmatrix} 0 & 4 \\ 1 & 0 \end{pmatrix}$ and $\begin{pmatrix} 0 & 2 \\ 2 & 0 \end{pmatrix}$ are conjugate. How do their eigenvalues and eigenvectors compare?

14. Prove that, if a matrix A is diagonalizable, then $p_A(A) = 0$. [This result, known as the Cayley-Hamilton Theorem, is true even if A is not diagonalizable, but that case is more difficult to prove.]

68 Chapter 4. An Introduction to Eigenvalues

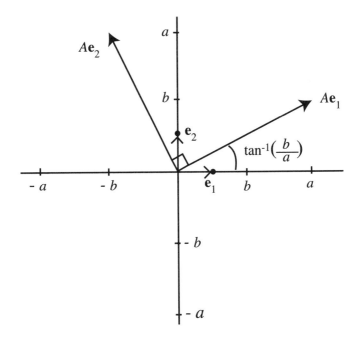

Figure 4.4: A has no real eigenvalues

4.4 The Need for Complex Eigenvalues

We have seen that the characteristic polynomial of an $n \times n$ real matrix is an n-th order real polynomial. Such a polynomial has exactly n roots (counting multiplicities), but some of the roots may be complex. We shall see that these complex eigenvalues can be treated exactly as before, with the added wrinkle that our eigenvectors will then also be complex. But there are no complex vectors in $\mathbb{R}^n$, so we are forced to expand our focus from $\mathbb{R}^n$ to $\mathbb{C}^n$.

As an example, consider the matrix

$$A = \begin{pmatrix} a & -b \\ b & a \end{pmatrix}, \qquad (4.24)$$

where a and b are real numbers. Multiplication by A sends $\mathbf{e}_1$ to $a\mathbf{e}_1 + b\mathbf{e}_2$, and sends $\mathbf{e}_2$ to $a\mathbf{e}_2 - b\mathbf{e}_1$, and so sends $(x, y)^T$ to $(ax - by, bx + ay)^T$. This is a rotation in the x-y plane by an angle $\tan^{-1}(b/a)$, followed by stretching by a factor of $\sqrt{a^2 + b^2}$. See Figure 4.4.

What about eigenvalues and eigenvectors? It is easy to see that A has no eigenvectors in $\mathbb{R}^2$ (unless $b = 0$). If $\mathbf{v}$ is a nonzero vector

4.4. The Need for Complex Eigenvalues

in $\mathbb{R}^2$, then $A\mathbf{v}$ makes an angle of $\tan^{-1}(b/a)$ with $\mathbf{v}$, and so cannot be a multiple of $\mathbf{v}$. However, we will soon see that the matrix A does have eigenvectors in $\mathbb{C}^2$.

We apply our 3-step method for diagonalizing A. The characteristic polynomial of A is

$$p_A(\lambda) = \begin{vmatrix} \lambda - a & b \\ -b & \lambda - a \end{vmatrix} = (\lambda - a)^2 + b^2, \qquad (4.25)$$

which has no real roots. However, $p_A(\lambda)$ has the complex roots $\lambda_1 = a + bi$ and $\lambda_2 = a - bi$. Next we find the eigenspaces. For E_{a+bi} we solve

$$\begin{pmatrix} bi & -b \\ b & bi \end{pmatrix} \mathbf{x} = 0. \qquad (4.26)$$

The solution is all multiples of $\mathbf{b}_1 = (1, -i)^T$ (or $(i, 1)^T$, if you prefer, since $i(1, -i)^T = (i, 1)^T$). Similarly E_{a-bi} is all multiples of $\mathbf{b}_2 = (1, i)^T$. It is easy to check that $\mathbf{b}_1$ and $\mathbf{b}_2$ are linearly independent, that $\mathbf{e}_1 = (\mathbf{b}_1 + \mathbf{b}_2)/2$, and that $\mathbf{e}_2 = i(\mathbf{b}_1 - \mathbf{b}_2)/2$.

In going from $\mathbf{e}_1$ and $\mathbf{e}_2$ to $\mathbf{b}_1$ and $\mathbf{b}_2$ we have made a subtle change in our space. While originally we only considered real linear combinations of $\mathbf{e}_1$ and $\mathbf{e}_2$, now we are considering complex linear combinations of $\mathbf{b}_1$ and $\mathbf{b}_2$, which is the same as taking complex linear combinations of $\mathbf{e}_1$ and $\mathbf{e}_2$. Our space has grown from $\mathbb{R}^2$ to $\mathbb{C}^2$. The matrix A does not have eigenvectors in $\mathbb{R}^2$ because the eigenvectors $\mathbf{b}_1$ and $\mathbf{b}_2$ happen to be complex. They live in $\mathbb{C}^2$, not $\mathbb{R}^2$.

A similar process applies to more general real vector spaces. If V is a real vector space with a basis $\mathcal{B}$, we define the space V_c, called the *complexification* of V, to be the set of all complex linear combinations of the elements of $\mathcal{B}$. V sits inside V_c the same way that $\mathbb{R}^n$ sits inside $\mathbb{C}^n$, as the set of vectors whose coefficients happen to all be real. Any vector $\mathbf{v}$ in V_c can be written as

$$\mathbf{v} = \mathbf{v}_R + i\mathbf{v}_I, \qquad (4.27)$$

where $\mathbf{v}_R$ and $\mathbf{v}_I$, called the real and imaginary parts of $\mathbf{v}$, are vectors in V. If L is an operator on V and λ_0 is a complex root of $p_L(\lambda)$, then the eigenvector corresponding to λ_0 is not in V, but is in V_c.

In practice, this expansion, from $\mathbb{R}^2$ to $\mathbb{C}^2$, or from V to V_c, makes very little difference. $\mathbb{R}^2$ is still there, sitting inside $\mathbb{C}^2$. Since A is a real matrix, multiplying a real vector by A gives another real vector. By working inside $\mathbb{C}^2$, however, we allow ourselves more eigenvectors, and therefore more calculational tools. In our example, $A^k\mathbf{e}_1$ is a real vector, but it is most easily computed using complex numbers:

$$\begin{aligned}A^k\mathbf{e}_1 &= A^k(\mathbf{b}_1+\mathbf{b}_2)/2 = ((a+bi)^k\mathbf{b}_1+(a-bi)^k\mathbf{b}_2)/2)\\ &= ((a+bi)^k+(a-bi)^k)\mathbf{e}_1/2 - i((a+bi)^k-(a-bi)^k)\mathbf{e}_2/2\\ &= Re[(a+bi)^k]\mathbf{e}_1 + Im[(a+bi)^k]\mathbf{e}_2.\end{aligned} \quad (4.28)$$

The real geometry of complex eigenvalues

We have seen that complex eigenvalues can be a useful computational tool, but that doesn't tell us what they mean. It turns out that complex eigenvalues have a simple interpretation in terms of 2-dimensional real geometry.

Let B be a real $n \times n$ matrix, and suppose that $a - bi$ is a complex root of $p_B(\lambda)$, hence an eigenvalue of B, with corresponding eigenvector $\mathbf{v} = \mathbf{v}_R + i\mathbf{v}_I$. Since complex roots of real polynomials come in pairs, $a+bi$ is also an eigenvalue of B, and the corresponding eigenvector is $\bar{\mathbf{v}} = \mathbf{v}_R - i\mathbf{v}_I$. Then

$$B\mathbf{v}_R = B(\mathbf{v}+\bar{\mathbf{v}})/2 = ((a-bi)\mathbf{v}+(a+bi)\bar{\mathbf{v}})/2 = a\mathbf{v}_R + b\mathbf{v}_I, \quad (4.29)$$

and

$$B\mathbf{v}_I = -iB(\mathbf{v}-\bar{\mathbf{v}})/2 = -i((a-bi)\mathbf{v}-(a+bi)\bar{\mathbf{v}})/2 = -b\mathbf{v}_R + a\mathbf{v}_I. \quad (4.30)$$

In other words B, acting on the $\mathbf{v}_R$-$\mathbf{v}_I$ plane, looks just like our matrix A acting on $\mathbb{R}^2$ (see Figure 4.5). Indeed, the matrix of the linear operator B, restricted to this plane and expressed in the $\{\mathbf{v}_R, \mathbf{v}_I\}$ basis, is precisely $\begin{pmatrix} a & -b \\ b & a \end{pmatrix}$. So it turns out that our matrix A isn't just a convenient example. It's at the heart of *every* example of a real matrix with complex eigenvalues.

To sum up, the geometry of real and complex eigenvalues is simple. If λ is a real eigenvalue of a linear operator L with eigenvector ξ, then L just stretches the line spanned by ξ by a factor λ. If $a \mp bi$ are a pair of complex eigenvalues with eigenvectors $\mathbf{v}_R \pm i\mathbf{v}_I$, there is a plane, spanned by $\mathbf{v}_R$ and $\mathbf{v}_I$, where L just stretches by $\sqrt{a^2+b^2}$ and rotates by an angle $\tan^{-1}(b/a)$. Remarkably, the action of every real diagonalizable matrix can be decomposed into these simple stretches and rotations.

4.4. The Need for Complex Eigenvalues

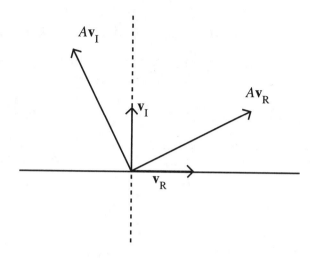

Figure 4.5: A rotates and stretches the $\mathbf{v}_R$-$\mathbf{v}_I$ plane.

Exercises

1. Check explicitly that the matrix A of (4.24) can be written as

$$A = \begin{pmatrix} 1 & 1 \\ -i & i \end{pmatrix} \begin{pmatrix} a+bi & 0 \\ 0 & a-bi \end{pmatrix} \begin{pmatrix} 1 & 1 \\ -i & i \end{pmatrix}^{-1}, \qquad (4.31)$$

as predicted by Theorem 4.5.

2. In the x-y plane, let L be rotation counterclockwise by a fixed angle θ. Find the matrix of L with respect to the standard basis, then find the eigenvalues and eigenvectors of L.

3. On $\mathbb{R}^3$, let $L\mathbf{x} = A\mathbf{x}$, where $\begin{pmatrix} 0 & 1 & 0 \\ 0 & 0 & 1 \\ 1 & 0 & 0 \end{pmatrix}$. Diagonalize L. Describe geometrically what L does to a vector in $\mathbb{R}^3$.

4. In problem 4.3, let $\mathbf{d}_1$ be the real eigenvector of A, and let $\mathbf{d}_2$ and $\mathbf{d}_3$ be the real and imaginary parts of one of the complex eigenvectors. Compute $[L]_\mathcal{D}$.

5. Diagonalize the matrix $\begin{pmatrix} 2 & 5 \\ -1 & 0 \end{pmatrix}$.

6. On $\mathbb{R}^2$, let $L\mathbf{x} = A\mathbf{x}$, where $A = \begin{pmatrix} 3 & 4 \\ -1 & 3 \end{pmatrix}$. Diagonalize L. Now let $\mathbf{d}_1$ and $\mathbf{d}_2$ be the real and imaginary parts of one of the eigenvectors of A. Compute $[L]_\mathcal{D}$.

7. Consider the matrices $\begin{pmatrix} 1 & -2 \\ 2 & 1 \end{pmatrix}$, $\begin{pmatrix} 1 & -4 \\ 1 & 1 \end{pmatrix}$ and $\begin{pmatrix} 1 & -1 \\ 4 & 1 \end{pmatrix}$.
Show that these matrices are all conjugate. How do the eigenvalues compare? How do the eigenvectors compare?

8. Is there a real matrix conjugate to $\begin{pmatrix} 3+2i & 0 \\ 0 & 3+2i \end{pmatrix}$?

9. Is there a real matrix conjugate to $\begin{pmatrix} 3+2i & 0 \\ 0 & 3-2i \end{pmatrix}$?

10. Diagonalize the matrix $\begin{pmatrix} 0 & 1 & 1 \\ -1 & 0 & 1 \\ -1 & -1 & 0 \end{pmatrix}$.

11. Diagonalize the matrix $\begin{pmatrix} -7 & 2 & -5 \\ 4 & -3 & 6 \\ 7 & 8 & -6 \end{pmatrix}$.

12. Diagonalize the matrix $\begin{pmatrix} 2 & 1 & 2 \\ -1 & 2 & 2 \\ -2 & -2 & 2 \end{pmatrix}$.

4.5 When is an Operator Diagonalizable?

We have seen that an operator L on a vector space V can be expressed as a diagonal matrix, or *diagonalized*, when we can find a basis of V consisting of eigenvectors of L. In this section we consider when that can be done.

To develop our intuition, we begin with an example of a matrix that cannot be diagonalized. Let

$$A = \begin{pmatrix} 0 & 1 \\ 0 & 0 \end{pmatrix}. \tag{4.32}$$

It is easy to see that $p_A(\lambda) = \lambda^2$, so the only eigenvalue is 0. However, solving $A\mathbf{v} = 0$ we see that E_0 is 1-dimensional, namely all multiples of $(1,0)^T$. Similarly the matrix $A+aI$, where a is any scalar, has a as its only eigenvalue, with multiples of $(1,0)^T$ as its only eigenvectors.

Such behavior almost never happens by accident. If you perturb a nondiagonalizable matrix, even slightly, the perturbed matrix is almost always diagonalizable. For example, let ϵ be a very small nonzero number (say, 10^{-3}). Then the matrix

$$A_\epsilon = \begin{pmatrix} 0 & 1 \\ \epsilon^2 & 0 \end{pmatrix} \tag{4.33}$$

is diagonalizable. See Figure 4.6 and Exercise 1.

4.5. When is an Operator Diagonalizable?

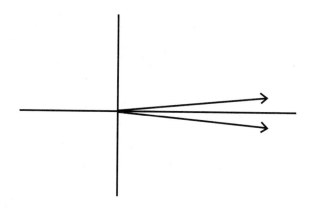

Figure 4.6: Eigenvectors of A_ϵ

A matrix that is very close to being nondiagonalizable is said to be *ill-conditioned*. When doing numerical analysis, ill-conditioned matrices can be big trouble, as small roundoff errors can cause big changes in coefficients.

In general, a matrix (or a linear operator) is diagonalizable if we can find enough linearly independent eigenvectors to span the vector space. It turns out that finding enough eigenvectors is sometimes hard, as in the example above, but that establishing linear independence is easy.

Theorem 4.6 *Let $b_1, \ldots, b_m$ be eigenvectors of L, corresponding to the eigenvalues $\lambda_1, \ldots, \lambda_m$, all of which are different. Then the b_i's are linearly independent.*

Proof: Suppose that the b_i's were linearly dependent. Let r be the smallest integer such that the set $\{b_1, \ldots, b_r\}$ is linearly dependent. That is, suppose we can write 0 as a nontrivial linear combination of $b_1, \ldots, b_r$, but not as a nontrivial combination of $b_1, \ldots, b_{r-1}$. Under these assumptions we will show how to write 0 as a nontrivial combination of $b_1, \ldots, b_{r-1}$, which is a contradiction that shows that r cannot exist, and thus that the b_i's are linearly independent.

By assumption, we can find a set of coefficients $\{a_i\}$, not all zero, such that

$$0 = a_1 b_1 + \cdots + a_r b_r. \tag{4.34}$$

If $a_r = 0$ we are done. Otherwise, at least one of the other coefficients is nonzero, since $a_r \mathbf{b}_r \neq 0$. But then

$$\begin{aligned} 0 &= (L - \lambda_r I) \sum_{i=1}^{r} a_i \mathbf{b}_i = \sum_{i=1}^{r} a_i (L - \lambda_r I) \mathbf{b}_i \\ &= \sum_{i=1}^{r} a_i (\lambda_i - \lambda_r) \mathbf{b}_i = \sum_{i=1}^{r-1} a_i (\lambda_i - \lambda_r) \mathbf{b}_i, \end{aligned} \quad (4.35)$$

since $a_r(\lambda_r - \lambda_r) = 0$. Since, for $i \neq r$, $\lambda_i - \lambda_r \neq 0$, at least one of the terms in the last sum is nonzero, which means we have written 0 as a nontrivial linear combination of $\mathbf{b}_1, \ldots, \mathbf{b}_{r-1}$. ■

Corollary 4.7 *If the characteristic polynomial $p_L(\lambda)$ has n distinct roots, where n is the dimension of the vector space V, then L is diagonalizable.*

Proof: If $p_L(\lambda)$ has n distinct roots, then we can find n eigenvectors, one for each eigenvalue. By Theorem 4.6, these eigenvectors are linearly independent and so form a basis. ■

Of course, $p_L(\lambda)$ is an nth order polynomial, and nth order polynomials always have n (possibly complex) roots, counting multiplicities. Typically these roots are all different. Corollary 4.7 tells us that trouble can only occur when two or more roots are the same.

Here we see the difference between the nondiagonalizable matrix A of (4.26) and the diagonalizable matrices A_ϵ. The polynomial $p_A(\lambda)$ had a double root at $\lambda = 0$. By perturbing A, we split the double root into two distinct roots $\lambda = \pm \epsilon$. By Corollary 4.7, the matrices A_ϵ are then diagonalizable.

Definition *The* algebraic multiplicity $m_a(\lambda_0)$ *of an eigenvalue λ_0 is the multiplicity of λ_0 as a root of $p_L(\lambda)$. That is, $p_L(\lambda)$ is divisible by exactly $m_a(\lambda_0)$ powers of $(\lambda - \lambda_0)$. If $m_a(\lambda_0) > 1$, then λ_0 is said to be a* degenerate *eigenvalue. Note that $n = \sum_i (m_a(\lambda_i))$. The* geometric multiplicity $m_g(\lambda_0)$ *is the dimension of E_{λ_0}. The* deficiency *of λ_0 is $m_a(\lambda_0) - m_g(\lambda_0)$. If an eigenvalue has nonzero deficiency, it is said to be* deficient. *An operator is said to be degenerate if one (or more) of its eigenvalues is degenerate, and is said to be deficient if one (or more) of its eigenvalues is deficient.*[2]

---

[2] Warning: Our usage of the term "degenerate" is standard in quantum mechanics. However, some mathematicians use the terms "multiple" or "repeated" instead, and use the term "degenerate" or "defective" to mean what we call "deficient".

4.5. When is an Operator Diagonalizable?

Theorem 4.8 *If λ_0 is an eigenvalue of L, then*

$$1 \leq m_g(\lambda_0) \leq m_a(\lambda_0). \tag{4.36}$$

Proof: Since λ_0 is an eigenvalue, it has an eigenvector, and E_{λ_0} is at least 1-dimensional. Now pick a basis $\mathbf{d}_1, \ldots, \mathbf{d}_{m_g(\lambda_0)}$ for E_{λ_0} and extend it to a basis $\mathbf{d}_1, \ldots, \mathbf{d}_n$ of V. Since, for $i \leq m_g(\lambda_0)$, $L\mathbf{d}_i = \lambda_0 \mathbf{d}_i$, $[L]_\mathcal{D}$ takes the form

$$[L]_\mathcal{D} = \begin{pmatrix} \lambda_0 I_{m_g} & A \\ 0 & B \end{pmatrix}, \tag{4.37}$$

where A is some $m_g(\lambda_0) \times (n - m_g(\lambda_0))$ matrix and B is some $(n - m_g(\lambda_0)) \times (n - m_g(\lambda_0))$ matrix. But then

$$p_L(\lambda) = p_{[L]_\mathcal{D}}(\lambda) = (\lambda - \lambda_0)^{m_g(\lambda_0)} p_B(\lambda). \tag{4.38}$$

Since $p_L(\lambda_0)$ is divisible by $(\lambda - \lambda_0)^{m_g(\lambda_0)}$, $m_a(\lambda_0)$ is at least $m_g(\lambda_0)$. ∎

We return to the general question of when a matrix (or operator) is diagonalizable. The complete answer is:

Theorem 4.9 *An operator L on a complex vector space is diagonalizable if, and only if, all of its eigenvalues have $m_a = m_g$.*

Proof: If all the eigenvalues have $m_a = m_g$, then let $\mathbf{b}_1, \ldots, \mathbf{b}_{m_g(\lambda_1)}$ be a basis for E_{λ_1}, let $\mathbf{b}_{m_g(\lambda_1)+1}, \ldots, \mathbf{b}_{m_g(\lambda_1)+m_g(\lambda_2)}$ be a basis for E_{λ_2}, and so on. I claim that the $\mathbf{b}_i$'s are linearly independent. Since there are $\sum m_g = \sum m_a = n$ of them, they then form a basis.

To see linear independence, suppose that

$$0 = a_1 \mathbf{b}_1 + \cdots + a_n \mathbf{b}_n. \tag{4.39}$$

Let $\mathbf{d}_1$ be the sum of the first $m_g(\lambda_1)$ terms, $\mathbf{d}_2$ the sum of the next $m_g(\lambda_2)$ terms, and so on. We then have $\sum_i \mathbf{d}_i = 0$. But by Theorem 4.6, the nonzero $\mathbf{d}_i$'s are linearly independent, and so cannot sum to zero. Thus all the $\mathbf{d}_i$'s must be zero. But if

$$0 = \mathbf{d}_1 = \sum_{i=1}^{m_g(\lambda_1)} a_i \mathbf{b}_i, \tag{4.40}$$

then $a_1, \ldots, a_{m_g(\lambda_1)}$ must all be zero, since the $\mathbf{b}_i$'s form a basis for E_{λ_1}. Similarly all the other a_i's must be zero, and we have shown that $\mathbf{b}_1, \ldots, \mathbf{b}_n$ are linearly independent vectors.

For the converse, note that we can only find $m_g(\lambda_i)$ linearly independent eigenvectors with eigenvalue λ_i. Thus a collection of linearly independent eigenvectors can have at most

$$\sum_i m_g(\lambda_i) = n - \sum_i (m_a(\lambda_i) - m_g(\lambda_i)) \qquad (4.41)$$

elements in it. If any eigenvalues are deficient, then this is less than n, and one cannot form a basis of eigenvectors. ∎

To sum up, most operators have distinct eigenvalues, and so are diagonalizable (Corollary 4.7). A few operators have degenerate eigenvalues. Of these, some have deficiencies and are nondiagonalizable, while others have $m_g = m_a$ and are diagonalizable (Theorem 4.9). We will see that many classes of operators (e.g., self-adjoint operators, or unitary operators) are always diagonalizable, whether they are degenerate or not. Operators that are not diagonalizable are treated in Section 4.9.

Exercises

1. Find the eigenvalues and eigenvectors of the matrix A_ϵ of (4.33). What happens to the eigenvectors as $\epsilon \to 0$? Decompose the vectors $(1,0)^T$ and $(0,1)^T$ as linear combinations of the eigenvectors of A_ϵ. What happens to the coefficients as $\epsilon \to 0$?

2. Ask MATLAB (or another computer program) to find the eigenvalues and eigenvectors of the matrix A (4.32). How do you interpret the result? Ask MATLAB to diagonalize A_ϵ for a very small value of ϵ. What is MATLAB trying to do with A? (This example should serve as a caution on the use of computers.)

For each of the following matrices, find the eigenvalues, and for each eigenvalue determine the algebraic and geometric multiplicities. Which matrices are diagonalizable?

3. $\begin{pmatrix} 2 & 2 \\ 0 & 2 \end{pmatrix}.$

4. $\begin{pmatrix} 3 & -1 \\ 1 & 1 \end{pmatrix}.$

5. $\begin{pmatrix} 3 & -1 \\ 0 & 1 \end{pmatrix}.$

6. $\begin{pmatrix} 2 & 0 \\ 0 & 2 \end{pmatrix}.$

7. $\begin{pmatrix} 3 & -1 & 0 \\ -1 & 3 & 0 \\ 1 & 1 & 4 \end{pmatrix}.$

8. $\begin{pmatrix} 2 & -1 & -1 \\ 0 & 3 & 1 \\ 2 & 1 & 5 \end{pmatrix}.$

9. $\begin{pmatrix} 4 & 1 & -1 \\ 0 & 3 & 1 \\ 2 & 1 & 5 \end{pmatrix}.$

4.6 Traces, Determinants, and Tricks of the Trade

In this section we cover some standard tricks for determining the eigenvalues of a linear operator. Each of these tricks has a limited range of applicability but, taken together, they can be very powerful. For the most part these tricks do not directly get you the eigenvectors. However, once the eigenvalues are known the eigenvectors can be found by solving $(\lambda I - A)\mathbf{x} = 0$.

Trick 1: The trace is the sum of the eigenvalues

Definition *The* trace *of a matrix A is the sum of its diagonal entries, and is denoted $Tr(A)$.*

A key property of the trace is that, if A and B are both $n \times n$ matrices, then $Tr(AB) = Tr(BA)$. This is called the *cyclic property of traces*. To see that, we compute

$$\begin{aligned}Tr(AB) &= \sum_i (AB)_{ii} = \sum_i \sum_j A_{ij} B_{ji} \\ &= \sum_{i,j} B_{ji} A_{ij} = \sum_j (BA)_{jj} = Tr(BA). \end{aligned} \quad (4.42)$$

(Actually, the formula $Tr(AB) = Tr(BA)$ is true even when A and B are not square, say if A is $n \times m$ and B is $m \times n$. However, this formula does *not* necessarily hold if A and B are of infinite size. See Section 3.4 for an example of the problems that may arise.)

We next consider the trace of a linear operator. If L is a linear operator on a vector space V, and if $\mathcal{B}$ and $\mathcal{D}$ are both bases for V, then we note that

$$\begin{aligned} Tr([L]_{\mathcal{D}}) &= Tr(P_{\mathcal{DB}}[L]_{\mathcal{B}} P_{\mathcal{DB}}^{-1}) = Tr((P_{\mathcal{DB}}[L]_{\mathcal{B}})(P_{\mathcal{DB}}^{-1})) \\ &= Tr((P_{\mathcal{DB}}^{-1})(P_{\mathcal{DB}}[L]_{\mathcal{B}})) = Tr([L]_{\mathcal{B}}). \end{aligned} \quad (4.43)$$

That is, the trace of a matrix of a linear operator does not depend on what basis we use to compute that matrix. Thus the following definition makes sense:

Definition *The trace of a linear operator L on a finite dimensional vector space V is the sum of the diagonal entries of $[L]_{\mathcal{B}}$, where $\mathcal{B}$ is any basis of V.*

Now suppose that $\mathcal{B}$ is a basis of eigenvectors. Then $[L]_\mathcal{B}$ is a diagonal matrix, whose diagonal entries are the eigenvalues $\lambda_1, \ldots, \lambda_n$. Thus the trace of $[L]_\mathcal{B}$, and hence the trace of L in *any* basis, is the sum of the eigenvalues.

Example: Consider the matrix $\begin{pmatrix} 1 & 2 \\ 2 & 4 \end{pmatrix}$. The trace of this matrix is $1 + 4 = 5$, so the two eigenvalues add up to 5. In fact, the eigenvalues are 0 and 5. Once you figure out that the first eigenvalue is 0 (say, by noticing that the matrix is not invertible), then the trace tells you the second eigenvalue.

Trick 2: The determinant is the product of the eigenvalues

Determinants follow a simple product rule: $\det(AB) = \det(A)\det(B)$. As a result, if $B = PAP^{-1}$, then

$$\det(B) = \det(P)\det(A)\det(P)^{-1} = \det(A). \qquad (4.44)$$

It follows that, if $[L]_\mathcal{B}$ and $[L]_\mathcal{D}$ are matrix representations of the same operator in different bases, then $\det([L]_\mathcal{B}) = \det([L]_\mathcal{D})$. Thus it makes sense to define the determinant of a linear operator.

Definition *The determinant of a linear operator L on a finite dimensional vector space A is the determinant of the matrix $[L]_\mathcal{B}$, where $\mathcal{B}$ is any basis of V.*

But now suppose, as with traces, that $\mathcal{B}$ is a basis of eigenvectors. Then $[L]_\mathcal{B}$ is a diagonal matrix, whose diagonal entries are the eigenvalues, and the determinant is the product of the eigenvalues.

This trick is most powerful when combined with the trick involving traces. Together they allow you to compute the last two eigenvalues once you have computed the first $n-2$.

Example: The matrix $\begin{pmatrix} 1 & 2 \\ 4 & -1 \end{pmatrix}$ has trace 0 and determinant -9.

Since the trace is zero, the sum of the eigenvalues is zero, so the eigenvalues must be negatives of each other. Their product is -9, so the eigenvalues must be 3 and -3.

Trick 3: Scaling and adding multiples of the identity

If A is an operator with eigenvalues $\lambda_1, \ldots, \lambda_n$ and eigenvectors $\mathbf{b}_1, \ldots, \mathbf{b}_n$, then $aA + cI$ is an operator with the exact same eigenvectors, and with eigenvalues $a\lambda_i + c$. For $(aA+cI)\mathbf{b}_i = aA\mathbf{b}_i + cI\mathbf{b}_i = a\lambda_i\mathbf{b}_i + c\mathbf{b}_i = (a\lambda_i + c)\mathbf{b}_i$.

Example: Once we understand that the eigenvalues of $\begin{pmatrix} 0 & 1 \\ 1 & 0 \end{pmatrix}$ are ± 1, with eigenvectors $(1, \pm 1)^T$, we can easily diagonalize the matrix $\begin{pmatrix} 5 & 3 \\ 3 & 5 \end{pmatrix}$. The eigenvalues are $5 + 3 = 8$ and $5 - 3 = 2$, and the eigenvectors are still $(1,1)^T$ and $(1,-1)^T$, respectively.

Trick 4: Block diagonal and block triangular matrices

Suppose a matrix M takes the block triangular form

$$M = \begin{pmatrix} A & B \\ 0 & C \end{pmatrix}, \tag{4.45}$$

where A and C are square matrices, B is a rectangular matrix, and 0 represents a rectangular block. Then

$$p_\lambda(M) = p_\lambda(A)p_\lambda(C). \tag{4.46}$$

The characteristic equation does not depend on B at all, so the eigenvalues of M are the eigenvalues of A and the eigenvalues of C. This can simplify your work considerably, as diagonalizing two smaller matrices (say, a 2×2 and a 3×3) is much easier than diagonalizing a big (say, 5×5) matrix.

Finding the eigenvectors is a little trickier. If $\mathbf{b}_i$ is an eigenvector of A with eigenvalue λ_i, then $\begin{pmatrix} \mathbf{b}_i \\ 0 \end{pmatrix}$ is an eigenvector of M with eigenvalue λ_i. However, if $\mathbf{d}_j$ is an eigenvector of C with eigenvalue λ_j, then $\begin{pmatrix} 0 \\ \mathbf{d}_j \end{pmatrix}$ is typically *not* an eigenvector of M. The eigenvector of M with eigenvalue λ_j is instead $\begin{pmatrix} \boldsymbol{\xi}_j \\ \mathbf{d}_j \end{pmatrix}$, where $\boldsymbol{\xi}_j$ and $\mathbf{d}_j$ satisfy

$$(\lambda_j I - A)\boldsymbol{\xi}_j = B\mathbf{d}_j. \tag{4.47}$$

If λ_j is not an eigenvalue of A, then $\lambda_j I - A$ is invertible and $\boldsymbol{\xi}_j$ is uniquely determined.

If λ_j is an eigenvalue of A, however, then $B\mathbf{d}_j$ may not lie in the range of $\lambda_j I - A$. If $B\mathbf{d}_j$ is not in the range, then $\boldsymbol{\xi}_j$ cannot exist, and the matrix M is not diagonalizable. In that case, λ_j is deficient.

Next, suppose that the matrix M' takes the form

$$M' = \begin{pmatrix} A' & 0 \\ B' & C' \end{pmatrix}. \tag{4.48}$$

In this case the analog to (4.46) still applies, and the eigenvalues of M' are the eigenvalues of A' and the eigenvalues of C'. In this case the eigenvectors corresponding to the eigenvalues of C' are easy to find; they are of the form $\begin{pmatrix} 0 \\ \mathbf{d}_j \end{pmatrix}$. However, the eigenvectors corresponding to the eigenvalues of A' are harder.

The easiest case is when our matrix is block diagonal, not just block triangular. That is, suppose that $\hat{M}$ takes the form

$$\hat{M} = \begin{pmatrix} \hat{A} & 0 \\ 0 & \hat{C} \end{pmatrix}. \tag{4.49}$$

Then, not only are the eigenvalues just those of $\hat{A}$ and $\hat{C}$, but the eigenvectors are all of the form $\begin{pmatrix} \mathbf{b}_i \\ 0 \end{pmatrix}$ or $\begin{pmatrix} 0 \\ \mathbf{d}_j \end{pmatrix}$. Our vector space splits as the direct sum of two pieces, with $\hat{A}$ acting on the first piece and $\hat{C}$ acting on the second piece. The zero matrices in the upper right and lower left indicate that neither piece influences the other.

Trick 5: Probability matrices and transposes

Suppose that the rows of a matrix A all add up to a certain number, say c. Then c is an eigenvalue of A, with eigenvector $(1, 1, \ldots, 1)^T$. This happens occasionally, and it is a good idea to look out for this.

Next, suppose that the *columns* of A all add up to a certain number, say c. Then c must be an eigenvalue of A^T. The characteristic polynomial of A^T is the same as that of A, so c must also be an eigenvalue of A. However, we do not have an immediate formula for the corresponding eigenvector.

This happens fairly often in probability theory. Suppose a system can be in one of several states (e.g. the weather might be rainy, cloudy, or sunny). If it is in state j today there is a probability A_{ij} of being in state i tomorrow. The importance of the matrix A is as follows. If we denote the probability of being in state i on day n as $x_i(n)$ and combine these into a vector $\mathbf{x}(n) = (x_1(n), \ldots, x_m(n))^T$, then the probability distribution evolves in time as $\mathbf{x}(n) = A\mathbf{x}(n-1)$. (We will see in Chapter 5 how the solution to such an equation

4.6. Traces, Determinants, and Tricks of the Trade

depends on the eigenvalues and eigenvectors of A.) Now suppose we are in state j today. Since we must be in *some* state tomorrow, we must have $\sum_i A_{ij} = 1$. In other words, the sum of the entries in the j-th column of A add up to 1, and so 1 is an eigenvalue of A. For this reason, a matrix whose columns all add up to 1 is called a *probabilty matrix*. Probability matrices are discussed further in Section 5.6.

Example: We combine the five tricks to diagonalize the matrix

$$M = \begin{pmatrix} 5 & 3 & 3 & 0 & 4 \\ -3 & -5 & -2 & 7 & 1 \\ 0 & 0 & 1 & 0 & 1 \\ 0 & 0 & 1 & 2 & -1 \\ 0 & 0 & 1 & 1 & 3 \end{pmatrix}. \tag{4.50}$$

This is a block triangular matrix of the form (4.45) with

$$A = \begin{pmatrix} 5 & 3 \\ -3 & -5 \end{pmatrix}; \quad B = \begin{pmatrix} 3 & 0 & 4 \\ -2 & 7 & 1 \end{pmatrix}; \quad C = \begin{pmatrix} 1 & 0 & 1 \\ 1 & 2 & -1 \\ 1 & 1 & 3 \end{pmatrix}.$$

$$\tag{4.51}$$

The matrix A has trace 0 and determinant -16, so its eigenvalues add up to 0 and have -16 as their product. They must be 4 and -4. The matrix C has trace 6 and determinant 6, so the eigenvalues add up to 6 and have product 6. Each column adds up to 3, so C is 3 times a probability matrix, and so has 3 as an eigenvalue. The remaining two eigenvalues must then add up to 3 and have product 2, so they must be 1 and 2. Putting this all together, the eigenvalues of M must be 4, -4, 3, 1, and 2.

As for the eigenvectors, note that the eigenvectors of A are $(3, -1)^T$ and $(1, -3)^T$, so $(3, -1, 0, 0, 0)$ and $(1, -3, 0, 0, 0)^T$ are eigenvectors of M (with eigenvalue 4 and -4, respectively). The eigenvectors of C are $(1, -1, 2)^T$, $(1, -1, 0)^T$ and $(1, -2, 1)^T$ (with eigenvalues 3, 1 and 2, respectively). These are extended by (4.47) to eigenvectors $\left(-\frac{67}{7}, \frac{19}{7}, 1, -1, 2\right)^T$, $\left(\frac{3}{5}, \frac{-9}{5}, 1, -1, 0\right)^T$ and $\left(\frac{-1}{3}, -2, 1, -2, 1\right)^T$ of M.

Exercises

1. Derive a formula for the eigenvectors of a lower triangular matrix M' of the form (4.48) corresponding to eigenvalues of A'.

Find the eigenvalues of the following matrices. You do not need to find the eigenvectors.

82   Chapter 4. An Introduction to Eigenvalues

2. $\begin{pmatrix} 5 & 11 \\ 1 & -5 \end{pmatrix}.$
3. $\begin{pmatrix} 3 & -1 \\ 6 & -2 \end{pmatrix}.$

4. $\begin{pmatrix} 2 & 1 \\ 5 & 0 \end{pmatrix}.$
5. $\begin{pmatrix} 1005 & 11 \\ 1 & 995 \end{pmatrix}.$

6. $\begin{pmatrix} 1 & 1 & 1 \\ 1 & 1 & 1 \\ 1 & 1 & 1 \end{pmatrix}.$
7. $\begin{pmatrix} 6 & 2 & 2 \\ 2 & 6 & 2 \\ 2 & 2 & 6 \end{pmatrix}.$

8. $\begin{pmatrix} 1 & 0 & 2 \\ 2 & 3 & -2 \\ 2 & 2 & 5 \end{pmatrix}.$

Find the eigenvalues and eigenvectors of the following matrices.

9. $\begin{pmatrix} 1 & 2 & 2 \\ 2 & 1 & -3 \\ 0 & 0 & 1 \end{pmatrix}.$
10. $\begin{pmatrix} 1 & 2 & 0 \\ 2 & 1 & 0 \\ 2 & -3 & 1 \end{pmatrix}$

11. $\begin{pmatrix} 1 & 2 & 1 & 2 \\ 1 & 0 & 3 & 4 \\ 0 & 0 & 2 & -1 \\ 0 & 0 & 2 & -1 \end{pmatrix}.$

4.7 Simultaneous Diagonalization of Two Operators

In many applications we need to diagonalize two different operators, call them A and B, on the same vector space V. The question is whether we can use the same set of eigenvectors for both operators. The answer is quite simple.

Definition *Finding a basis of vectors that are eigenvectors both of A and B is called* simultaneously diagonalizing A and B. *If this can be done the operators A and B are said to be* simultaneously diagonalizable. *Two operators A and B are said to* commute *if $AB = BA$.*

Theorem 4.10 *Two diagonalizable operators on a finite dimensional vector space are simultaneously diagonalizable if and only if they commute.*

Proof: One implication is easy to prove, the other more involved. Let A and B be diagonalizable operators on an n-dimensional vector

4.7. Simultaneous Diagonalization of Two Operators

space V, and suppose first that A and B are simultaneously diagonalizable. Then there exists a basis $\mathcal{B} = \{\mathbf{b}_i\}$ with $A\mathbf{b}_i = \lambda_i$ and $B\mathbf{b}_i = \mu_i$. Then

$$AB\mathbf{b}_i = A\mu_i\mathbf{b}_i = \mu_i A\mathbf{b}_i = \mu_i\lambda_i\mathbf{b}_i = \lambda_i\mu_i\mathbf{b}_i = \lambda_i B\mathbf{b}_i = B\lambda_i\mathbf{b}_i = BA\mathbf{b}_i. \tag{4.52}$$

Since AB acting on any basis vector is the same as BA acting on that basis vector, AB acting on any linear combination of basis vectors (i.e. on any vector at all) is the same as BA acting on that linear combination. So $AB = BA$.

For the converse, we begin with a lemma:

Lemma 4.11 *If A and B commute, then B maps each eigenspace of A to itself.*

Proof of Lemma: If $\mathbf{v}$ is an eigenvector of A with eigenvalue λ, then

$$A(B\mathbf{v}) = (AB)\mathbf{v} = BA\mathbf{v} = B\lambda\mathbf{v} = \lambda B\mathbf{v}, \tag{4.53}$$

so $B\mathbf{v}$ is also in the eigenspace E_λ. ∎

Returning to the proof of Theorem 4.10, we suppose that A and B are diagonalizable and commute, and that the eigenvalues of A are $\lambda_1, \ldots, \lambda_m$, with λ_i having multiplicity $m_g(\lambda_i) = m_a(\lambda_i)$. (The algebraic and geometric multiplicities are the same since A is diagonalizable.) To simultaneously diagonalize A and B, we pick a preliminary basis $\mathbf{d}_1, \ldots, \mathbf{d}_n$ of eigenvectors of A, with the first $m_g(\lambda_1)$ eigenvectors corresponding to eigenvalue λ_1, the next $m_g(\lambda_2)$ eigenvectors corresponding to λ_2, and so on. This need *not* be a basis of eigenvectors of B. However, by the lemma, the matrix of B in this basis takes the form

$$[B]_\mathcal{D} = \begin{pmatrix} B_1 & 0 & \cdots & 0 \\ 0 & B_2 & \cdots & 0 \\ \vdots & \vdots & \ddots & \vdots \\ 0 & 0 & \cdots & B_m \end{pmatrix}, \tag{4.54}$$

where each B_i is an $m_g(\lambda_i) \times m_g(\lambda_i)$ block.

Now recall Trick 4 of Section 6. Diagonalizing every block B_i is tantamount to diagonalizing $[B]_\mathcal{D}$, and hence to diagonalizing B. Moreover, since B *is* diagonalizable, that means each block B_i is diagonalizable. But diagonalizing B_i means finding $m_g(\lambda_i)$ linearly independent vectors, all in the eigenspace E_{λ_i}, that are eigenvectors of B. Those vectors are eigenvectors both of A (with eigenvalue

λ_i) and of B (with unspecified eigenvalue). All together, we have $m_g(\lambda_1) + \cdots + m_g(\lambda_m) = n$ such vectors, which together make a basis for V. ∎

Example 1: Consider the 4×4 matrices

$$A = \begin{pmatrix} 0 & 4 & 0 & 0 \\ 1 & 0 & 0 & 0 \\ 0 & 0 & 0 & 4 \\ 0 & 0 & 1 & 0 \end{pmatrix}; \quad B = \begin{pmatrix} 0 & 0 & 1 & 0 \\ 0 & 0 & 0 & 1 \\ 1 & 0 & 0 & 0 \\ 0 & 1 & 0 & 0 \end{pmatrix}. \quad (4.55)$$

It is not hard to check that $AB = BA$, so A and B should be simultaneously diagonalizable. We diagonalize A first. The eigenvalues of A are 2 and -2, and we pick eigenvectors $\mathbf{d}_1 = (2, 1, 0, 0)^T$ and $\mathbf{d}_2 = (0, 0, 2, 1)^T$ with eigenvalue 2, and $\mathbf{d}_3 = (2, -1, 0, 0)^T$ and $\mathbf{d}_4 = (0, 0, 2, -1)^T$ with eigenvalue -2. Relative to the $\mathcal{D}$ basis, B takes the block-diagonal form

$$[B]_\mathcal{D} = \begin{pmatrix} 0 & 1 & 0 & 0 \\ 1 & 0 & 0 & 0 \\ 0 & 0 & 0 & 1 \\ 0 & 0 & 1 & 0 \end{pmatrix}. \quad (4.56)$$

The eigenvectors of the block B_1 are $(1, 1)^T$ and $(1, -1)^T$, indicating that $\mathbf{d}_1 \pm \mathbf{d}_2$ are eigenvectors of B with eigenvalues ± 1. Similarly, $(1, 1)^T$ and $(1, -1)^T$ are eigenvectors of B_2, so $\mathbf{d}_3 \pm \mathbf{d}_4$ are eigenvectors of B (with eigenvalues ± 1). Thus we can take our basis for $\mathbb{R}^4$ to be

$$\begin{aligned}
\mathbf{b}_1 &= \mathbf{d}_1 + \mathbf{d}_2 = (2, 1, 2, 1)^T; & A\mathbf{b}_1 &= 2\mathbf{b}_1; & B\mathbf{b}_1 &= \mathbf{b}_1 \\
\mathbf{b}_2 &= \mathbf{d}_1 - \mathbf{d}_2 = (2, 1, -2, -1)^T; & A\mathbf{b}_2 &= 2\mathbf{b}_2; & B\mathbf{b}_2 &= -\mathbf{b}_2 \\
\mathbf{b}_3 &= \mathbf{d}_3 + \mathbf{d}_4 = (2, -1, 2, -1)^T; & A\mathbf{b}_3 &= -2\mathbf{b}_3; & B\mathbf{b}_3 &= \mathbf{b}_3 \\
\mathbf{b}_4 &= \mathbf{d}_3 - \mathbf{d}_4 = (2, -1, -2, 1)^T; & A\mathbf{b}_4 &= -2\mathbf{b}_4; & B\mathbf{b}_4 &= -\mathbf{b}_4.
\end{aligned} \quad (4.57)$$

Each of the $\mathbf{b}_i$'s is an eigenvector both of A and of B, so we have simultaneously diagonalized A and B.

Returning to the general case, once we have simultaneously diagonalized A and B, we also have diagonalized $A + B$, AB, and any other reasonable function of A and B. For example, if $\mathbf{b}_i$ is an eigenvector of A with eigenvalue λ_i and an eigenvector of B with eigenvalue μ_i, then it is also an eigenvector of $A^2 + AB$ with eigenvalue $\lambda_i^2 + \lambda_i \mu_i$ and an eigenvector of $(A + B)^3$ with eigenvalue $(\lambda_i + \mu_i)^3$.

This has important applications to differential equations, where we often have to solve $C\mathbf{v} = 0$, where C is some differential operator

4.7. Simultaneous Diagonalization of Two Operators

and **v** is a function. If we can factor $C = AB$, where $AB = BA$, then we are just looking for simultaneous eigenvectors of A and B for which $\lambda_i \mu_i = 0$. But that means that $\lambda_i = 0$ or $\mu_i = 0$. So the space of solutions to $AB\mathbf{v} = 0$ is just the sum of the space of solutions to $A\mathbf{v} = 0$ and the space of solutions to $B\mathbf{v} = 0$. To put it another way, if $AB\mathbf{v} = 0$, then **v** can be written as $\mathbf{x} + \mathbf{y}$, where $A\mathbf{x} = B\mathbf{y} = 0$.

Example 2: Consider the ordinary differential equation

$$\frac{d^2 f}{dt^2} + 5\frac{df}{dt} + 6f = 0. \tag{4.58}$$

The left hand side can be rewritten as $(D+3)(D+2)f$, where $D = d/dt$. Since $(D+3)(D+2) = (D+2)(D+3)$, all we need to do is to solve $(D+2)f = 0$ and to solve $(D+3)f = 0$. These are first-order equations, and are easy to solve. The solutions to the first equation are of the form $c_1 e^{-2t}$, the solutions to the second are of the form $c_2 e^{-3t}$, so the most general solution to (4.58) is

$$f(t) = c_1 e^{-2t} + c_2 e^{-3t}. \tag{4.59}$$

Example 3: Consider the partial differential equation

$$\frac{\partial^2 f}{\partial x \partial y} + \frac{\partial f}{\partial y} + \frac{\partial f}{\partial x} + f = 0. \tag{4.60}$$

This can be written as $ABf = 0$, where $A = \frac{\partial}{\partial y} + 1$ and $B = \frac{\partial}{\partial x} + 1$ are commuting operators. The solutions to $Af = 0$ are of the form $f(x,y) = e^{-y} g(x)$, where g is an arbitrary function of x, and the solutions to $Bf = 0$ are of the form $f(x,y) = e^{-x} h(y)$, where h is an arbitrary function of y. Thus the general solution to (4.60) is

$$f(x,y) = e^{-y} g(x) + e^{-x} h(y), \tag{4.61}$$

where g and h are arbitrary functions.

Strictly speaking, examples 2 and 3 are not applications of Theorem 4.10, since Theorem 4.10 as stated only applies to finite dimensional vector spaces. However, as with most theorems in this course, there is a natural extension of this theorem to infinite dimensions.

Exercises

Check if each pair of matrices commute. If so, simultaneously diagonalize them.

1. $A = \begin{pmatrix} 1 & 0 & 0 \\ 0 & 1 & 0 \\ 0 & 0 & 2 \end{pmatrix}$, $\quad B = \begin{pmatrix} 1 & 1 & 0 \\ 1 & 1 & 0 \\ 0 & 0 & 2 \end{pmatrix}$.

2. $A = \begin{pmatrix} 2 & 0 & -1 \\ 0 & 1 & 0 \\ -1 & 0 & 2 \end{pmatrix}$, $\quad B = \begin{pmatrix} 1 & 2 & -1 \\ 1 & 0 & 1 \\ -1 & 2 & 1 \end{pmatrix}$.

3. $A = \begin{pmatrix} 1 & 1 & 1 \\ 0 & 2 & 1 \\ 1 & -1 & 3 \end{pmatrix}$, $\quad B = \begin{pmatrix} 0 & 1 & 4 \\ 1 & 0 & 4 \\ 3 & -3 & 5 \end{pmatrix}$.

4. $A = \begin{pmatrix} 1 & 2 & 3 \\ 2 & 3 & 1 \\ 3 & 1 & 2 \end{pmatrix}$, $\quad B = \begin{pmatrix} 0 & 1 & 0 \\ 1 & 0 & 0 \\ 0 & 0 & 1 \end{pmatrix}$.

5. $A = \begin{pmatrix} -1 & -2 \\ 1 & 2 \end{pmatrix}$, $\quad B = \begin{pmatrix} 2-e & 2-2e \\ e-1 & 2e-1 \end{pmatrix}$.

6. $A = \begin{pmatrix} 1 & 2 \\ 3 & 4 \end{pmatrix}$, $\quad B = \begin{pmatrix} 3 & 1 \\ 4 & 0 \end{pmatrix}$.

7. Prove the following extension of Theorem 4.10: Three diagonalizable operators on a finite-dimensional vector space are simultaneously diagonalizable if and only if each pair of operators commutes.

8. Extend Exercise 7 to cover an arbitrary number of operators.

4.8 Exponentials of Complex Numbers and Matrices

In Chapter 5 and beyond we will see that solutions to differential equations involving matrices often involve the exponentials of the eigenvalues. Since eigenvalues can be complex, we must make sense of exponentials of complex numbers. In the process we will also make sense of the exponential of a matrix, *any* function of a diagonalizable matrix, and any *analytic* function of an arbitrary matrix.

When x is a real number, there are three standard definitions of e^x:

4.8. Exponentials of Complex Numbers and Matrices

- e^x equals the number $e = 2.71828\ldots$ multiplied by itself x times.

- e^x is the sum of the infinite power series

$$1 + x + x^2/2 + \cdots + x^n/n! + \cdots \quad (4.62)$$

- The exponential is the inverse of the natural log, so $y = e^x$ means $x = \ln(y) = \int_1^y dt/t$. This is equivalent to saying that $dx/dy = 1/y$, so $y = e^x$ is the unique solution to the differential equation $dy/dx = y$ with initial condition $y(0) = 1$.

The first definition is easy to understand, at least when x is an integer, and quickly yields the important property

$$e^{a+b} = e^a e^b. \quad (4.63)$$

However, this definition does not make sense when x isn't an integer. It is possible to extend the definition to cover rational exponentials — $e^{p/q}$ is the positive q-th root of e^p — but this still leaves us wondering what $e^{\sqrt{2}}$ might mean.

The second definition applies to all real numbers, and it gives a practical means of computing e^x. The factorials in the denominators grow so quickly that the power series converges for arbitrarily large values of x, and converges extremely quickly for moderate values of x. However, it is not immediately clear that e^2 is the same thing as e times e.

Our theoretical understanding of the exponential, and in particular the property (4.63) for all real numbers, comes from the third definition. Taking the derivative of each term in (4.62), we see that the power series satisfies the differential equation, and therefore the third definition is equivalent to the second.

Next we turn our attention to the exponential of a complex number. The first definition ("multiply e by itself $a + bi$ times") makes no sense at all, but the second and third definitions carry over with only minor modification. Let z be a complex number and let t be a real number. We define

- $e^z = \sum_{n=0}^{\infty} z^n/n!$.

- $y = e^{zt}$ is the unique solution to the differential equation $dy/dt = zy$ with initial condition $y(0) = 1$ (so in particular $e^z = y(1)$.)

Taking the derivative of the power series for e^{zt}, term by term, we get a solution to the differential equation, so these definitions once again are equivalent. From the differential equation, one can again check that (4.63) holds.

Using trigonometric functions and real exponentials, we already know a solution to $dy/dt = (a+bi)y$. You should check that the derivative of

$$y = e^{at}(\cos(bt) + i\sin(bt)) \qquad (4.64)$$

with respect to t is

$$dy/dt = (a+bi)e^{at}(\cos(bt) + i\sin(bt)), \qquad (4.65)$$

and that $e^0(\cos(0) + i\sin(0)) = 1$. Thus

$$\exp((a+bi)t) = e^{at}(\cos(bt) + i\sin(bt)). \qquad (4.66)$$

Setting $t = 1$, we then have the important formula

$$e^{a+bi} = e^a(\cos(b) + i\sin(b)), \qquad (4.67)$$

and in particular Euler's formula

$$e^{i\theta} = \cos(\theta) + i\sin(\theta). \qquad (4.68)$$

This last result may also be derived directly from the power series:

$$\begin{aligned} e^{i\theta} &= 1 + i\theta + (i\theta)^2/2 + \cdots \\ &= (1 - \theta^2/2 + \theta^4/4! - \theta^6/6! + \cdots) + i(\theta - \theta^3/3! + \theta^5/5! - \cdots) \\ &= \cos(\theta) + i\sin(\theta). \end{aligned} \qquad (4.69)$$

Now we proceed to exponentials of matrices. If A is a square matrix, we know what integer powers of A mean: A^n is A multiplied by itself n times. From that we can define

$$e^A = \sum_{n=0}^{\infty} \frac{A^n}{n!} \qquad (4.70)$$

We can also consider the unique solution to the matrix-valued differential equation

$$dY/dt = AY; \qquad Y(0) = I, \qquad (4.71)$$

and define $e^{At} = Y(t)$. Once again, plugging the power series for e^{At} into the differential equation shows that these two definitions are equivalent.

4.8. Exponentials of Complex Numbers and Matrices

Although the exponentials of matrices make sense, property (4.63) does not hold in general. Because matrices do not generally commute, $e^A e^B$ and $e^B e^A$ are typically different matrices, with neither one equal to e^{A+B}.

Fact: If $AB = BA$, then $e^A e^B = e^B e^A = e^{A+B}$. Otherwise these three terms are generally all different.

If two matrices are conjugates, then their powers, and their exponentials, are also conjugates. If $B = PAP^{-1}$, then

$$B^n = PAP^{-1}PAP^{-1}PAP^{-1}\cdots PAP^{-1} = PA^n P^{-1}, \quad (4.72)$$

since all powers of P except the first P and the last P^{-1} cancel in pairs. Similarly,

$$e^B = \sum_{n=0}^{\infty} \frac{B^n}{n!} = \sum_{n=0}^{\infty} \frac{PA^n P^{-1}}{n!} = Pe^A P^{-1}. \quad (4.73)$$

This, together with Exercise 2, below, gives us yet another way to compute matrix exponentials. Suppose that A is diagonalizable. That is, $A = PDP^{-1}$, where D is diagonal with entries $\lambda_1, \ldots, \lambda_n$. Then

$$e^A = Pe^D P^{-1} = P \begin{pmatrix} e^{\lambda_1} & \cdots & & 0 \\ 0 & e^{\lambda_2} & \cdots & 0 \\ \vdots & \vdots & \ddots & 0 \\ 0 & 0 & \cdots & e^{\lambda_n} \end{pmatrix} P^{-1}. \quad (4.74)$$

That is, the eigenvectors of e^A are the same as the eigenvectors of A, while the eigenvalues of e^A are the exponentials of the eigenvalues of A.

The exponential of a linear operator is defined almost identically. If L is a linear operator, then L^2 is the operator L applied twice, L^3 is L applied 3 times, and so on. As before, we can define e^L by a power series,

$$e^L = \sum_{n=0}^{\infty} \frac{L^n}{n!} \quad (4.75)$$

and we can define e^{Lt} as the solution to the operator-valued differential equation

$$dY/dt = L \circ Y; \quad Y(0) = I. \quad (4.76)$$

Once again, plugging the power series into the differential equation shows that these definitions are equivalent. The operator e^L can also be described in terms of eigenvalues and eigenvectors. The eigenvectors of e^L are the same as the eigenvectors of L, while the eigenvalues of e^L are the exponentials of the eigenvalues of L.

These results suggest a means of defining almost any function of a diagonalizable matrix (or operator). Let $f(z)$ be any function of a complex variable z. Then, for $A = PDP^{-1}$ diagonalizable, we define

$$f(A) = P \begin{pmatrix} f(\lambda_1) & \cdots & & 0 \\ 0 & f(\lambda_2) & \cdots & 0 \\ \vdots & \vdots & \ddots & 0 \\ 0 & 0 & \cdots & f(\lambda_n) \end{pmatrix} P^{-1}. \qquad (4.77)$$

If $f(z)$ happens to be an analytic function, that is, a function that can be expressed as a convergent power series in z,

$$f(z) = a_0 + a_1 z + a_2 z^2 + \cdots, \qquad (4.78)$$

then one can define $f(A)$ by the same power series,

$$f(A) = a_0 + a_1 A + a_2 A^2 + \cdots. \qquad (4.79)$$

The first definition (4.77) of $f(A)$ allowed $f(z)$ to be any function, but required A to be diagonalizable. The second definition (4.79) requires $f(z)$ to be analytic, but puts no restrictions of A. When A is diagonalizable and f is analytic the two definitions agree, as can be seen by plugging $A = PDP^{-1}$ into (4.79) and making use of (4.72). In practice, both definitions are important.

Exercises

1. Let $A = \begin{pmatrix} 0 & -1 \\ 1 & 0 \end{pmatrix}$. Compute e^{At} from the power series. (You may need to remember the Taylor series for sine and cosine). Check that it satisfies the differential equation.

2. If D is a diagonal matrix with diagonal entries $\lambda_1, \ldots, \lambda_n$, show that e^{Dt} is a diagonal matrix with entries $e^{\lambda_1 t}, \ldots, e^{\lambda_n t}$.

3. Let $\mathbf{v}$ be an eigenvector of an operator L, with eigenvalue λ. Show that $\mathbf{v}$ is an eigenvector of e^L with eigenvalue e^λ.

4. Let $\mathcal{B}$ be a basis for a vector space and let L be an operator on that space. Let $A = [L]_\mathcal{B}$. Show that $[e^L]_\mathcal{B} = e^A$.

5. Compute the exponential of $\begin{pmatrix} 0 & -t \\ t & 0 \end{pmatrix}$ from equation (4.74), and compare to the results of Exercise 1.

6. Using (4.74), compute the exponential of $\begin{pmatrix} a & -b \\ b & a \end{pmatrix}$.

7. Let $A = \begin{pmatrix} 0 & 4 \\ 1 & 0 \end{pmatrix}$. Compute the absolute value of A. Notice that this is not at all the same thing as taking the absolute value of each entry of A.

8. Compute $\exp\begin{pmatrix} 0 & 1 \\ 0 & 0 \end{pmatrix}$. Since the matrix is not diagonalizable, you must use the power series.

9. Compute $\exp(A_\epsilon)$ by diagonalization, where A_ϵ is given by (4.33). Take the limit as $\epsilon \to 0$ and compare your result to the answer to Exercise 8.

10. Compute $\cos(A)$, where $A = \begin{pmatrix} \pi/4 & \pi/4 \\ \pi & \pi/4 \end{pmatrix}$.

11. Show that, if A is diagonalizable, then $\sin^2(A) + \cos^2(A) = I$. Does this identity also hold for nondiagonalizable matrices?

12. Show that there does not exist a matrix A such that $\sin(A) = \begin{pmatrix} 1 & 1 \\ 0 & 1 \end{pmatrix}$. [Hint: use the result of Exercise 11.]

4.9 Power Vectors and Jordan Canonical Form

Although most operators are diagonalizable, some are not. In this section we show how to analyze such nondiagonalizable operators. We define a generalization of eigenvectors, called *power vectors*.[3]. For most applications (e.g. differential equations), power vectors are almost as handy as eigenvectors, and every operator on a finite dimensional complex vector space V, diagonalizable or not, has enough power vectors to form a basis for V.

Definition *If a nonzero vector $\boldsymbol{\xi}$ has the property that*

$$(L - \lambda I)^p \boldsymbol{\xi} = 0, \qquad (4.80)$$

for some scalar λ and some positive integer p, then we say that $\boldsymbol{\xi}$ is a power vector *corresponding to the eigenvalue λ. The subspace*

---

[3] Power vectors are more frequently called *generalized eigenvectors* However, in Chapter 9 we will encounter a very different generalization of eigenvectors. To avoid confusion, we will use "power vector" in the present context and save "generalized eigenvector" for Chapter 9.

spanned by the power vectors corresponding to λ is called the power space *of* λ *and is denoted* $\tilde{E}_\lambda$. *If* $(L - \lambda I)^p \boldsymbol{\xi} = 0$ *but* $(L - \lambda I)^{p-1} \neq 0$, *we say that the power vector* $\boldsymbol{\xi}$ *has order p.*

Notice that a power vector of order 1 is the same thing as an eigenvector. Also notice that, if $\boldsymbol{\xi}$ is a power vector of order p, then $(L - \lambda I)\boldsymbol{\xi}$ is a power vector of order $p - 1$, $(L - \lambda I)^2 \boldsymbol{\xi}$ is a power vector of order $p - 2$, and so on through $(L - \lambda I)^{p-1}\boldsymbol{\xi}$, which is an eigenvector.

Example 1: Let
$$A = \begin{pmatrix} 2 & 1 \\ 0 & 2 \end{pmatrix}. \tag{4.81}$$

The characteristic polynomial is $(\lambda - 2)^2$, so the only eigenvalue is $\lambda = 2$. The vector $(1, 0)^T$ is an eigenvector of eigenvalue 2, while the vector $(0, 1)^T$ is a power vector of order 2 corresponding to the eigenvalue 2.

Example 2: The matrix
$$B = \begin{pmatrix} 2 & 1 \\ -1 & 0 \end{pmatrix} \tag{4.82}$$

has characteristic polynomial $(\lambda - 1)^2$. While $(1, -1)^T$ is an eigenvector with eigenvalue 1, $(1, 1)^T$ is a power vector of order 2 corresponding to the eigenvalue 1. Notice that $(B - I)\begin{pmatrix} 1 \\ 1 \end{pmatrix} = \begin{pmatrix} 2 \\ -2 \end{pmatrix}$, which is indeed an eigenvector.

Power vectors are important in large part because the action of an operator (or a function of an operator) on a power vector is almost as simple as the action of the operator on an eigenvectors. For example, if $\mathbf{v}$ is a eigenvector of L with eigenvalue λ, then $e^{Lt}\mathbf{v} = e^{\lambda t}\mathbf{v}$. If, however, $\boldsymbol{\xi}$ is a power vector of eigenvalue λ and order p, then

$$\begin{aligned} e^{Lt}\boldsymbol{\xi} &= e^{(L-\lambda I)t + \lambda t I}\boldsymbol{\xi} \\ &= e^{\lambda t} e^{(L-\lambda I)t}\boldsymbol{\xi} \\ &= e^{\lambda t} \sum_{k=0}^{\infty} \frac{t^k (L - \lambda I)^k \boldsymbol{\xi}}{k!} \\ &= e^{\lambda t} \sum_{k=0}^{p-1} \frac{t^k (L - \lambda I)^k \boldsymbol{\xi}}{k!}, \end{aligned} \tag{4.83}$$

since, for $k \geq p$, $(L - \lambda I)^k \boldsymbol{\xi} = 0$. The final expression is the sum of p terms, each proportional to $e^{\lambda t}$ times a power of t. Thus, for B as

4.9. Power Vectors and Jordan Canonical Form

in (4.82), we have

$$e^{Bt}\begin{pmatrix}1\\-1\end{pmatrix} = e^t\begin{pmatrix}1\\-1\end{pmatrix},$$
$$e^{Bt}\begin{pmatrix}1\\1\end{pmatrix} = e^t\begin{pmatrix}1\\1\end{pmatrix} + 2te^t\begin{pmatrix}1\\-1\end{pmatrix}. \quad (4.84)$$

Since every vector in $\mathbb{R}^2$ is a linear combination of $(1,-1)^T$ and $(1,1)^T$, it is fairly easy to compute the effect of e^{Bt} on an arbitrary vector in $\mathbb{R}^2$, which is tantamount to computing the matrix e^{Bt} itself.

Large powers of an operator also act (relatively) simply on power vectors. If $\boldsymbol{\xi}$ is a power vector of order p, then

$$\begin{aligned}L^n\boldsymbol{\xi} &= [(L-\lambda I)+\lambda I]^n\boldsymbol{\xi}\\ &= \sum_{k=0}^{n}\binom{n}{k}\lambda^{n-k}(L-\lambda I)^k\boldsymbol{\xi}\\ &= \sum_{k=0}^{p-1}\binom{n}{k}\lambda^{n-k}(L-\lambda I)^k\boldsymbol{\xi}.\end{aligned} \quad (4.85)$$

Indeed, it is not hard to compute the effect of *any* analytic function of an operator on a power vector. Suppose $f(x)$ is an analytic function with Taylor series

$$f(x) = \sum_{k=0}^{\infty}a_k(x-\lambda)^k \quad (4.86)$$

around $x=\lambda$. Then the linear operator $f(L)$ has the expansion

$$f(L) = \sum_{k=0}^{\infty}a_k(L-\lambda I)^k, \quad (4.87)$$

so, if $\boldsymbol{\xi}$ is a power vector of eigenvalue λ and order p, then

$$f(L)\boldsymbol{\xi} = \sum_{k=0}^{p-1}a_k(L-\lambda I)^k\boldsymbol{\xi}. \quad (4.88)$$

Since power vectors are easy to work with, analyzing problems involving nondiagonalizable matrices (or operators) just boils down to finding enough power vectors to form a basis. This can always be done, although the proof is somewhat lengthly. Theorems 4.12 and 4.13 summarize the essential theory of power vectors. The proofs are left to the exercises.

Theorem 4.12 *Let L be an operator acting on a finite dimensional complex vector space V.*

- *A collection of power vectors corresponding to different eigenvalues is linearly independent.*

- *Every power vector of L has order less than or equal to the algebraic multiplicity $m_a(\lambda)$ of its eigenvalue.*

- *The power space corresponding to λ is the same as the kernel of $(L - \lambda I)^{m_a(\lambda)}$.*

- *The dimension of the power space corresponding to λ is exactly equal to $m_a(\lambda)$.*

- *There exists a basis of V consisting of power vectors of L.*

Theorem 4.12 gives us a definite procedure for constructing a basis of power vectors. From the characteristic polynomial we can determine the eigenvalues and the algebraic multiplicities. For each eigenvalue λ, find a basis for the kernel of $(L-\lambda I)^{m_a(\lambda)}$. Concatenate these bases to get a basis for the vector space V.

Not all such bases are equally useful, however. In the exercises we sketch how to construct a particularly nice basis for each power space, relative to which the matrix of the linear operator is particularly simple:

Theorem 4.13 (Jordan Form Theorem) *If L is an operator on a finite dimensional complex vector space V, then there exists a basis $\mathcal{B}$ of V such that $[L]_\mathcal{B}$ is block-diagonal, with each block (called a Jordan block) of the form*

$$\begin{pmatrix} \lambda & 1 & 0 & \cdots & 0 & 0 \\ 0 & \lambda & 1 & \cdots & 0 & 0 \\ 0 & 0 & \lambda & \cdots & 0 & 0 \\ \vdots & \vdots & \vdots & \ddots & \vdots & \vdots \\ 0 & 0 & 0 & \cdots & \lambda & 1 \\ 0 & 0 & 0 & \cdots & 0 & \lambda \end{pmatrix}. \qquad (4.89)$$

That is, each Jordan block is a multiple of the identity plus a series of 1's just above the diagonal. The geometric multiplicity of an eigenvalue λ_0 is the number of blocks featuring λ_0. The algebraic multiplicity is the total dimension of these blocks. An operator

4.9. Power Vectors and Jordan Canonical Form

is thus diagonalizable if and only if its Jordan blocks are all 1×1. The matrix $[L]_\mathcal{B}$ is called the *Jordan canonical form* of the operator L. Two operators are conjugate if and only if their canonical forms agree, up to interchange of the order of the blocks.

Exercises

1. Show that L, and therefore any polynomial in L, maps $\tilde{E}_\lambda$ to itself.

2. Show that, if $\mu \neq \lambda$, that $L - \mu I$ maps $\tilde{E}_\lambda$ onto itself.

3. Prove the first statement of Theorem 4.12: A collection of power vectors corresponding to different eigenvalues is linearly independent.

This shows that concatenating the bases for the different power spaces yields a linearly independent set. The next 4 exercises show that this set spans our vector space.

4. Show that there exists a polynomial $h(x)$ such that $H(L)$ is identically zero. [Hint: if L is an $n \times n$ matrix, can $I, L, L^2, \ldots, L^{n^2}$ be linearly independent elements of $M_{n,n}$?]

5. Factorize the polynomial h of the previous exercise as $h(x) = (x-\lambda_1)^{n_1} \cdots (x-\lambda_k)^{n_k}$, and let $g_i(x)$ be the product of all the factors except $(x - \lambda_i)^{n_i}$. Show that the range of $g_i(L)$ is contained in $\tilde{E}_{\lambda_i}$.

6. Show that there exist polynomials $f_i(x)$ such that $\sum_i f_i(x) g_i(x)$ equals 1. [Hint: what is the greatest common factor of the g_i's?]

7. Show that any vector can be written as a linear combination of power vectors whose eigenvalues are roots of $h(x)$.

Next we proceed to prove the Jordan Form Theorem. The statements in Theorem 4.12 about the algebraic multiplicity of λ then follow.

8. If $\boldsymbol{\xi}$ is a power vector of eigenvalue λ and order p, show that $\mathbf{b}_p = \boldsymbol{\xi}, \mathbf{b}_{p-1} = (L-\lambda I)\boldsymbol{\xi}, \mathbf{b}_{p-2} = (L-\lambda I)^2 \boldsymbol{\xi}, \ldots, \mathbf{b}_1 = (L-\lambda I)^{p-1}\boldsymbol{\xi}$ is a linearly independent set of elements of $\tilde{E}_\lambda$. (This shows, by the way, that p is bounded by the dimension of $\tilde{E}_\lambda$.)

9. Let W be the span of the $\mathbf{b}_i$'s of the previous exercise. Show that L maps W to itself. What is the matrix $[L|_W]_\mathcal{B}$?

10. Suppose that $\boldsymbol{\xi}$ is a power vector of the largest possible order and W is constructed as above. Let W^c be the set of vectors in $\tilde{E}_\lambda$ with the following property: None of the vectors $(L - \lambda I)^i \mathbf{v}$, as i

ranges from zero to the order of **v**, are nonzero elements of W. Show that W^c is a subspace of $\tilde{E}_\lambda$, and that $\tilde{E}_\lambda = W \oplus W^c$. [Hint: you will need to use the fact that p was the largest possible order, and thus that there is no vector **w** with $(L - \lambda I)\mathbf{w} = \boldsymbol{\xi}$.]

11. Show that L maps W^c to itself.

12. Prove that any operator has a Jordan canonical form.

13. Compute the characteristic polynomial of a matrix in Jordan form. What does this imply about the dimension of $\tilde{E}_\lambda$ and the possible orders of elements of $\tilde{E}_\lambda$?

14. Show that two operators are conjugate if and only if they have the same Jordan form. [Hint: consider the dimension of the kernel of $(L - \lambda I)^i$ for various values of i.]

15. Prove the Cayley-Hamilton Theorem: If $p_A(\lambda)$ is the characteristic polynomial of the matrix A, then $p_A(A) = 0$.

Exercises 16–20 refer to the matrix $A = \begin{pmatrix} 1 & -3 & -1 \\ 1 & 5 & 1 \\ -2 & -6 & 0 \end{pmatrix}$.

16. Find the eigenvalues of A and a basis for each eigenspace.

17. Find a power vector $\boldsymbol{\xi}$ that is not an eigenvector. What is the order of $\boldsymbol{\xi}$?

18. Construct the space W as in Exercise 9. Find a vector that is neither in W nor in W^c.

19. Find a basis that puts A in Jordan canonical form.

20. Compute e^{At} for an arbitrary scalar t.

21. Consider the operator on $\mathbb{R}_2[t]$ that translates polynomials a distance 1, namely $(T_1\mathbf{p})(t) = \mathbf{p}(t - 1)$. Compute the matrix of T_1 relative to the standard basis of $\mathbb{R}_2[t]$. T_1 is not diagonalizable. Find a basis that puts T_1 in Jordan form.

Chapter 5

Some Crucial Applications

In this chapter[1] we start to apply eigenvalues and eigenvectors to problems ranging from physics to population dynamics. The applications differ mainly in whether the time evolution is discrete or continuous, and whether the equations are first-order or second-order. In all cases, a basis $\mathcal{B}$ of eigenvectors gives coordinates in which problems decouple, and we can employ the strategy of Figure 5.1. To convert an initial condition $\mathbf{x}(0)$ into a final state $\mathbf{x}(n)$, we first find the coordinates $[\mathbf{x}(0)]_\mathcal{B}$, then solve the decoupled evolution equations for $[\mathbf{x}(n)]_\mathcal{B}$, and finally convert back to our original basis.

5.1 Discrete-Time Evolution: $\mathbf{x}(n) = A\mathbf{x}(n-1)$

In Chapter 1 we considered the population of owls and mice in a certain forest. Since owls eat mice, the more mice there are, the more the owls have to eat, and the more owls there will be next year. But the more owls there are, the more mice get eaten, and the fewer mice there will be next year. Let $x_1(n)$ and $x_2(n)$ describe the owl population (in hundreds) and mouse populations (in tens of thousands) in year n, and assume the equations governing population growth are

$$\begin{aligned} x_1(n) &= 0.4x_1(n-1) + 0.6x_2(n-1), \\ x_2(n) &= -0.3x_1(n-1) + 1.3x_2(n-1). \end{aligned} \quad (5.1)$$

---

[1] In this chapter, and only in this chapter, the dimension of a typical vector space will be denoted m, rather than n, since the letter n already designates the number of time steps in discrete evolution problems.

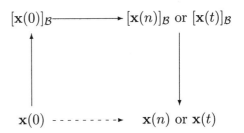

Figure 5.1: The strategy for solving coupled evolution problems

If in year 0 the populations are $x_1 = 2$, $x_2 = 3$, what will the populations be in year n?

In general, we consider equations

$$\mathbf{x}(n) = A\mathbf{x}(n-1), \tag{5.2}$$

where A is a square matrix of appropriate size. In the owls-mice example, we have

$$A = \begin{pmatrix} 0.4 & 0.6 \\ -0.3 & 1.3 \end{pmatrix}. \tag{5.3}$$

Since $\mathbf{x}(1) = A\mathbf{x}(0)$ and $\mathbf{x}(2) = A\mathbf{x}(1) = A^2\mathbf{x}(0)$, etc., we generally have

$$\mathbf{x}(n) = A^n\mathbf{x}(0). \tag{5.4}$$

The hard thing is computing $A^n\mathbf{x}(0)$ from A and $\mathbf{x}(0)$.

Suppose A is diagonalizable, with eigenvalues $\lambda_1, \ldots, \lambda_m$ and corresponding eigenvectors $\mathbf{b}_1, \ldots, \mathbf{b}_m$. We expand $\mathbf{x}$ as

$$\mathbf{x}(n) = a_1(n)\mathbf{b}_1 + \cdots + a_m(n)\mathbf{b}_m, \tag{5.5}$$

and rewrite equation (5.2) in terms of the coefficients a_i. The result is

$$a_1(n)\mathbf{b}_1 + \cdots + a_m(n)\mathbf{b}_m = a_1(n-1)\lambda_1\mathbf{b}_1 + \cdots + a_m(n-1)\lambda_m\mathbf{b}_m. \tag{5.6}$$

In other words,

$$a_i(n) = \lambda_i a_i(n-1), \tag{5.7}$$

and therefore

$$a_i(n) = \lambda_i^n a_i(0), \tag{5.8}$$

5.1. Discrete-Time Evolution: $\mathbf{x}(n) = A\mathbf{x}(n-1)$

hence
$$\mathbf{x}(n) = \sum_{i=1}^{m} \lambda_i^n a_i(0) \mathbf{b}_i. \tag{5.9}$$

Returning to the owls-mice example, the characteristic polynomial of the matrix A is

$$p_A(\lambda) = \begin{vmatrix} \lambda - 0.4 & -0.6 \\ 0.3 & \lambda - 1.3 \end{vmatrix} = (\lambda - 0.4)(\lambda - 1.3) + 0.18 = \lambda^2 - 1.7\lambda + 0.7, \tag{5.10}$$

with roots $\lambda_1 = 1$ and $\lambda_2 = 0.7$. Solving $(\lambda I - A)\mathbf{x} = 0$ in the two cases we find the eigenvectors $\mathbf{b}_1 = (1,1)^T$ and $\mathbf{b}_2 = (2,1)^T$. Our initial condition is

$$\mathbf{x}(0) = \begin{pmatrix} 2 \\ 3 \end{pmatrix} = 4\mathbf{b}_1 - \mathbf{b}_2. \tag{5.11}$$

Plugging into equation (5.9) we have that

$$\mathbf{x}(n) = 4(1)^n \mathbf{b}_1 - (0.7)^n \mathbf{b}_2 = 4 \begin{pmatrix} 1 \\ 1 \end{pmatrix} - (0.7)^n \begin{pmatrix} 2 \\ 1 \end{pmatrix}. \tag{5.12}$$

The individual populations are then

$$x_1(n) = 4 - 2(0.7)^n; \qquad x_2(n) = 4 - (0.7)^n. \tag{5.13}$$

As $n \to \infty$, both x_1 and x_2 approach 4, and our forest approaches a limiting population of 400 owls and 40,000 mice.

Discrete-time evolution equations may also describe populations of different groups within a single species. The Rennaissance Italian mathematician Fibonacci considered the following model of rabbit population growth. (See Figure 5.2.) Female rabbits are either juvenile or adult. Each month, each juvenile grows to adulthood, and each adult female gives birth (on average) to one juvenile female. Starting with one juvenile female rabbit, how many female rabbits will there be after n months?

Let $x_1(n)$ and $x_2(n)$ be the number of juvenile and adult female rabbits, respectively, after n months. Our evolution equations are

$$x_1(n) = x_2(n-1); \qquad x_2(n) = x_1(n-1) + x_2(n-1), \tag{5.14}$$

or, in matrix form,

$$\mathbf{x}(n) = A\mathbf{x} = \begin{pmatrix} 0 & 1 \\ 1 & 1 \end{pmatrix} \mathbf{x}(n-1). \tag{5.15}$$

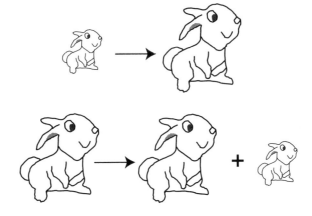

Figure 5.2: Fibonacci's rabbit model

These equations follow from the fact that each of last month's juveniles grew into one adult, and each of last month's adults remained an adult while giving rise to one juvenile.

To solve the problem we find the eigenvalues and eigenvectors of A. The characteristic polynomial of A is

$$p_A(\lambda) = \lambda(\lambda - 1) - 1 = \lambda^2 - \lambda - 1, \tag{5.16}$$

with roots $\lambda_1 = (1 + \sqrt{5})/2 \approx 1.618$ and $\lambda_2 = (1 - \sqrt{5})/2 \approx -0.618$. λ_1 is sometimes called the *golden mean*. The corresponding eigenvectors are $\mathbf{b}_1 = (\sqrt{5} - 1, 2)^T$ and $\mathbf{b}_2 = (-\sqrt{5} - 1, 2)^T$. The initial condition is

$$\mathbf{x}(0) = \begin{pmatrix} 1 \\ 0 \end{pmatrix} = \frac{1}{2\sqrt{5}}\mathbf{b}_1 - \frac{1}{2\sqrt{5}}\mathbf{b}_2. \tag{5.17}$$

Thus the population distribution after n months is

$$\begin{aligned}\mathbf{x}(n) &= \frac{1}{2\sqrt{5}}\left(\frac{1+\sqrt{5}}{2}\right)^n \mathbf{b}_1 - \frac{1}{2\sqrt{5}}\left(\frac{1-\sqrt{5}}{2}\right)^n \mathbf{b}_2 \\ &= \frac{1}{\sqrt{5}}\begin{pmatrix} \left(\frac{1+\sqrt{5}}{2}\right)^{n-1} - \left(\frac{1-\sqrt{5}}{2}\right)^{n-1} \\ \left(\frac{1+\sqrt{5}}{2}\right)^n - \left(\frac{1-\sqrt{5}}{2}\right)^n \end{pmatrix},\end{aligned} \tag{5.18}$$

and the total population is

$$x_1(n) + x_2(n) = \frac{1}{\sqrt{5}}\left(\left(\frac{1+\sqrt{5}}{2}\right)^{n+1} - \left(\frac{1-\sqrt{5}}{2}\right)^{n+1}\right). \tag{5.19}$$

5.1. Discrete-Time Evolution: $\mathbf{x}(n) = A\mathbf{x}(n-1)$

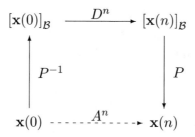

Figure 5.3: The strategy for discrete-time evolution

Since $|\lambda_2| < 1$, eventually λ_2^n gets very small, and our population is described extremely well by the approximation

$$\mathbf{x}(n) \approx \frac{1}{2\sqrt{5}} \left(\frac{1+\sqrt{5}}{2}\right)^n \mathbf{b}_1 = \frac{1}{\sqrt{5}} \begin{pmatrix} \left(\frac{1+\sqrt{5}}{2}\right)^{n-1} \\ \left(\frac{1+\sqrt{5}}{2}\right)^n \end{pmatrix}. \tag{5.20}$$

The population growth is asymptotically exponential, multiplying by approximately 1.618 each month, and always with approximately 1.618 times more adults than juveniles.

We close this section with a comment on the functional form (5.4) of the solution to the general equation (5.2). If A is diagonalizable, with eigenvalues $\lambda_1, \ldots, \lambda_m$ and eigenvectors $\mathbf{b}_1, \ldots \mathbf{b}_m$, then we have seen (Theorem 4.5) that

$$A = PDP^{-1}, \tag{5.21}$$

where D is a diagonal matrix, whose diagonal entries are the eigenvalues λ_i, and P is a matrix whose columns are the eigenvectors $\mathbf{b}_i$. From equation (4.72) we then have

$$A^n = PD^nP^{-1}, \tag{5.22}$$

and so

$$\mathbf{x}(n) = PD^nP^{-1}\mathbf{x}(0). \tag{5.23}$$

This equation is easy to understand on its own merits. See Figure 5.3. If $\mathbf{x}(0) = \sum_i a_i(0)\mathbf{b}_i$, we have

$$P^{-1}\mathbf{x}(0) = \begin{pmatrix} a_1(0) \\ \vdots \\ a_m(0) \end{pmatrix} = [\mathbf{x}]_{\mathcal{B}}. \tag{5.24}$$

In other words, P^{-1} converts from the coordinates $(x_1, \ldots, x_n)^T$, in which the problem appears coupled, to the coordinates $(a_1, \ldots, a_n)^T$, in which the problem decouples. Next we have

$$D^n P^{-1} \mathbf{x}(0) = \begin{pmatrix} \lambda_1^n a_1(0) \\ \vdots \\ \lambda_m^n a_m(0) \end{pmatrix} = [\mathbf{x}(n)]_\mathcal{B}. \tag{5.25}$$

That is, D^n just multiplies each coordinate a_i by λ_i^n. Finally,

$$PD^n P^{-1} \mathbf{x}(0) = \lambda_1^n a_1(0) \mathbf{b}_1 + \cdots + \lambda_m^n a_m(0) \mathbf{b}_m, \tag{5.26}$$

which is precisely equation (5.9).

Exercises

1. In the Fibonacci rabbit problem, let $y(n) = x_1(n) + x_2(n)$ be the total population of (female) rabbits. Show that, for $n \geq 2$, $y(n) = y(n-1) + y(n-2)$. The progression of $y(n)$'s, namely 1, 1, 2, 3, 5, 8, 13, 21,..., is called the Fibonacci sequence.

2. Imagine Fibonacci's rabbit problem with each adult giving birth to two juveniles per month. Starting with one juvenile rabbit, how many juveniles and how many adults will there be in n months?

3. Let $A = \frac{1}{\sqrt{2}} \begin{pmatrix} 1 & -1 \\ 1 & 1 \end{pmatrix}$ and let $\mathbf{x}(0) = \begin{pmatrix} 3 \\ 4 \end{pmatrix}$. Find $\mathbf{x}(n)$ for all n. Describe qualitatively what is happening.

4. Let $A = \begin{pmatrix} 3 & 1 & 1 \\ 1 & 2 & 2 \\ 1 & 2 & 2 \end{pmatrix}$ and let $\mathbf{x}(0) = \begin{pmatrix} 3 \\ 1 \\ -1 \end{pmatrix}$. Find $\mathbf{x}(n)$ for all n. Describe qualitatively what is happening.

5. Let $A = \begin{pmatrix} 0.3 & 0.2 & 0.4 \\ 0.4 & 0.6 & 0.2 \\ 0.3 & 0.2 & 0.4 \end{pmatrix}$ and let $\mathbf{x}(0) = \begin{pmatrix} 0.5 \\ 0.2 \\ 0.3 \end{pmatrix}$. Find $\mathbf{x}(n)$ for all n. Describe what happens as $n \to \infty$.

6. Let $A = \begin{pmatrix} 0.2 & 1.1 & 2.3 \\ 1.1 & 0.5 & -1.7 \\ 2.3 & -1.7 & 0.7 \end{pmatrix}$ and let $\mathbf{x}(0) = \begin{pmatrix} 1 \\ 0 \\ 0 \end{pmatrix}$. Find $\mathbf{x}(n)$ for all n. You may wish to use technology to diagonalize A and to invert P.

7. Let $A = \begin{pmatrix} 0.5738 & 0.1775 & -0.2218 & -0.1584 & 0.4994 \\ 0.3051 & 0.0387 & -0.0336 & -0.3644 & 0.0491 \\ -0.3462 & -0.2716 & -0.0465 & 0.4543 & 0.1453 \\ -0.0079 & -0.8545 & 0.0851 & 0.8213 & 0.0217 \\ 0.9806 & 0.2944 & 0.7080 & 0.2595 & 0.7346 \end{pmatrix}$

and let $\mathbf{x}(0) = (1,0,0,0,0)^T$. Using technology, compute $\mathbf{x}(n)$ for several values of n. Diagonalize A (using technology) and describe the asymptotic behavior of $\mathbf{x}(n)$ for n large. How does $x_1(100)$ compare to $x_1(99)$? What about x_2? How do the proportions $x_1 : x_2 : x_3 : x_4 : x_5$ depend on n?

8. Griffins and Dragons live in the Enchanted Forest. The number of dragons D and griffins G in the forest each year is determined by the populations the previous year, according to the formulas:

$$D(n) = 1.5D(n-1) + G(n-1),$$
$$G(n) = D(n-1)$$

If in year 0 there are 25 dragons and no griffins, what will the populations be in year k? (Don't worry about your answers being fractional. Mythical animals don't have to come in whole units.) In the long run, will the populations grow, shrink, or approach a nonzero equilibrium value? After a long time, approximately what will the ratio of dragons to griffins be?

9. Deer can be classified as juvenile or adult. Every year 50% of the juveniles survive to become adults, while the other 50% die. Every year the adults give birth to an average of 0.6 juveniles each (that is, 1.2 births per mating pair), and 70% of the adults survive the year. If there are 100 adults and 100 juveniles one year, how many adults and how many juveniles can be expected 10 years later?

10. Repeat the previous problem with an adult survival rate of 80% instead of 70%.

5.2 First-Order Continuous-Time Evolution: $d\mathbf{x}/dt = A\mathbf{x}$

The discrete-time models of Section 5.1 are appropriate when events occur in definite cycles. Animal populations, for example, always fluctuate with the seasons, so it makes little sense to compare the number of owls in January to the number in July. It does make sense to compare July of one year to July of the next.

Some events, however, happen evenly and continuously, or at least so fast that they can be modeled as continuous. The population of bacteria in a Petri dish, or the number of atoms of a radioactive sample are, strictly speaking, integers. However, they vary so quickly and so much that they are well described by a differential equation

$$\frac{dx}{dt} = f(x), \tag{5.27}$$

where x is the quantity of interest, and $f(x)$ is some function that describes the rate of growth. We are interested in problems where $f(x)$ is linear in x. When x is a single variable, this means

$$f(x) = rx, \tag{5.28}$$

for some constant r. In such cases the solution is

$$x(t) = e^{rt}x(0). \tag{5.29}$$

When $r > 0$ (exponential growth), r is called the growth rate, or the interest rate if the problem is financial. The *time scale* is $1/r$, and the *doubling time* is $\ln(2)/r \approx 0.69/r$. When $r < 0$ (exponential decay), $|r|$ is called the decay rate, $1/|r|$ is called the *mean lifetime* or *time scale*, and $\ln(2)/|r|$ is called the *half-life*. Examples of exponential growth include money earning interest and populations of bacteria with unlimited food. Examples of exponential decay include radiactive samples and the excess temperature of a hot object in cool surroundings.

We are interested in coupled linear first-order differential equations, that is

$$\frac{d\mathbf{x}}{dt} = A\mathbf{x}, \tag{5.30}$$

where A is a fixed matrix. In such cases the solution is

$$\mathbf{x}(t) = e^{At}\mathbf{x}(0), \tag{5.31}$$

as can be seen by taking the derivative of the right hand side. (Recall that the derivative of e^{At} is Ae^{At}. This is one of the definitions of e^{At}.) However, this expression does us little good until we learn to efficiently compute e^{At}.

As in the previous section, we assume that A is diagonalizable, with eigenvalues $\lambda_1, \ldots, \lambda_m$ and corresponding eigenvectors $\mathbf{b}_1, \ldots, \mathbf{b}_m$. As usual, we decompose

$$\mathbf{x}(t) = a_1(t)\mathbf{b}_1 + \cdots + a_m(t)\mathbf{b}_m, \tag{5.32}$$

5.2. First-Order Continuous-Time Evolution: $d\mathbf{x}/dt = A\mathbf{x}$

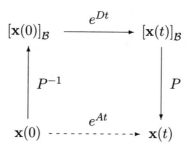

Figure 5.4: The strategy for continuous-time evolution

and plug this into the equation (5.30). The result is

$$\frac{da_1(t)}{dt}\mathbf{b}_1 + \cdots + \frac{da_m(t)}{dt}\mathbf{b}_m = a_1(t)\lambda_1\mathbf{b}_1 + \cdots + a_m(t)\lambda_m\mathbf{b}_m, \quad (5.33)$$

which implies

$$\frac{da_i(t)}{dt} = \lambda_i a_i(t). \quad (5.34)$$

The equations have decoupled, and we solve the scalar equations to get

$$a_i(t) = e^{\lambda_i t} a_i(0), \quad (5.35)$$

or

$$\mathbf{x}(t) = \sum_{i=1}^{m} e^{\lambda_i t} a_i(0) \mathbf{b}_i. \quad (5.36)$$

Notice the similarity between equations (5.32–5.36) and (5.5–5.9). Although the form of the evolution equation has changed, the method of decoupling has not.

If we write $A = PDP^{-1}$, then $e^{At} = Pe^{Dt}P^{-1}$, and each factor corresponds to one step in our strategy for determining time evolution. P^{-1} converts $\mathbf{x}(0)$ to $[\mathbf{x}(0)]_\mathcal{B}$, e^{Dt} converts $[\mathbf{x}(0)]_\mathcal{B}$ to $[\mathbf{x}(t)]_\mathcal{B}$, and P converts $[\mathbf{x}(t)]_\mathcal{B}$ to $\mathbf{x}(t)$. See Figure 5.4.

Example: Suppose that radioactive material X decays into material Y at a rate r_1. Y is itself radioactive, and decays into material Z at a rate r_2. Material Z is stable. If we start at time 0 with 1 kilogram of pure material X, how much X, and how much Y, will we have at time t?

Let x_1 and x_2 denote the quantities of material X and Y, respectively. These quantities satisfy the coupled differential equations

$$dx_1/dt = -r_1 x_1,$$

$$dx_2/dt = r_1 x_1 - r_2 x_2. \tag{5.37}$$

This is of the form (5.30) with $\mathbf{x} = (x_1, x_2)^T$ and

$$A = \begin{pmatrix} -r_1 & 0 \\ r_1 & -r_2 \end{pmatrix}. \tag{5.38}$$

The eigenvalues of A are $-r_1$ and $-r_2$, with eigenvectors $\mathbf{b}_1 = \begin{pmatrix} r_2 - r_1 \\ r_1 \end{pmatrix}$ and $\mathbf{b}_2 = \begin{pmatrix} 0 \\ 1 \end{pmatrix}$. The initial condition is

$$\mathbf{x}(0) = \begin{pmatrix} 1 \\ 0 \end{pmatrix} = \frac{1}{r_2 - r_1} \mathbf{b}_1 - \frac{r_1}{r_2 - r_1} \mathbf{b}_2. \tag{5.39}$$

By (5.36), our solution is then

$$\mathbf{x}(t) = \frac{e^{-r_1 t}}{r_2 - r_1} \mathbf{b}_1 - \frac{r_1 e^{-r_2 t}}{r_2 - r_1} \mathbf{b}_2. \tag{5.40}$$

Expanding this we get

$$\begin{aligned} x_1(t) &= e^{-r_1 t}, \\ x_2(t) &= \frac{r_1}{r_2 - r_1}(e^{-r_1 t} - e^{-r_2 t}). \end{aligned} \tag{5.41}$$

So far we have assumed that A is diagonalizable. What if it isn't? For example, what happens in our radioactive decay example if $r_1 = r_2$? There are two practical approaches. The choice of which to use depends on your ultimate goal and upon the details of the situation. (Both methods are explored further in the exercises.)

One approach is to perturb the matrix. Since almost every matrix is diagonalizable, a nondiagonalizable matrix can always be written as the limit, as $\epsilon \to 0$, of a family A_ϵ of diagonalizable matrices. We solve the equation $d\mathbf{x}/dt = A_\epsilon \mathbf{x}$ by diagonalization, and then take the limit as $\epsilon \to 0$. For example, in our radioactive decay problem we can take $r_2 = r_1 + \epsilon$. This approach is particularly useful if we can obtain the solution for $\epsilon \neq 0$ without too much extra work (especially if we have already done that case) and the expressions allow for easy limits.

A second approach is to use power vectors (see Section 4.9). Recall that, if $\mathbf{b}$ is a power vector of eigenvalue λ and order p, then $(A - \lambda I)^p \mathbf{b} = 0$, so

$$e^{At} \mathbf{b} = e^{(A - \lambda I)t + \lambda t I} \mathbf{b}$$

5.2. First-Order Continuous-Time Evolution: $d\mathbf{x}/dt = A\mathbf{x}$

$$= e^{\lambda t} e^{(A-\lambda I)t} \mathbf{b}$$

$$= e^{\lambda t} \sum_{k=0}^{\infty} \frac{t^k (A-\lambda I)^k \mathbf{b}}{k!}$$

$$= e^{\lambda t} \sum_{k=0}^{p-1} \frac{t^k (A-\lambda I)^k \mathbf{b}}{k!}. \tag{5.42}$$

If we write $\mathbf{x}(0)$ as a linear combination of power vectors, we compute the action of e^{At} on each power vector by (5.42), and recombine to get $\mathbf{x}(t)$. The advantage of this approach is that the functional form of the solution, as a polynomial in t times $e^{\lambda t}$, with the order of the polynomial one less than the order of the power vector, is clear. The disadvantage is that one has completely different formulas for the diagonalizable and nondiagonalizable cases, obscuring the fact that the solution for any fixed t depends smoothly on the entries of A.

Exercises

In Exercises 1–5, find the solution $\mathbf{x}(t)$ to the coupled equations $d\mathbf{x}/dt = A\mathbf{x}$ for all time t.

1. Let $A = \begin{pmatrix} 0 & 1 \\ 1 & 0 \end{pmatrix}$ and let $\mathbf{x}(0) = \begin{pmatrix} 1 \\ 0 \end{pmatrix}$.

2. Let $A = \begin{pmatrix} 0 & 2 \\ -2 & 0 \end{pmatrix}$ and let $\mathbf{x}(0) = \begin{pmatrix} 1 \\ 0 \end{pmatrix}$. Repeat for $\mathbf{x}(0) = \begin{pmatrix} 0 \\ 1 \end{pmatrix}$.

3. Let $A = \begin{pmatrix} 0 & 4 \\ -1 & 0 \end{pmatrix}$ and let $\mathbf{x}(0) = \begin{pmatrix} 1 \\ 0 \end{pmatrix}$. Repeat for $\mathbf{x}(0) = \begin{pmatrix} 0 \\ 1 \end{pmatrix}$. Compare to the results of Exercise 2.

4. Let $A = \begin{pmatrix} -0.7 & 0.2 & 0.4 \\ 0.4 & -0.4 & 0.2 \\ 0.3 & 0.2 & -0.6 \end{pmatrix}$ and let $\mathbf{x}(0) = \begin{pmatrix} 0.5 \\ 0.2 \\ 0.3 \end{pmatrix}$. Describe what happens as $t \to \infty$.

5. Let $A = \begin{pmatrix} 0.2 & 1.1 & 2.3 \\ 1.1 & 0.5 & -1.7 \\ 2.3 & -1.7 & 0.7 \end{pmatrix}$ and let $\mathbf{x}(0) = \begin{pmatrix} 1 \\ 0 \\ 0 \end{pmatrix}$. You may wish to use technology to diagonalize A.

Exercises 6–8 refer to the radioactive decay example, with Exercises 7 and 8 contrasting the two approaches to nondiagonalizable matrices:

6. If $r_1 < r_2$, find the time when x_2 achieves its maximum. Describe qualitatively what is going on. Repeat for $r_1 > r_2$.

7. If $r_1 = r_2$, the solution (5.31) is still correct, but it can no longer be computed by diagonalization, since in that case A is not diagonalizable. Instead, take the limit of (5.41) as $r_2 \to r_1$. [Hint: Use L'Hopital's rule with r_2 as your variable.] Check your answers by plugging them back into the differential equations (5.37).

8. For $r_1 = r_2$, compute $e^{At} \begin{pmatrix} 1 \\ 0 \end{pmatrix}$ directly, using the fact that $(1,0)^T$ is a power vector. Compare to the result of Exercise 7.

9. Let A be a single $p \times p$ Jordan block (see Section 4.9). Find the general solution to $dx/dt = A\mathbf{x}$.

10. Let $A = \begin{pmatrix} 1 & 1 \\ -1 & -1 \end{pmatrix}$ and let $\mathbf{x}(0) = \begin{pmatrix} 1 \\ 0 \end{pmatrix}$. Solve for $\mathbf{x}(t)$ for all t.

5.3 Second-order Continuous-Time Evolution: $d^2\mathbf{x}/dt^2 = A\mathbf{x}$

Thanks to Newton's Third Law, $\mathbf{F} = m\mathbf{a}$, second-order continuous-time evolution problems are extremely common. The positions of a mass on a spring, a pendulum, and a ball rolling near the bottom of a hill are all described by the equation

$$\frac{d^2x}{dt^2} = ax, \qquad (5.43)$$

with a a negative constant. The positions of a ball near the top of a hill or an upside-down pendulum are also described by (5.43), except that in these cases the constant a is positive. We are going to consider coupled systems, governed by equations

$$\frac{d^2\mathbf{x}}{dt^2} = A\mathbf{x}. \qquad (5.44)$$

As usual, we assume that A is diagonalizable, and for now we assume that the eigenvalues are real. (In Chapter 7 we will prove that these assumptions hold whenever A is real and symmetric, as is typical in problems of this sort.) However, some of the eigenvalues of A may be positive, while others may be negative or zero. To understand the solutions to (5.44) we must understand the solutions to (5.43) for all possible values of a.

5.3. Second-Order Continuous-Time Evolution: $d^2x/dt^2 = Ax$

If $a > 0$, let $\kappa = \sqrt{a}$. The solutions to (5.43) are all linear combinations of $e^{\kappa t}$ and $e^{-\kappa t}$. We can just as easily say that the solutions are all linear combinations of the hyperbolic cosine and sine. These are defined to be

$$\cosh(\kappa t) = \frac{e^{\kappa t} + e^{-\kappa t}}{2},$$

$$\sinh(\kappa t) = \frac{e^{\kappa t} - e^{-\kappa t}}{2}. \quad (5.45)$$

The inverse relations are

$$e^{\kappa t} = \cosh(\kappa t) + \sinh(\kappa t),$$
$$e^{-\kappa t} = \cosh(\kappa t) - \sinh(\kappa t). \quad (5.46)$$

Notice that $\cosh(0) = \sinh'(0) = 1$, while $\sinh(0) = \cosh'(0) = 0$. Thus if

$$x(t) = c_1 \cosh(\kappa t) + c_2 \sinh(\kappa t), \quad (5.47)$$

we must have

$$x(0) = c_1; \quad \dot{x}(0) = \kappa c_2, \quad (5.48)$$

where $\dot{x}$ denotes dx/dt, and hence

$$c_1 = x(0); \quad c_2 = \dot{x}(0)/\kappa. \quad (5.49)$$

In terms of the original exponentials, we can write

$$x(t) = \tilde{c}_1 e^{\kappa t} + \tilde{c}_2 e^{-\kappa t}, \quad (5.50)$$

where

$$\tilde{c}_1 = (c_1 + c_2)/2 = (\kappa x(0) + \dot{x}(0))/2\kappa,$$
$$\tilde{c}_2 = (c_1 - c_2)/2 = (\kappa x(0) - \dot{x}(0))/2\kappa. \quad (5.51)$$

The analysis for $a < 0$ is completely analogous. Let $\omega = \sqrt{-a}$. The solutions to (5.43) are all linear combinations of $\sin(\omega t)$ and $\cos(\omega t)$. By Euler's formula (4.68), we can just as easily say that the solutions are all linear combinations of $e^{i\omega t}$ and $e^{-i\omega t}$. These are related by

$$\cos(\omega t) = \frac{e^{i\omega t} + e^{-i\omega t}}{2},$$

$$\sin(\omega t) = \frac{e^{i\omega t} - e^{-i\omega t}}{2i}, \quad (5.52)$$

and the inverse relations
$$\begin{aligned} e^{i\omega t} &= \cos(\omega t) + i\sin(\omega t), \\ e^{-i\omega t} &= \cos(\omega t) - i\sin(\omega t). \end{aligned} \quad (5.53)$$

Notice that $\cos(0) = \sin'(0) = 1$, while $\sin(0) = \cos'(0) = 0$, so if
$$x(t) = c_1 \cos(\omega t) + c_2 \sin(\omega t), \quad (5.54)$$
we must have
$$x(0) = c_1; \qquad \dot{x}(0) = \omega c_2, \quad (5.55)$$
and so
$$c_1 = x(0); \qquad c_2 = \dot{x}(0)/\omega. \quad (5.56)$$

We can also express our solution in terms of the complex exponentials
$$x(t) = \tilde{c}_1 e^{i\omega t} + \tilde{c}_2 e^{-i\omega t}, \quad (5.57)$$
where
$$\begin{aligned} \tilde{c}_1 &= (c_1 - ic_2)/2 = (\omega x(0) - i\dot{x}(0))/2\omega, \\ \tilde{c}_2 &= (c_1 + ic_2)/2 = (\omega x(0) + i\dot{x}(0))/2\omega. \end{aligned} \quad (5.58)$$

To summarize, regardless of whether a is positive or negative, we can express any solution as a linear combination of $e^{\pm\sqrt{a}\,t}$, or repackage these exponentials as (possibly hyperbolic) sines and cosines. If $a > 0$ we have
$$x(t) = x(0) \cosh(\sqrt{a}\,t) + (\dot{x}(0)/\sqrt{a}) \sinh(\sqrt{a}\,t), \quad (5.59)$$
while if $a < 0$ we have
$$x(t) = x(0) \cos(\sqrt{-a}\,t) + (\dot{x}(0)/\sqrt{-a}) \sin(\sqrt{-a}\,t). \quad (5.60)$$

Finally, if $a = 0$, then the general solution is $x(t) = c_1 + c_2 t$. Plugging in our initial conditions we see that $c_1 = x(0)$ and $c_2 = \dot{x}(0)$, so
$$x(t) = x(0) + \dot{x}(0)t. \quad (5.61)$$
Notice that the result (5.61) can be obtained as the limit of either (5.59) or (5.60) as $a \to 0$.

We are now able to solve (5.44). We expand $\mathbf{x}(t)$ in a basis of eigenvectors of A, as in (5.32). Plugging (5.32) into the equation (5.44) we get
$$\frac{d^2 a_1(t)}{dt^2}\mathbf{b}_1 + \cdots + \frac{d^2 a_m(t)}{dt^2}\mathbf{b}_m = a_1(t)\lambda_1 \mathbf{b}_1 + \cdots + a_m(t)\lambda_m \mathbf{b}_m, \quad (5.62)$$

5.3. Second-Order Continuous-Time Evolution: $d^2\mathbf{x}/dt^2 = A\mathbf{x}$

which is tantamount to the m decoupled scalar equations

$$\frac{d^2 a_i(t)}{dt^2} = \lambda_i a_i(t) \tag{5.63}$$

that we have just learned how to solve. If $\lambda_i > 0$ we have

$$a_i(t) = \cosh(\sqrt{\lambda_i}\,t)a_i(0) + \dot{a}_i(0)\sinh(\sqrt{\lambda_i}\,t)/\sqrt{\lambda_i}, \tag{5.64}$$

if $\lambda_i < 0$ we have

$$a_i(t) = \cos(\sqrt{-\lambda_i}\,t)a_i(0) + \dot{a}_i(0)\sin(\sqrt{-\lambda_i}\,t)/\sqrt{-\lambda_i}, \tag{5.65}$$

while if $\lambda_i = 0$ we have

$$a_i(t) = a_i(0) + \dot{a}_i(0)t. \tag{5.66}$$

In any case, once we have computed the $a_i(t)$'s, we can plug them into the expansion (5.32) to get $\mathbf{x}(t)$.

The eigenvectors of A are called the *normal modes* of the system. The ones with $\lambda < 0$ are called *stable modes*, as they are associated with oscillatory (also called *sinusoidal*) behavior, while the ones with $\lambda > 0$ are called *unstable modes*, and are associated with exponential behavior.

As an example, consider the two blocks of Figure 5.5, connected by springs. Each block is assumed to have mass m, while each spring is assumed to have spring constant k. Suppose we push the left block one inch to the right of equilibrium and then let go of both blocks. What will the positions of both blocks be at time t?

Let x_1 and x_2 be the positions of the two blocks, relative to equilibrium. The first (leftmost) spring is stretched a distance x_1, and so applies a force $-kx_1$ to the first block. The second spring is stretched a distance $x_2 - x_1$. It applies a force $kx_2 - kx_1$ to the first block and $kx_1 - kx_2$ to the second block. The third spring is stretched a distance $-x_2$ and applies a force $-kx_2$ to the second block. Acceleration equals force divided by mass, so

$$\frac{d^2 x_1}{dt^2} = \frac{-2k}{m}x_1 + \frac{k}{m}x_2,$$
$$\frac{d^2 x_2}{dt^2} = \frac{-k}{m}x_1 - \frac{2k}{m}x_2. \tag{5.67}$$

Putting this in matrix form gives (5.44) with

$$A = \frac{k}{m}\begin{pmatrix} -2 & 1 \\ 1 & -2 \end{pmatrix}. \tag{5.68}$$

112   Chapter 5. Some Crucial Applications

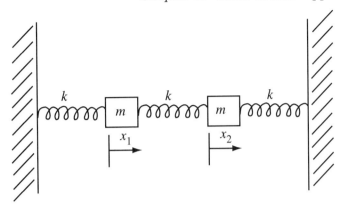

Figure 5.5: Coupled oscillators

The eigenvalues of A are $\lambda_1 = -k/m$ and $\lambda_2 = -3k/m$, with eigenvectors $\mathbf{b}_1 = (1,1)^T$ and $\mathbf{b}_2 = (1,-1)^T$. The natural frequencies of the system are $\omega_1 = \sqrt{-\lambda_1} = \sqrt{k/m}$ and $\omega_2 = \sqrt{-\lambda_2} = \sqrt{3k/m}$.

The two normal modes are not difficult to visualize physically. See Figure 5.6. In the first mode ($\mathbf{b}_1$), the two blocks move together, and the middle spring never gets stretched or compressed. The middle spring exerts no force, and may as well not even be there. Each block then moves with its own natural frequency, $\sqrt{k/m}$, and the two blocks remain forever in phase. The second mode ($\mathbf{b}_2$) has the two blocks moving completely out of phase. The middle spring gets compressed by twice the amplitude of the motion and provides a restoring force of $-2kx_i$ to the i-th block. Added to the effect of the outside springs, each block feels a restoring force of $-3kx_i$, and so oscillates with frequency $\sqrt{3k/m}$. Although the two normal modes are easy to visualize, it is still amazing that an arbitrary motion of the two blocks can be decomposed as a linear combination of the two modes.

In our problem, the initial conditions are

$$\mathbf{x}(0) = \begin{pmatrix} 1 \\ 0 \end{pmatrix} = \frac{1}{2}(\mathbf{b}_1 + \mathbf{b}_2); \quad \dot{\mathbf{x}}(0) = \begin{pmatrix} 0 \\ 0 \end{pmatrix} = 0\mathbf{b}_1 + 0\mathbf{b}_2, \quad (5.69)$$

so
$$a_1(0) = a_2(0) = 1/2; \quad \dot{a}_1(0) = \dot{a}_2(0) = 0. \quad (5.70)$$

By equation (5.65),

$$\begin{pmatrix} a_1(t) \\ a_2(t) \end{pmatrix} = \frac{1}{2} \begin{pmatrix} \cos(\omega_1 t) \\ \cos(\omega_2 t) \end{pmatrix}, \quad (5.71)$$

5.3. Second-Order Continuous-Time Evolution: $d^2\mathbf{x}/dt^2 = A\mathbf{x}$

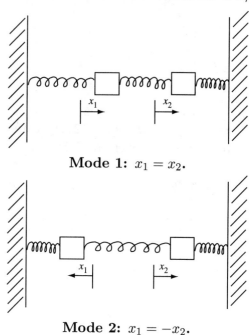

Mode 1: $x_1 = x_2$.

Mode 2: $x_1 = -x_2$.

Figure 5.6: Two normal modes of oscillation

so

$$\mathbf{x}(t) = a_1(t)\mathbf{b}_1 + a_2(t)\mathbf{b}_2 = \frac{1}{2}\begin{pmatrix} \cos(\omega_1 t) + \cos(\omega_2 t) \\ \cos(\omega_1 t) - \cos(\omega_2 t) \end{pmatrix}. \quad (5.72)$$

A small variant of this problem allows for a striking laboratory demonstration. Suppose the three springs are not of the same strength. Rather, suppose that the first and third spring have spring constant k_1, while the middle spring has spring constant $k_2 \ll k_1$. If k_2 were zero, we would have two independent oscillators. Since k_2 is merely small, we have two oscillating blocks that are weakly coupled by the middle spring. As before, we take the initial conditions $\mathbf{x}(0) = (1,0)^T$, $\dot{\mathbf{x}}(0) = (0,0)^T$ and examine how the system behaves.

In this case, we adjust equation (5.67) to account for the different strengths of the springs:

$$\frac{d^2 x_1}{dt^2} = \frac{-(k_1 + k_2)}{m} x_1 + \frac{k_2}{m} x_2,$$

$$\frac{d^2 x_2}{dt^2} = \frac{k_2}{m} x_1 - \frac{k_1 + k_2}{m} x_2. \quad (5.73)$$

This is still of the form (5.44), only now with

$$A = \begin{pmatrix} -(k_1+k_2)/m & k_2/m \\ k_2/m & -(k_1+k_2)/m \end{pmatrix}. \quad (5.74)$$

The eigenvectors are still $\mathbf{b}_1 = (1,1)^T$ and $\mathbf{b}_2 = (1,-1)^T$, but the eigenvalues are now $-k_1/m$ and $-(k_1+2k_2)/m$, respectively. The solution to our problem is still given by (5.72), only now with

$$\omega_1 = \sqrt{k_1/m}; \quad \omega_2 = \sqrt{(k_1+2k_2)/m}. \quad (5.75)$$

Since $k_2 \ll k_1$, ω_2 is very close to ω_1. Let $\nu = \omega_2 - \omega_1$. Then, using the identity

$$\cos(\omega_2 t) = \cos(\omega_1 t + \nu t) = \cos(\omega_1 t)\cos(\nu t) - \sin(\omega_1 t)\sin(\nu t), \quad (5.76)$$

we can rewrite (5.72) as

$$\mathbf{x}(t) = \frac{1}{2}\begin{pmatrix} \cos(\omega_1 t)(1+\cos(\nu t)) - \sin(\omega_1 t)\sin(\nu t) \\ \cos(\omega_1 t)(1-\cos(\nu t)) + \sin(\omega_1 t)\sin(\nu t) \end{pmatrix}. \quad (5.77)$$

As long as $t \ll \pi/(2\nu)$, $1-\cos(\nu t)$ and $\sin(\nu t)$ are both small, so the second block is hardly moving at all. This is to be expected, since we started with only the first block disturbed, and it takes time for the weak middle spring to get the second block moving. Once t gets close to $\pi/(2\nu)$, both blocks are moving in roughly equal amounts. When t gets close to π/ν, then $1+\cos(\nu t)$ and $\sin(\nu t)$ are both small, and the *first* block is hardly moving at all. It has transfered *all* of its energy to the second block. When $t \approx 3\pi/(2\nu)$ both blocks are moving again, and when $t \approx 2\pi/\nu$ the second block has stopped, and the first block is doing all the oscillating. The pattern repeats itself, with energy being gradually passed back and forth between the two blocks, with frequency ν. See Figure 5.7.

5.3. Second-Order Continuous-Time Evolution: $d^2\mathbf{x}/dt^2 = A\mathbf{x}$

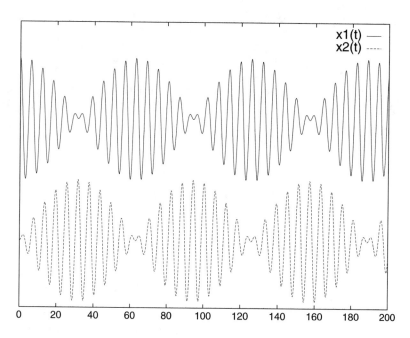

Figure 5.7: The effect of weak coupling, with $\omega_1 = 1$ and $\nu = 0.1$

Exercises

1. Let $A = \begin{pmatrix} 0 & 2 \\ 2 & 0 \end{pmatrix}$, let $\mathbf{x}(0) = \begin{pmatrix} 1 \\ 0 \end{pmatrix}$ and let $\dot{\mathbf{x}}(0) = \begin{pmatrix} 0 \\ 1 \end{pmatrix}$. Solve for $\mathbf{x}(t)$ for all t.

2. Let $A = \begin{pmatrix} 0 & 4 \\ 1 & 0 \end{pmatrix}$, let $\mathbf{x}(0) = \begin{pmatrix} 1 \\ 0 \end{pmatrix}$ and let $\dot{\mathbf{x}}(0) = \begin{pmatrix} 0 \\ 1 \end{pmatrix}$. Solve for $\mathbf{x}(t)$ for all t. Compare to Exercise 1.

3. Let $A = \begin{pmatrix} 0.2 & 1.1 & 2.3 \\ 1.1 & 0.5 & -1.7 \\ 2.3 & -1.7 & 0.7 \end{pmatrix}$, let $\mathbf{x}(0) = \begin{pmatrix} 1 \\ 0 \\ 0 \end{pmatrix}$ and let $\dot{\mathbf{x}}(0) = \begin{pmatrix} 0 \\ 1 \\ 0 \end{pmatrix}$. Find $\mathbf{x}(t)$ for all t. You may wish to use technology to diagonalize A.

Exercises 4–7 refer to generalizations of the coupled spring problem of Figure 5.1. Exercises 8 and 9 require a prior knowledge of inductors and capacitors.

4. Consider the equations for two weakly coupled oscillators (5.73), with $m = k_1 = 1$ and $k_2 = 0.01$. If you start the system with

$\mathbf{x}(0) = \begin{pmatrix} 1 \\ 0 \end{pmatrix}$ and $\dot{\mathbf{x}}(0) = 0$, how long will it be before x_1 and $\dot{x}_1$ are both essentially zero, and (almost) all the energy of motion is transfered to the second block?

5. Consider the three coupled blocks illustrated below, with each block having mass $m = 1$, and each spring having spring constant $k = 1$. Write down the equations of motion for this coupled system. Find the normal modes and oscillation frequencies.

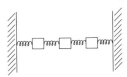

6. Repeat Exercise 5, only with four coupled blocks.

7. Consider an array of N coupled blocks, all of mass m and all springs of constant k. Show that, for any integer j, the vector $\mathbf{x} = (\sin(j\pi/(N+1)), \sin(2j\pi/(N+1)), \ldots, \sin(Nj\pi/(N+1)))^T$ is an eigenvector of the coefficient matrix A. What is the corresponding eigenvalue?

8. Consider the pictured LC circuit with all capacitors having equal capacitance C, and all inductors having equal inductivity L. If the currents around the two loops are I_1 and I_2, show that $\mathbf{x} = (I_1, I_2)^T$ satisfies the differential equation (5.44) for an appropriate coefficient matrix A. Find that matrix, find the normal modes, and find the oscillation frequencies. What happens if the middle capacitance is made different from the other two?

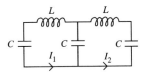

9. Consider the pictured LC circuit with all capacitors having equal capacitance C, and all inductors having equal inductivity L. Set up the differential equations for $\mathbf{x} = (I_1, I_2, I_3)^T$, and solve for initial conditions $\mathbf{x}(0) = (1, 0, 0)^T$, and $\dot{\mathbf{x}}(0) = 0$.

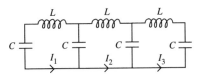

5.4 Reducing Second-Order Problems to First-Order

In Section 5.3 we discussed how to solve $d^2\mathbf{x}/dt^2 = A\mathbf{x}$ if the eigenvalues of A are real. However, there are cases where A has complex eigenvalues, and cases where $d^2\mathbf{x}/dt^2$ depends not only on $\mathbf{x}$ but also on $\dot{\mathbf{x}}$. We therefore need a procedure to solve the system of equations

$$\frac{d^2\mathbf{x}}{dt^2} = A\mathbf{x} + B\frac{d\mathbf{x}}{dt}, \qquad (5.78)$$

with A and B as general as possible. As usual, before solving the problem when $\mathbf{x} \in \mathbb{R}^m$, we consider the case where x is a single variable.

Suppose that

$$\frac{d^2x}{dt^2} = ax + b\frac{dx}{dt}. \qquad (5.79)$$

If we define a new variable $y = dx/dt$, then $dx/dt = y$ (by definition), and $dy/dt = ax + by$. Therefore x and y together satisfy the coupled first-order system

$$\frac{d}{dt}\begin{pmatrix} x \\ y \end{pmatrix} = \begin{pmatrix} 0 & 1 \\ a & b \end{pmatrix}\begin{pmatrix} x \\ y \end{pmatrix}, \qquad (5.80)$$

which may be solved by the method of Section 5.2. The characteristic polynomial is $p_A(\lambda) = \lambda^2 - b\lambda - a$, and so the eigenvalues are

$$\lambda_{1,2} = \frac{b \pm \sqrt{b^2 + 4a}}{2}. \qquad (5.81)$$

The corresponding eigenvectors take the form

$$\mathbf{b}_i = \begin{pmatrix} 1 \\ \lambda_i \end{pmatrix}, \qquad (5.82)$$

and our general solution is of the form

$$x(t) = c_1 e^{\lambda_1 t} + c_2 e^{\lambda_2 t}; \qquad y(t) = c_1 \lambda_1 e^{\lambda_1 t} + c_2 \lambda_2 e^{\lambda_2 t}. \qquad (5.83)$$

A problem arises if $b^2 + 4a = 0$ (for example, if $a = b = 0$). In that case, the two eigenvalues are the same, and it turns out that the matrix is not diagonalizable. However, we can analyze this case as a limit of the cases where $\lambda_1 \ne \lambda_2$. We use the fact that, regardless of the values of a and b, the set of solutions to (5.79) is a 2-dimensional subspace of the space of functions of t. If λ_2 is not equal to λ_1, then we have shown that a basis for this space of solutions is $e^{\lambda_1 t}$ and $e^{\lambda_2 t}$. However, the functions

$$f_1(t) = e^{\lambda_1 t}; \qquad f_2(t) = \frac{e^{\lambda_2 t} - e^{\lambda_1 t}}{\lambda_2 - \lambda_1} \qquad (5.84)$$

also form a basis. Taking the limit of these functions as $\lambda_2 \to \lambda_1$ gives the two functions $e^{\lambda_1 t}$ and $te^{\lambda_1 t}$, which do indeed solve (5.79) in the case that $b^2 + 4a = 0$. In this case the general solution is

$$x(t) = c_1 e^{\lambda_1 t} + c_2 t e^{\lambda_1 t}; \qquad y(t) = c_1 \lambda_1 e^{\lambda_1 t} + c_2 (1 + \lambda_1 t) e^{\lambda_1 t}. \qquad (5.85)$$

The case $b^2 + 4a = 0$ can also be solved using power vectors. Here $\lambda = b/2$ is an eigenvalue with algebraic multiplicity two, and geometric multiplicity one, $\mathbf{b}_1 = (1, \lambda)^T$ is an eigenvector, and $\mathbf{b}_2 = (0, 1)^T$ is a power vector of order 2, with $(A - \lambda I)\mathbf{b}_2 = \mathbf{b}_1$, and $(A - \lambda I)^2 \mathbf{b}_2 = 0$. As before,

$$e^{At} \mathbf{b}_1 = e^{\lambda t} \mathbf{b}_1 = \begin{pmatrix} e^{\lambda t} \\ \lambda e^{\lambda t} \end{pmatrix}, \qquad (5.86)$$

corresponding to the solution $x = e^{\lambda t}$, $y = \dot{x} = \lambda e^{\lambda t}$. We also compute

$$\begin{aligned} e^{At} \mathbf{b}_2 &= e^{\lambda t} e^{(A-\lambda I)t} \mathbf{b}_2 = e^{\lambda t}(\mathbf{b}_2 + t(A - \lambda I)\mathbf{b}_2) \\ &= e^{\lambda t}(\mathbf{b}_2 + t\mathbf{b}_1) = \begin{pmatrix} t e^{\lambda t} \\ (1 + \lambda t) e^{\lambda t} \end{pmatrix}, \end{aligned} \qquad (5.87)$$

corresponding to the solution $x = te^{\lambda t}$, $y = \dot{x} = (1 + \lambda t)e^{\lambda t}$. Since every possible initial condition $(x(0), y(0))^T$ is a linear combination of $\mathbf{b}_1$ and $\mathbf{b}_2$, the general solution to our equations is again given by (5.85).

We now turn to the coupled system of equations (5.78). If we define new variables $y_i = dx_i/dt$, then $d\mathbf{x}/dt = \mathbf{y}$, and $d\mathbf{y}/dt = A\mathbf{x} + B\mathbf{y}$, so

$$\frac{d}{dt} \begin{pmatrix} \mathbf{x} \\ \mathbf{y} \end{pmatrix} = \begin{pmatrix} 0 & I \\ A & B \end{pmatrix} \begin{pmatrix} \mathbf{x} \\ \mathbf{y} \end{pmatrix}. \qquad (5.88)$$

The behavior of the system is governed by the eigenvalues and eigenvectors of the $2m \times 2m$ matrix in (5.88). Unfortunately, these cannot be computed simply from the eigenvalues of A and the eigenvalues of B, since A and B may have different eigenvectors. In general, we have no alternative but to diagonalize the entire $2m \times 2m$ matrix $\begin{pmatrix} 0 & I \\ A & B \end{pmatrix}$.

5.4. Reducing Second-Order Problems to First-Order

Once we have diagonalized this matrix, however, we can read off the solutions to (5.78). To each eigenvalue λ there is a solution $\mathbf{x}(t) = \mathbf{v}e^{\lambda t}$, where $\begin{pmatrix} \mathbf{v} \\ \lambda \mathbf{v} \end{pmatrix}$ is the corresponding eigenvector. Sometimes there are multiple eigenvalues, in which case there may be more than one mode that grows at the same rate.

If an eigenvalue happens to be deficient, with deficiency $m_a(\lambda) - m_g(\lambda)$, then, in addition to solutions that grow as $e^{\lambda t}$, there are solutions that grow as powers of t times $e^{\lambda t}$. The number of linearly independent solutions of this sort is precisely the deficiency of λ, with each such solution coming from a power vector of order greater than 1.

Higher order discrete-time evolution problems, also called difference equations, can be approached in the same way. Consider the seconnd-order difference equation

$$\mathbf{x}(n) = A\mathbf{x}(n-2) + B\mathbf{x}(n-1). \qquad (5.89)$$

Defining the new variable $\mathbf{y}(n) = \mathbf{x}(n-1)$, we obtain

$$\begin{pmatrix} \mathbf{y} \\ \mathbf{x} \end{pmatrix}(n) = \begin{pmatrix} 0 & I \\ A & B \end{pmatrix} \begin{pmatrix} \mathbf{y} \\ \mathbf{x} \end{pmatrix}(n-1). \qquad (5.90)$$

That is, we have converted a second-order system in m variables into a first-order system in $2m$ variables, which can then be solved by the methods of Section 5.1.

Exercises

1. Convert the third-order system of ODEs $d^3\mathbf{x}/dt^3 = A\mathbf{x} + B d\mathbf{x}/dt + C d^2\mathbf{x}/dt^2$ in m variables into a first-order system in $3m$ variables. What matrix governs this first-order system? Given an eigenvalue to this matrix, how do you go about constructing a solution to the original third order system?

2. Generalize the situation of Exercise 1 to cover k-th order systems of ODEs.

3. Suppose that $x(0) = 0$, $x(1) = 1$, and for $n \geq 2$, $x(n) = x(n-1) + 2x(n-2)$. Convert this to a 2×2 first-order matrix problem, and solve to get a closed-form expression for $x(n)$.

4. Consider the $m \times m$ second-order system $d^2\mathbf{x}/dt^2 = A\mathbf{x}$, for a diagonalizable matrix with real eigenvalues λ_i (some positive and some negative) and corresponding eigenvectors $\mathbf{b}_i$. Convert this to a

$2m \times 2m$ first-order system. Solve the first-order system, and show how the solutions of the first-order system correspond to solutions to the original second-order system.

5. Find the general solution to (5.78) if $A = \begin{pmatrix} -1 & 0 \\ 0 & -1 \end{pmatrix}$ and $B = \begin{pmatrix} 0 & 1 \\ 1 & 0 \end{pmatrix}$.

6. Find the general solution to (5.78) if $A = \begin{pmatrix} 0 & 1 \\ 1 & 0 \end{pmatrix}$ and $B = \begin{pmatrix} 0 & 1 \\ -1 & 0 \end{pmatrix}$.

7. Find the general solution to (5.78) if $A = \begin{pmatrix} -1 & 0 \\ 0 & -1 \end{pmatrix}$ and $B = \begin{pmatrix} 0 & 1 \\ -1 & 0 \end{pmatrix}$. This describes the motion of a charged particle in a quadratic potential and a constant magnetic field.

5.5 Long-Time Behavior and Stability

If A is a diagonalizable matrix with eigenvalues $\lambda_1, \ldots, \lambda_m$ and eigenvectors $\mathbf{b}_1, \ldots, \mathbf{b}_m$, then we have seen that the solutions to the discrete-time evolution equation

$$\mathbf{x}(n) = A\mathbf{x}(n-1) \tag{5.91}$$

take the form

$$\mathbf{x}(n) = \sum_{i=1}^{m} c_i \lambda_i^n \mathbf{b}_i. \tag{5.92}$$

If we are given initial conditions, then we may compute the constants c_i from

$$\mathbf{x}(0) = \sum_{i=1}^{m} c_i \mathbf{b}_i. \tag{5.93}$$

Similarly, the solutions to the first-order system of differential equations

$$\frac{d\mathbf{x}}{dt} = A\mathbf{x} \tag{5.94}$$

are all of the form

$$\mathbf{x}(t) = \sum_{i=1}^{m} c_i e^{\lambda_i t} \mathbf{b}_i, \tag{5.95}$$

5.5. Long-Time Behavior and Stability

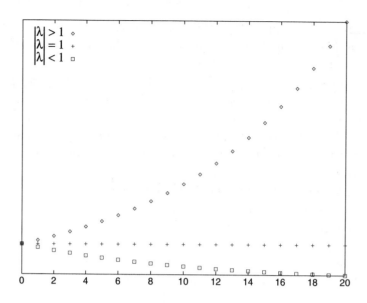

Figure 5.8: Stability for first-order discrete-time evolution

while the solutions to the second-order system

$$\frac{d^2\mathbf{x}}{dt^2} = A\mathbf{x} \qquad (5.96)$$

are of the form

$$\mathbf{x}(t) = \sum_{i=1}^{m}(c_i e^{\sqrt{\lambda_i}\,t} + d_i e^{-\sqrt{\lambda_i}\,t})\mathbf{b}_i, \qquad (5.97)$$

where the constants c_i and d_i can be computed from initial conditions. We wish to understand how such solutions behave as $n \to \infty$ or $t \to \infty$.

For the discrete problem we consider each term in the sum (5.92). See Figure 5.8. If $|\lambda_i| < 1$, then $|c_i \lambda_i^n| \to 0$ as $n \to \infty$. The term is shrinking exponentially in time, and we say the mode is *stable*. Stable modes cannot contribute to the limiting form of $\mathbf{x}(n)$. If $|\lambda_i| > 1$, then, as long as $c_i \neq 0$, $|c_i \lambda_i^n|$ grows exponentially to infinity. We say such modes are *unstable*. If $|\lambda_i| = 1$, then $|c_i \lambda_i^n|$ neither grows nor shrinks, and we say the mode is *neutrally stable*.

Now suppose that the largest eigenvalue, in absolute value, is λ_i. This is called the *dominant eigenvalue*, and the corresponding eigenvector $\mathbf{b}_i$ is called the *dominant mode*. If $c_i \neq 0$, then $c_i \lambda_i^n \mathbf{b}_i$ will eventually be larger than any other term in the sum (5.92), indeed larger than the sum of all the other terms. Regardless of whether $|\lambda_i|$ is bigger, smaller, or equal to one, the *direction* of $\mathbf{x}(n)$ will approach that of $\mathbf{b}_i$. In other words, the asymptotic behavior of a solution is controlled by its largest eigenvalue and the corresponding eigenvector, as long as the corresponding coefficient is nonzero. In order for a system to be stable, *all* of its eigenvalues must have $|\lambda_i| \leq 1$. If even one eigenvalue has magnitude greater than 1, that mode will grow in time, and the system will eventually grow without bounds.

In the owls-mice example, the largest eigenvalue was 1, indicating that there was a steady-state solution. The other eigenvalue was 0.7, which is smaller than 1. The second eigenvector represented deviations from steady-state, and those deviations shrank with time. The limiting distribution was proportional to $\mathbf{b}_1$.

We next turn our attention to first-order continuous-time problems. Here the $\mathbf{b}_i$ term in the solution (5.95) grows as $e^{\lambda_i t}$. Remember that λ_i may be complex. Since

$$|e^{\lambda t}| = e^{\lambda_R t}, \quad (5.98)$$

where λ_R is the real part of λ, the long-time behavior depends on whether λ_R is positive, negative, or zero. Modes whose eigenvalues have positive real parts grow exponentially with time, and are called unstable. Modes whose eigenvalues have negative real parts shrink exponentially, and are called stable. Modes with pure imaginary eigenvalues are neutrally stable; while their phases may oscillate, their amplitudes neither grow nor shrink in time. See Figure 5.9.

Just as in the discrete-time problem, for large time, the sum (5.95) will typically be dominated by a single term. In this case the dominant eigenvalue is the one with the largest real part. To understand the long-time behavior of a system, you just have to understand the dominant eigenvector, as long as its coefficient is nonzero. A system is stable only if *all* its eigenvalues have nonpositive real parts.

For the second-order systems, the eigenvalue λ_i is associated to $e^{\pm\sqrt{\lambda_i}\,t}$ behavior. What matters is the real part of $\pm\sqrt{\lambda_i}$. However, if the real part of $\sqrt{\lambda_i}$ is negative, then the real part of $-\sqrt{\lambda_i}$ is positive, and vice-versa. Thus the only way a mode can be stable is if the eigenvalue has pure imaginary square roots. In other words, the eigenvalue itself must be real and negative. In such cases, the mode neither grows nor shrinks with time, but merely oscillates. The

5.5. Long-Time Behavior and Stability

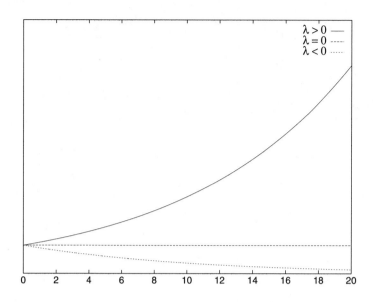

Figure 5.9: Stability for first-order continuous-time evolution

only way to have a stable system of solutions to (5.44) is if all the eigenvalues of A are real and negative. See Figure 5.10.

In some sense, we already know that. If λ_i is negative, then the corresponding solutions involve sines and cosines of $\omega_i t$, where $\omega_i^2 = -\lambda_i$. Sines and cosines neither grow nor shrink with time. However, if λ_i is positive, or complex, then one of the two modes $e^{\pm\sqrt{\lambda_i}\,t}$ is exponentially growing, while the other is exponentially shrinking. Unless the coefficient of the growing mode happens to be zero, the system will eventually grow without bound.

To sum up, for discrete-time problems a mode is stable if $|\lambda_i| < 1$, unstable if $|\lambda_i| > 1$, and neutral if $|\lambda_i| = 1$. For first-order differential equations, a mode is stable if the real part of λ_i is negative, unstable if it is positive, and neutrally stable if λ is pure imaginary. For second-order differential equations of the form (5.96), a mode is unstable unless λ_i is real and negative. In that case, the mode is neutrally stable, and represents oscillations whose amplitude neither grows nor shrinks with time. For a system to be stable, all of its modes must be stable. Even a single unstable mode (with a nonzero coefficient) will cause the solution to grow without bound.

One can sometimes discover the existence of an unstable mode (or a stable mode) by looking at the determinant or trace of the coefficient matrix A. For discrete-time evolution problems, look at

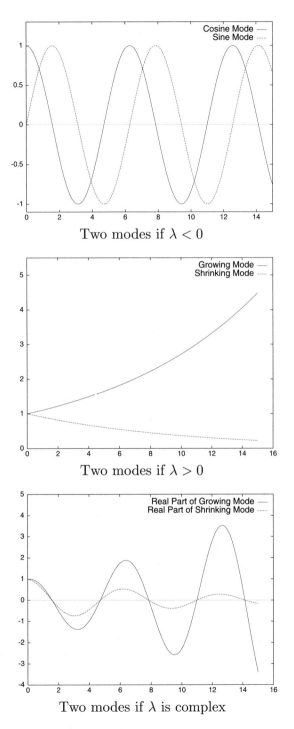

Figure 5.10: Stability for second-order continuous-time evolution

5.5. Long-Time Behavior and Stability

the determinant. If the determinant of A has magnitude greater than one, then at least one of the eigenvalues must be greater than one, so the system is unstable. If the determinant is less than one, then there is at least one stable mode, but there may be some unstable modes, so we cannot say whether the system as a whole is stable. For first-order continuous-time problems, look at the trace of A. If the real part of the trace of A is positive, then at least one of the eigenvalues has positive real part, and the system is unstable. If the real part of the trace is negative, then there is at least one stable mode, but we cannot immediately tell whether any unstable modes exist.

Exercises

For Exercises 1–6, identify the stable, unstable, and neutrally stable modes, and indicate whether or not the system as a whole is stable.

1. $\mathbf{x}(n) = A\mathbf{x}(n-1)$, with $A = \begin{pmatrix} 0 & 1 \\ 1 & 1 \end{pmatrix}$.

2. $\mathbf{x}(n) = A\mathbf{x}(n-1)$, with $A = \begin{pmatrix} 0 & 2 \\ 1 & 1 \end{pmatrix}$.

3. $d\mathbf{x}/dt = A\mathbf{x}$, with $A = \begin{pmatrix} 1 & 1 \\ -1 & 0 \end{pmatrix}$.

4. $d^2\mathbf{x}/dt^2 = A\mathbf{x}$, with $A = \begin{pmatrix} -1 & 4 \\ 1 & -1 \end{pmatrix}$.

5. $d\mathbf{x}/dt = A\mathbf{x}$, with $A = \begin{pmatrix} 0 & 1 & -1 \\ -1 & 0 & 1 \\ 1 & -1 & 0 \end{pmatrix}$.

6. $d^2\mathbf{x}/dt^2 = A\mathbf{x}$, with $A = \begin{pmatrix} 0 & 1 & 0 \\ 1 & 0 & 1 \\ 0 & 1 & 0 \end{pmatrix}$.

7. Let $\mathbf{x}(n) = A\mathbf{x}(n-1)$, with $A = \begin{pmatrix} 0 & k \\ 1 & 1 \end{pmatrix}$, where k is an unspecified real constant. For what values of k is the system stable? For what values is there one unstable mode? For what values are there two unstable modes?

8. Let $\mathbf{x}(n) = A\mathbf{x}(n-1)$, with A a real 2×2 matrix whose determinant equals one. Show that the system is neutrally stable if $-2 < Tr(A) < 2$, but is unstable if $|Tr(A)| > 2$.

9. In the deer-growth model of Exercises 9 and 10 of Section 5.1, suppose that the survival rate for juveniles is p_1, the survival rate for adults is p_2, and the number of births per adult per year is k. Find a criterion involving p_1, p_2, and k that will determine whether the deer population eventually grows, stabilizes, or shrinks to zero.

5.6 Markov Chains and Probability Matrices[2]

Many systems in the real world have a memory; where they are tomorrow depends not only on where they are today, but also on where they were yesterday. These systems are difficult to model mathematically, and the delayed action of yesterday's news can lead to dramatic feedback effects.

In this section we consider a simpler possibility. We imagine a system that can be in a finite number of states, and assume that the state of the system tomorrow depends on the state of the system today, and on a roll of the dice, but not on any past history. Such a system is called a *Markov chain*. We will see that the probabilities in such systems evolve according to equation (5.2), with the matrix A taking a particular form.

For example, the weather in a certain southwestern city is (almost) always either sunny (state 1), cloudy (state 2), or rainy (state 3). If it is sunny today, there is a 70% chance of it being sunny tomorrow, a 20% chance of clouds, and a 10% chance of rain. If it's cloudy today, there is 50% chance of sun tomorrow, a 30% chance of clouds, and a 20% chance of rain. If it rains today, then tomorrow has a 30% chance of sun, a 30% chance of clouds, and a 40% chance of rain. If it's raining today, what is the weather likely to be 3 days from now? 10 days from now? A year from now?

Let a_{ij} be the probability of being in state i tomorrow, given that we're in state j today. a_{ij} is called a *transition probability*, and the matrix

$$A = \begin{pmatrix} a_{11} & a_{12} & a_{13} \\ a_{21} & a_{22} & a_{23} \\ a_{31} & a_{32} & a_{33} \end{pmatrix} = \begin{pmatrix} 0.7 & 0.5 & 0.3 \\ 0.2 & 0.3 & 0.3 \\ 0.1 & 0.2 & 0.4 \end{pmatrix} \quad (5.99)$$

is called the *transition matrix*. Notice that every entry of A is between 0 and 1, and that each column of A adds up to 1. This is because probabilities are always between 0 and 1, and, no matter

---

[2] This material is not used in the remainder of the book, and can be skipped on first reading.

5.6. Markov Chains and Probability Matrices

what state we're in today, there is a 100% chance of being in *some* state tomorrow. These properties occur frequently enough to deserve a name:

Definition *A probability matrix is a square matrix with all nonnegative entries such that each column adds up to 1. A regular probability matrix is a probability matrix with strictly positive entries.*

Let $x_i(n)$ be the probability of being in state i on day n. Now

$$\begin{aligned} x_1(n+1) &= P(\text{sun tomorrow}) \\ &= P(\text{sun today and sun tomorrow}) \\ &\quad + P(\text{clouds today and sun tomorrow}) \\ &\quad + P(\text{rain today and sun tomorrow}) \\ &= a_{11}P(\text{sun today}) + a_{12}P(\text{clouds today}) \\ &\quad + a_{13}P(\text{rain today}) \\ &= a_{11}x_1(n) + a_{12}x_2(n) + a_{13}x_3(n). \end{aligned} \quad (5.100)$$

The other probabilities evolve by

$$x_i(n+1) = \sum_j a_{ij} x_j(n), \quad (5.101)$$

or, in matrix form,

$$\mathbf{x}(n+1) = A\mathbf{x}(n). \quad (5.102)$$

This is something we know how to solve! We just find the eigenvalues and eigenvectors of A, decompose the initial condition $\mathbf{x}(0) = (0, 0, 1)^T$ (i.e., rain on day zero) as a linear combination of eigenvectors, multiply each coefficient by the n-th power of the corresponding eigenvalue, and recombine to get $\mathbf{x}(n)$.

In this example, the eigenvalues are $\lambda_1 = 1$, $\lambda_2 = (2+\sqrt{2})/10 \approx 0.3414$ and $\lambda_3 = (2-\sqrt{2})/10 \approx 0.0586$, and the eigenvectors are

$$\mathbf{b}_1 = \begin{pmatrix} 18/31 \\ 15/62 \\ 11/62 \end{pmatrix} \approx \begin{pmatrix} 0.5806 \\ 0.2419 \\ 0.1774 \end{pmatrix}, \mathbf{b}_2 = \begin{pmatrix} -\sqrt{2}-2 \\ 1 \\ \sqrt{2}+1 \end{pmatrix} \approx \begin{pmatrix} -3.414 \\ 1 \\ 2.414 \end{pmatrix}$$

and $\mathbf{b}_3 = \begin{pmatrix} \sqrt{2} \\ -1-\sqrt{2} \\ 1 \end{pmatrix} \approx \begin{pmatrix} 1.414 \\ -2.414 \\ 1 \end{pmatrix}$. The initial condition decomposes as

$$\mathbf{x}(0) = (0, 0, 1)^T = \mathbf{b}_1 + 0.2553\mathbf{b}_2 + 0.2060\mathbf{b}_3, \quad (5.103)$$

so

$$\mathbf{x}(n) = 1^n \mathbf{b}_1 + 0.2553(0.3414)^n \mathbf{b}_2 + 0.2060(0.0586)^n \mathbf{b}_3. \quad (5.104)$$

128   Chapter 5. Some Crucial Applications

Plugging in $n = 2$, $n = 3$ and $n = 10$ we get

$$\mathbf{x}(2) = \begin{pmatrix} 0.48 \\ 0.27 \\ 0.25 \end{pmatrix}; \quad \mathbf{x}(3) = \begin{pmatrix} 0.546 \\ 0.252 \\ 0.202 \end{pmatrix}; \quad \mathbf{x}(10) \approx \begin{pmatrix} 0.580626 \\ 0.241941 \\ 0.177433 \end{pmatrix}.$$
(5.105)

In other words, if it is raining today, the day after tomorrow has a 48% chance of sun, a 27% chance of clouds, and a 25% chance of rain; 3 days from now there is about a 55% chance of sun, a 25% chance of clouds, and a 20% chance of rain. 10 days from now there is about a 58% chance of sun, a 24% chance of clouds, and an 18% chance of rain. When n is large, λ_2^n and λ_3^n both go to zero, and we are left with

$$\lim_{n \to \infty} \mathbf{x}(n) = \mathbf{b}_1 \approx \begin{pmatrix} 0.580645 \\ 0.241936 \\ 0.177419 \end{pmatrix}.$$
(5.106)

This limit is approached quite quickly; $\mathbf{x}(10)$ is already practically equal to $\mathbf{b}_1$.

This convergence to a steady-state limiting probability distribution is typical of Markov chains. The general properties of Markov chains and probability matrices are listed in the following theorem:

Theorem 5.1 *If A is a probability matrix, then*

1. $\lambda_1 = 1$ *is an eigenvalue of A. The entries of the corresponding eigenvector $\mathbf{b}_1$ are all non-negative. We can normalize $\mathbf{b}_1$ so that the sum of its entries is 1.*

2. *The sum of the entries of $\mathbf{x}(n) = A^n \mathbf{x}(0)$ is the same as the sum of the entries of $\mathbf{x}(0)$. If the entries of $\mathbf{x}(0)$ are all non-negative, then the entries of $\mathbf{x}(n)$ are all non-negative.*

3. *If ξ is an eigenvector with eigenvalue not equal to 1, then the sum of the entries of ξ is zero.*

4. *A has no eigenvalues of magnitude greater than 1 (i.e., there are no unstable modes).*

5. *If A is a regular probability matrix, then the only neutrally stable mode is $\mathbf{b}_1$. In this case, $\mathbf{x}(n)$ converges exponentially to $c\mathbf{b}_1$, where c is the sum of the entries of $\mathbf{x}(0)$. Also, in this case A^n converges exponentially to a matrix whose columns are all equal to $\mathbf{b}_1$.*

5.6. Markov Chains and Probability Matrices

Proof: We prove the statements in the order listed. Let $\mathbf{r} = (1, \ldots, 1)$. Since each column of A adds up to 1, we have

$$\mathbf{r}A = \mathbf{r}, \qquad (5.107)$$

and consequently also

$$\mathbf{r}A^n = \mathbf{r} \qquad (5.108)$$

for all n. This implies that $\mathbf{r}^T$ is an eigenvector of A^T with eigenvalue equal to 1. But the characteristic polynomials of A and A^T are the same, so $\lambda_1 = 1$ must also be an eigenvalue of A. The fact that the entries of $\mathbf{b}_1$ all can be chosen non-negative will appear later.

To get the sum of the entries of a vector, you just multiply $\mathbf{r}$ by that vector. Thus

$$\sum_i x_i(n) = \mathbf{r}\mathbf{x}(n) = \mathbf{r}A^n\mathbf{x}(0) = \mathbf{r}\mathbf{x}(0) = \sum_i x_i(0). \qquad (5.109)$$

As for the entries of $\mathbf{x}(n)$ being non-negative, we know that the entries of A are non-negative. Since $x_i(n) = \sum_j a_{ij} x_j(n-1)$, and since the entries a_{ij} are all non-negative, the non-negativity of $\mathbf{x}(n)$ follows from that of $\mathbf{x}(n-1)$. By induction, the non-negativity of $\mathbf{x}(n)$ then follows from that of $\mathbf{x}(0)$.

Now suppose that $A\xi = \lambda\xi$, with $\lambda \neq 1$. Then

$$\sum \xi_i = \mathbf{r}\xi = (\mathbf{r}A)\xi = \mathbf{r}(A\xi) = \lambda\mathbf{r}\xi = \lambda \sum_i \xi_i. \qquad (5.110)$$

Since the sum of the entries of ξ equals λ times itself, it must equal zero. Since the vector $\mathbf{e}_1 = (1, 0, 0, \ldots, 0)^T$, whose entries add up to 1, can be expanded as a linear combination of eigenvectors, and since all the eigenvectors except $\mathbf{b}_1$ have entries that add up to zero, the entries for $\mathbf{b}_1$ cannot add up to zero. We normalize $\mathbf{b}_1$ so that its entries add up to 1.

Now let $\mathbf{x}(0)$ be a general vector with non-negative entries that add up to 1. We have shown that $\mathbf{x}(n)$ is then a vector with non-negative entries that add up to 1. In particular, each entry $x_i(n)$ remains between 0 and 1 for all n. We decompose $\mathbf{x}(0)$ as a linear combination of eigenvectors

$$\mathbf{x}(0) = \sum_{i=1}^m c_i \mathbf{b}_i, \qquad (5.111)$$

and we can choose $\mathbf{x}(0)$ so that all the coefficients c_i are nonzero. If there were any unstable modes, then as n grew large the sum

$$\mathbf{x}(n) = \sum_{i=1}^{m} c_i \lambda_i^n \mathbf{b}_i \qquad (5.112)$$

would diverge, and the various entries of $\mathbf{x}(n)$ would grow large or become negative. Since the entries remain between 0 and 1, such divergence is impossible, and there are no unstable modes.

For statement 5, suppose that A is regular. Let α equal the smallest entry. Then $A = (A - \alpha B) + \alpha B$, where

$$B = \begin{pmatrix} 1 & 1 & \cdots & 1 \\ 1 & 1 & \cdots & 1 \\ \vdots & \vdots & \ddots & \vdots \\ 1 & 1 & \cdots & 1 \end{pmatrix}. \qquad (5.113)$$

Now suppose $\mathbf{v}$ is an eigenvector of A, other than $\mathbf{b}_1$, with eigenvalue λ. Since the entries of $\mathbf{v}$ add up to zero, $B\mathbf{v} = 0$, so $\mathbf{v}$ is an eigenvector of $A - \alpha B$, also with eigenvalue λ. But $A - \alpha B$ is itself $(1 - m\alpha)$ times a probability matrix, since the entries of $A - \alpha B$ are all non-negative and each column adds up to $1 - m\alpha$. Since probability matrices have no eigenvalues larger than 1, λ cannot be larger, in magnitude, than $1 - m\alpha$, which is strictly less than 1. Thus $\mathbf{v}$ cannot be a neutrally stable mode.

Since the first eigenvalue of A is $\lambda_1 = 1$, we write

$$\mathbf{x}(n) = c_1 \mathbf{b}_1 + \sum_{i=2}^{m} c_i \lambda_i^n \mathbf{b}_i. \qquad (5.114)$$

Sum the entries on both sides. Since the sum of the entries of $\mathbf{b}_1$ is 1, and the sum of the entries of $\mathbf{b}_i$ is zero, c_1 must equal the sum of the entries of $\mathbf{x}(n)$, which equals the sum of the entries of $\mathbf{x}(0)$. Since $|\lambda_i| < 1$, the right hand side of (5.114) converges exponentially to $c_1 \mathbf{b}_1$, the difference being bounded by a constant times the n-th power of the second-largest eigenvalue, hence by a constant times $(1 - m\alpha)^n$. As for the convergence of A^n, note that the i-th column of A^n is $A^n \mathbf{e}_i$, which we have shown converges to $\mathbf{b}_1$, since $\mathbf{e}_i$ is a vector whose entries add up to 1.

Finally, since $\mathbf{x}(n)/c_1$ converges to $\mathbf{b}_1$, and since the entries of $\mathbf{x}(n)$ are all non-negative for all n, the entries of $\mathbf{b}_1$ must all be non-negative. ■

5.6. Markov Chains and Probability Matrices

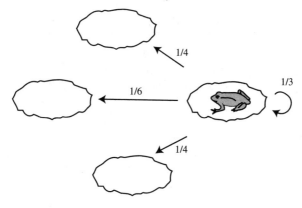

Figure 5.11: Where will the frog be after n jumps?

Corollary 5.2 *If A is a probability matrix and some power of A is regular, then A has no unstable modes, one neutrally stable mode with eigenvalue 1, and the rest stable modes.*

Proof: If A^k is a regular probability matrix, then all but one of the eigenvalues of A^k have norm strictly less than 1. But the eigenvalues of A^k are the k-th powers of the eigenvalues of A, so all but one of the eigenvalues of A must have norm strictly less than 1. ∎

Example: Imagine that a lily pond has four lily pads, arranged in a square. See Figure 5.11. A frog sits on one of the pads. Every time he jumps, the frog has a probability of 1/4 of jumping left, 1/4 of jumping right, 1/6 of jumping to the pad diagonally opposite, and 1/3 of jumping straight up and down, landing on the pad he started on. If the frog starts out on the first pad, what probability distribution describes his position after n jumps?

Numbering the pads clockwise around the square, the transition matrix,

$$A = \begin{pmatrix} 1/3 & 1/4 & 1/6 & 1/4 \\ 1/4 & 1/3 & 1/4 & 1/6 \\ 1/6 & 1/4 & 1/3 & 1/4 \\ 1/4 & 1/6 & 1/4 & 1/3 \end{pmatrix}, \qquad (5.115)$$

is a regular 4×4 probability matrix. Notice that the smallest entry of A has size 1/6. By Theorem 5.1, the largest eigenvalue of A is exactly 1, and the other three have norm at most $1 - 4/6 = 1/3$. Thus the probability distribution $\mathbf{x}(n)$ will converge to a limiting equilibrium at least as fast as $1/3^n$. Furthermore, the limiting distribution is an eigenvector of eigenvalue 1, so $A\mathbf{b}_1 = \mathbf{b}_1$, and it is not hard to see that $\mathbf{b}_1 = (1/4, 1/4, 1/4, 1/4)^T$. In the long run, the frog spends 1/4 of its time on each pad. Without diagonalizing A, we have already understood the qualitative behavior of the Markov chain.

To obtain more precise information, we diagonalize A. The eigenvalues are 1, 0, 1/6, and 1/6, with eigenvectors

$$\mathbf{b}_1 = \begin{pmatrix} 1/4 \\ 1/4 \\ 1/4 \\ 1/4 \end{pmatrix}, \quad \mathbf{b}_2 = \begin{pmatrix} 1 \\ -1 \\ 1 \\ -1 \end{pmatrix}, \quad \mathbf{b}_3 = \begin{pmatrix} 1 \\ 0 \\ -1 \\ 0 \end{pmatrix}, \quad \mathbf{b}_4 = \begin{pmatrix} 0 \\ 1 \\ 0 \\ -1 \end{pmatrix},$$
(5.116)

respectively. We decompose our initial condition as

$$\mathbf{x}(0) = (1, 0, 0, 0)^T = \mathbf{b}_1 + (1/4)\mathbf{b}_2 + (1/2)\mathbf{b}_3, \quad (5.117)$$

so our probability distribution after n jumps is

$$\mathbf{x}(n) = \mathbf{b}_1 + \frac{0^n}{4}\mathbf{b}_2 + \frac{6^{-n}}{2}\mathbf{b}_3 = \frac{1}{4}\begin{pmatrix} 1 + (2/6^n) \\ 1 \\ 1 - (2/6^n) \\ 1 \end{pmatrix}. \quad (5.118)$$

Thus, for example, the probability of our frog being on pad 3 after 2 jumps is $1/4 - 1/72 = 17/72$. As predicted, the probability distribution converges quickly to $\mathbf{b}_1$.

When a Markov chain has a transition matrix that is regular, Theorem 5.1 tells us that the probability distribution converges exponentially to the unique steady-state solution. We now consider what happens if A is not regular. In this case, $\mathbf{x}(n)$ need not converge to $\mathbf{b}_1$, but instead may oscillate around $\mathbf{b}_1$. However, the time average of $\mathbf{x}(n)$ still converges. Before stating the general theorem, we consider a simple example.

Suppose we have a stoplight that can be either red or green. Every minute it changes color. If it starts out red, what color will it be after n minutes? Obviously, if n is even the light will be red, while if n is odd it will be green; there is no convergence as $n \to \infty$. However, there is a well-defined limiting time-average. In the long run, the light will be red half the time, and green half the time. The time-average $(1/2, 1/2)^T$ is $\mathbf{b}_1$, the eigenvector of eigenvalue 1 for the transition matrix $A = \begin{pmatrix} 0 & 1 \\ 1 & 0 \end{pmatrix}$.

5.6. Markov Chains and Probability Matrices

Theorem 5.3 *If A is a probability matrix and $\mathbf{x}(0)$ is a vector whose entries add up to 1, then the time-average*

$$\lim_{N \to \infty} \frac{1}{N} \sum_{n=1}^{N} \mathbf{x}(n) \qquad (5.119)$$

converges to a solution to the steady-state equation $A\mathbf{x} = \mathbf{x}$. If the eigenvalue 1 is nondegenerate, then the limit is $\mathbf{b}_1$.

Proof: As usual, we decompose our initial vector $\mathbf{x}(0)$ as a sum (5.111) of eigenvectors, with $c_1 = 1$, from which we get $\mathbf{x}(n)$ as a sum of eigenvectors. Suppose that the first k eigenvalues are equal to 1 and the remaining $m - k$ eigenvalues are not (typically $k = 1$, but we are considering the general case). Then

$$\begin{aligned}
\frac{1}{N} \sum_{n=1}^{N} \mathbf{x}(n) &= \sum_{i=1}^{k} c_i \mathbf{b}_i + \frac{1}{N} \sum_{j=k+1}^{m} c_j \left(\sum_{n=1}^{N} \lambda_j^n \right) \mathbf{b}_j \\
&= \sum_{i=1}^{k} c_i \mathbf{b}_i + \frac{1}{N} \sum_{j=k+1}^{m} c_j \frac{\lambda_j - \lambda_j^{N+1}}{1 - \lambda_j} \mathbf{b}_j. \quad (5.120)
\end{aligned}$$

As $N \to \infty$, the sum over j is bounded, since each λ_j has magnitude at most 1. Dividing by N, the sum over j disappears in the limit, and we are left with the sum over i. But each $\mathbf{b}_i$ is an eigenvector with eigenvalue 1, so the sum over i gives an element of E_1. Finally, if the eigenvalue 1 is nondegenerate, then $k = 1$ and the sum over i consists of a single term, namely $c_1 \mathbf{b}_1 = \mathbf{b}_1$. ∎

Exercises

1. A certain metropolitan area has a population of 9 million, of whom 5 million live in the city, and 4 million live in the suburbs. Each year, only 20% of the suburbanites move to the city (and 80% stay in the suburbs), while 40% of the city dwellers move to the suburbs (and 60% stay in the city). Set up the time evolution equations and diagonalize the coefficient matrix. (You may ignore the effects of births, deaths, and migration into and out of the metropolitan area.)

2. Compute and graph the city and suburban populations for years 0–10. You may wish to use technology.

3. In the long run, will the population of the city decrease to zero, or stabilize? Give closed-form expressions for the city and suburban populations in n years, and for the limiting distribution.

4. In the previous example, suppose the initial populations were 9 million suburbanites and 3 million city dwellers. How would that affect the long-term distribution?

5. A certain football team derives confidence from each win but gets demoralized after each loss. After winning a game, it has a 90% chance of winning the next game, but after losing a game it has only a 20% chance of winning the next game. In the long run, what faction of the games will this streaky team win?

6. A very different football team gets overconfident after a win but works hard after a loss. After winning a game, it has only a 20% chance of winning the next game, but after losing a game it has a 90% chance of winning the next game. In the long run, what faction of the games will this contrary team win? Qualitatively, explain the difference between this result and the result of the previous problem.

In Exercises 7–10, for each of the following matrices, find the steady state, estimate the convergence rate from Theorem 5.1, and compare to the actual size of the second largest eigenvalue.

7. $\begin{pmatrix} 0.7 & 0.2 & 0.3 \\ 0.1 & 0.5 & 0.3 \\ 0.2 & 0.3 & 0.4 \end{pmatrix}$.

8. $\begin{pmatrix} 0.1 & 0.8 & 0.1 \\ 0.1 & 0.1 & 0.8 \\ 0.8 & 0.1 & 0.1 \end{pmatrix}$.

9. $\begin{pmatrix} 0.5 & 0.45 & 0.1 \\ 0.45 & 0.5 & 0.1 \\ 0.05 & 0.05 & 0.8 \end{pmatrix}$.

10. $\begin{pmatrix} 0.65 & 0.49 & 0.01 & 0.01 \\ 0.33 & 0.49 & 0.01 & 0.01 \\ 0.01 & 0.01 & 0.24 & 0.33 \\ 0.01 & 0.01 & 0.74 & 0.65 \end{pmatrix}$.

For Exercises 11–14, find the neutrally stable modes for the given matrices. Given the initial condition, compute $\mathbf{x}(n)$ for n up to 10, and compute the time-average $(\sum_{k=1}^{n} \mathbf{x}(k))/n$ for n up to 10 (you may wish to use technology for this).

11. $\begin{pmatrix} 0 & 0 & 1 \\ 1 & 0 & 0 \\ 0 & 1 & 0 \end{pmatrix}$, $\mathbf{x}(0) = \begin{pmatrix} 1 \\ 0 \\ 0 \end{pmatrix}$.

5.7. Linear Analysis near Fixed Points of Nonlinear Problems

12. $\begin{pmatrix} 0 & 0.4 & 0 & 0.6 \\ 0.6 & 0 & 0.4 & 0 \\ 0 & 0.6 & 0 & 0.4 \\ 0.4 & 0 & 0.6 & 0 \end{pmatrix}$, $\mathbf{x}(0) = \begin{pmatrix} 1 \\ 0 \\ 0 \\ 0 \end{pmatrix}$.

13. $\begin{pmatrix} 1/3 & 1/2 & 0 & 0 \\ 2/3 & 1/2 & 0 & 0 \\ 0 & 0 & 1/4 & 1/3 \\ 0 & 0 & 3/4 & 2/3 \end{pmatrix}$, $\mathbf{x}(0) = \begin{pmatrix} 1 \\ 0 \\ 0 \\ 0 \end{pmatrix}$.

14. $\begin{pmatrix} 1/3 & 1/2 & 0 & 0 \\ 2/3 & 1/2 & 0 & 0 \\ 0 & 0 & 1/4 & 1/3 \\ 0 & 0 & 3/4 & 2/3 \end{pmatrix}$, $\mathbf{x}(0) = \begin{pmatrix} 0 \\ 0 \\ 1 \\ 0 \end{pmatrix}$.

Compare to the results of Exercise 13.

15. Suppose a matrix L has non-negative off-diagonal entries and columns that add up to zero, show that e^{Lt} is a probability matrix for all t. If the entries of L are all nonzero, show that e^{Lt} is a regular probability matrix. (Matrices of this type come up in continuous-time Markov processes. The entries ℓ_{ij} describe the probability per unit time of jumping from state j to state i, and the probability distribution evolves according to $d\mathbf{x}/dt = L\mathbf{x}$.)

16. Consider a continuous version of the city-suburb problem. Every city dweller has a probability per unit time of 0.4 per year (or 0.4/12 per month or 0.4/365 per day) of moving to the suburbs, and every suburbanite has a probability per unit time of 0.2 per year of moving to the city. Set up the differential equation that governs the two populations and find the general solution. Then find the particular solution corresponding to the initial conditions of Exercise 1. How does the answer differ from the discrete-time problem?

5.7 Linear Analysis near Fixed Points of Nonlinear Problems

There are many linear systems in the real world, but there are far more nonlinear systems. Fortunately, every smooth system, no matter how nonlinear, is approximately linear near its steady state solutions, also known as fixed points. In this section we consider nonlinear discrete-time evolution problems, nonlinear systems of first-order ODE, and nonlinear systems of second-order ODEs, and show how they can be approximated by the linear problems that we have already learned how to solve, and whose qualitative behavior we understand.

Discrete-time problems

We begin with a discrete-time nonlinear evolution equation,

$$\mathbf{x}(n) = \mathbf{f}(\mathbf{x}(n-1)), \tag{5.121}$$

where $\mathbf{f}(\mathbf{x}) = (f_1(\mathbf{x}), \ldots, f_m(\mathbf{x}))^T$ is some vector-valued function. A *fixed point* of such a system is a point $\mathbf{x} = \mathbf{a}$ where

$$\mathbf{f}(\mathbf{a}) = \mathbf{a}. \tag{5.122}$$

If you start the system at $\mathbf{x}(0) = \mathbf{a}$, then you will have $\mathbf{x}(n) = \mathbf{a}$ for all n. However, in the real world it is impossible to specify an initial condition exactly. The best we can do is to make $\mathbf{x}(0)$ very close to $\mathbf{a}$. If $\mathbf{x}(0)$ starts close to $\mathbf{a}$, will $\mathbf{x}(n)$ stay close to $\mathbf{a}$, or will it run away? If it runs away, at what rate does it do so?

To analyze this problem, we set

$$\mathbf{y}(n) = \mathbf{x}(n) - \mathbf{a}. \tag{5.123}$$

The evolution equation for $\mathbf{y}$ is

$$\mathbf{y}(n) = \mathbf{f}(\mathbf{x}(n-1)) - \mathbf{a} = \mathbf{f}(\mathbf{a} + \mathbf{y}(n-1)) - \mathbf{f}(\mathbf{a}). \tag{5.124}$$

As long as the function $\mathbf{f}$ is at least twice differentiable, we can expand this last term in a Taylor series

$$\mathbf{f}(\mathbf{a} + \mathbf{y}) - \mathbf{f}(\mathbf{a}) = \sum_{i=1}^{m} \frac{\partial \mathbf{f}}{\partial x_i}\bigg|_{\mathbf{x}=\mathbf{a}} y_i + O(|\mathbf{y}|^2), \tag{5.125}$$

where "$O(|\mathbf{y}|^2)$", read "order y squared", means a term that is bounded in size by a multiple of $|\mathbf{y}|^2$. Defining a matrix A whose entries are

$$A_{ij} = \frac{\partial f_i}{\partial x_j}\bigg|_{\mathbf{x}=\mathbf{a}}, \tag{5.126}$$

the evolution equation for $\mathbf{y}$ becomes

$$\mathbf{y}(n) = A\mathbf{y}(n-1) + O(|\mathbf{y}(n-1)|^2) \approx A\mathbf{y}(n-1), \tag{5.127}$$

where the approximation is valid as long as $\mathbf{y}$ remains small. The approximation (5.127) is called the *linearization* of the equations (5.121) near the fixed point $\mathbf{a}$.

5.7. Linear Analysis near Fixed Points of Nonlinear Problems 137

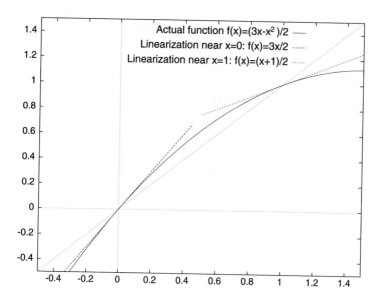

Figure 5.12: Linearization of $f(x) = (3x - x^2)/2$ near fixed points

Example 1: To illustrate the idea of linearization, we first consider a problem in just one dimension, using the function $f(x) = (3x - x^2)/2$ pictured in Figure 5.12. This function has two fixed points, $a = 0$ and $a = 1$. Since $m = 1$, A is just a 1×1 matrix, with the only entry being $f'(a)$. We need to do a separate analysis for each fixed point.

Working near the fixed point $a = 0$, we have $A = f'(0) = 3/2$, and $y = x - 0 = x$, so in this case (5.127) becomes

$$x(n+1) \approx 3x(n)/2. \qquad (5.128)$$

This is an unstable situation, with x growing by a factor of $3/2$ at each step until this approximation (5.128) is no longer valid.

The other fixed point is $a = 1$, where $A = f'(1) = 1/2$ and $y = x - 1$. Here the approximation (5.127) becomes

$$y(n+1) \approx y(n)/2. \qquad (5.129)$$

Thus if x starts out close to 1, $y = x - 1$ shrinks by a factor of two at each step, so x approaches 1 exponentially and the approximation (5.129) remains valid forever. In short, $a = 1$ is stable. Two trajectories, one starting at $x(0) = 0.01$ and the other at $x(0) = 1.1$, are shown in Table 5.1.

In m dimensions, A is an $m \times m$ matrix and we consider its eigenvalues. If the eigenvalues of A are all smaller than 1, then

	Near $a = 0$	Near $a = 1$
n	$x(n)$	$x(n)$
0	0.0100	1.1000
1	0.0149	1.0450
2	0.0223	1.0215
3	0.0332	1.0105
4	0.0493	1.0052
5	0.0727	1.0026
6	0.1064	1.0013
7	0.1540	1.0006
8	0.2191	1.0003
9	0.3046	1.0002
10	0.4105	1.0001

Table 5.1: Two trajectories for $x(n+1) = (3x(n) - x(n)^2)/2$

$\mathbf{y}(n)$ will shrink in size, approaching zero exponentially, with the decay rate determined by the size of the largest eigenvalue. In this case, we say that $\mathbf{a}$ is a *stable fixed point* of the equation (5.121). Moreover, since $\mathbf{y}$ remains small, the approximation (5.127) remains valid forever, justifying our calculation.

If some of the eigenvalues of A are larger than 1, however, then $\mathbf{y}(n)$ will eventually start to grow exponentially, with the growth rate determined by the largest eigenvalue. In this case, we say that $\mathbf{a}$ is an *unstable fixed point*. Eventually, $\mathbf{y}$ grows to the point that the approximation (5.127) ceases to be valid. To understand what happens beyond that point you have to study the full nonlinear equations (5.121), not the linearization (5.127).

If the largest eigenvalue of A has norm equal to 1, then we are on the borderline. $\mathbf{y}(n)$ will not grow or shrink exponentially, but it may grow (or shrink) more slowly, depending on the form of the $O(|\mathbf{y}|^2)$ terms in (5.125).

In summary, $\mathbf{a}$ is a stable fixed point of the nonlinear system of equations (5.121) if and only if the linearization (5.127) is stable. The linearization may not tell us *anything* about what happens when $\mathbf{x}$ is far from $\mathbf{a}$, but it tells us *everything* about what happens when $\mathbf{x}$ is close to $\mathbf{a}$.

Example 2: Consider the evolution

$$\begin{pmatrix} x_1(n) \\ x_2(n) \end{pmatrix} = \begin{pmatrix} x_2(n-1)^2 \\ x_1(n-1)^3 \end{pmatrix} \tag{5.130}$$

5.7. Linear Analysis near Fixed Points of Nonlinear Problems

in $\mathbb{R}^2$. There are two fixed points, namely $(0,0)^T$ and $(1,1)^T$. At $\mathbf{a} = (0,0)^T$, the matrix A of derivatives equals $\begin{pmatrix} 0 & 0 \\ 0 & 0 \end{pmatrix}$, both eigenvalues are equal to zero, so the origin is a stable fixed point. If $\mathbf{x}(0)$ is close to the origin, then $\mathbf{x}(n)$ will remain close to the origin for all n. At $\mathbf{a} = (1,1)^T$, however, we have

$$A = \begin{pmatrix} 0 & 2 \\ 3 & 0 \end{pmatrix}, \qquad (5.131)$$

with eigenvalues $\pm\sqrt{6}$. Both eigenvalues are (in magnitude) greater than one, so both modes are unstable, and $(1,1)^T$ is an unstable fixed point. If $\mathbf{x}(0)$ is close to (but not equal to) $(1,1)^T$, then $\mathbf{x}(n) - (1,1)^T$ will increase exponentially until $\mathbf{x}(n)$ is no longer close to $(1,1)^T$.

First-Order nonlinear ODEs

Linearization applies not only to discrete-time evolution problems, but also to continuous-time problems. Suppose that $\mathbf{x}(t)$ satisfies the first-order differential equation

$$d\mathbf{x}/dt = \mathbf{f}(\mathbf{x}). \qquad (5.132)$$

A fixed point is a point $\mathbf{a}$ where $\mathbf{f}(\mathbf{a}) = 0$. Thus, if $\mathbf{x}(0) = \mathbf{a}$, then $\mathbf{x}(t) = \mathbf{a}$ for all time. We wish to understand what happen if $\mathbf{x}(0)$ is merely close to $\mathbf{a}$, and not exactly equal.

Once again we define $\mathbf{y} = \mathbf{x} - \mathbf{a}$, rewrite the differential equation in terms of $\mathbf{y}$, and expand the right-hand side in a Taylor series around $\mathbf{y} = 0$. The result is

$$d\mathbf{y}/dt = A\mathbf{y} + O(|\mathbf{y}|^2) \approx A\mathbf{y}, \qquad (5.133)$$

where the matrix A is again given by (5.126). Again we look at the solutions to the linearization (in this case (5.133)) as a guide to what happens to solutions to the exact equations (5.132).

If the eigenvalues of A all have negative real part, then $\mathbf{y}$ shrinks exponentially, so $\mathbf{x}$ approaches $\mathbf{a}$ exponentially. The rate of approach is governed by the eigenvalue of A with greatest real part. In this case we say that $\mathbf{a}$ is a stable fixed point. If one or more eigenvalues have positive real part, then $\mathbf{y}$ will eventually grow exponentially, with rate governed by the eigenvalue with greatest real part, until $\mathbf{y}$ is so large that the approximation (5.133) ceases to be valid. In this case **a** is an unstable fixed point. If the eigenvalue with greatest real part is pure imaginary, then **y** will not grow or shrink exponentially, but may grow or shrink at a lesser rate, depending on the $O(|\mathbf{y}|^2)$ terms. Such borderline cases can be very delicate.

As an example, consider the differential equation

$$\frac{d}{dt}\begin{pmatrix} x_1 \\ x_2 \end{pmatrix} = \begin{pmatrix} x_1^2 - x_2 \\ x_1 - x_2 \end{pmatrix}. \tag{5.134}$$

The fixed points are $(0,0)^T$ and $(1,1)^T$, and

$$A = \begin{pmatrix} 2x_1 & -1 \\ 1 & -1 \end{pmatrix}. \tag{5.135}$$

At $\mathbf{a} = (0,0)^T$, this matrix has eigenvalues $(-1 \pm i\sqrt{3})/2$. Since both eigenvalues have negative real part, both modes are stable and $(0,0)^T$ is a stable fixed point. At $\mathbf{a} = (1,1)^T$ the eigenvalues are $(1 \pm \sqrt{5})/2$. One eigenvalue is positive, indicating an unstable mode, while the other eigenvalue is negative, indicating a stable mode. Since it only takes one unstable mode to make a fixed point unstable, $(1,1)^T$ is an unstable fixed point.

Nonlinear forces

Finally, we consider a case of great physical importance. Suppose a particle of mass m is subject to a force **F** that depends on its position **x**. In that case its acceleration will be $\mathbf{f}(\mathbf{x}) = \mathbf{F}(\mathbf{x})/m$. In most cases, **F** will be minus the gradient of a potential energy function $V(\mathbf{x})$, but the analysis below applies even if the force field is not conservative. We are considering the differential equation

$$\frac{d^2\mathbf{x}}{dt^2} = \mathbf{f}(\mathbf{x}). \tag{5.136}$$

A fixed point **a** is a point where the force $\mathbf{f}(\mathbf{a})$ is zero. Such a point might be a minimum of the potential energy, or a maximum, or a saddle point. If you start the system at $\mathbf{x} = \mathbf{a}$ with zero velocity, then it will stay there forever. What happens if you bump the system a little, that is starting with $\mathbf{x}(0)$ merely close to **a**, and with $\dot{\mathbf{x}}(0)$ merely close to zero?

Once again we define $\mathbf{y} = \mathbf{x} - \mathbf{a}$, rewrite the differential equation in terms of **y**, and expand the right-hand side in a Taylor series around $\mathbf{y} = 0$. The result is

$$d^2\mathbf{y}/dt^2 = A\mathbf{y} + O(|\mathbf{y}|^2) \approx A\mathbf{y}, \tag{5.137}$$

5.7. Linear Analysis near Fixed Points of Nonlinear Problems

where the matrix A is yet again given by (5.126).

A particularly important case is when the force comes from a potential energy function. In that case

$$A_{ij} = \frac{-1}{m} \frac{\partial^2 V}{\partial x_i \partial x_j}\bigg|_{\mathbf{x}=\mathbf{a}}. \tag{5.138}$$

Since mixed partials are equal, this means that A is a real symmetric matrix. We will see in Chapter 7 that such matrices are always diagonalizable and have real eigenvalues. The only question is whether the eigenvalues are positive or negative.

Whether or not $\mathbf{f}$ comes from a potential energy functional, A determines the behavior of the system. If A has any positive eigenvalues, or any complex eigenvalues, then $\mathbf{y}$ will grow exponentially until the approximation (5.137) ceases to be valid; the fixed point is unstable. If all of the eigenvalues are real and negative, then $\mathbf{y}$ will oscillate around 0. In other words, $\mathbf{x}$ will oscillate around the stable fixed point $\mathbf{a}$, and the frequency and pattern of those oscillations are determined by the eigenvalues and eigenvectors of A.

A word of caution is in order. Since the modes are at best neutrally stable, the nonlinear terms we have neglected can cause oscillations to grow over time, even in the absence of unstable modes. When $\mathbf{f}$ comes from a potential energy function this does not happen (because of conservation of energy), but when $\mathbf{f}$ is not a gradient this typically does happen. This sort of growth is much slower than the exponential growth associated with unstable modes, so our approximation (5.137) remains valid for a (relatively) long time, but may not be valid forever.

As an example, we consider the second-order differential equation

$$\frac{d^2}{dt^2} \begin{pmatrix} x_1 \\ x_2 \end{pmatrix} = \begin{pmatrix} x_1^2 - x_2 \\ x_1 - x_2 \end{pmatrix}. \tag{5.139}$$

This is the same right-hand side as in the nonlinear first-order system of differential equations (5.134). As before, the fixed points are $(0,0)^T$ and $(1,1)^T$,

$$A = \begin{pmatrix} 2x_1 & -1 \\ 1 & -1 \end{pmatrix}. \tag{5.140}$$

and at $\mathbf{a} = (0,0)^T$ the matrix has eigenvalues $(-1 \pm i\sqrt{3})/2$. Since both eigenvalues are complex, both modes are unstable and $(0,0)^T$ is an unstable fixed point. At $\mathbf{a} = (1,1)^T$ the eigenvalues are $(1 \pm \sqrt{5})/2$. The positive eigenvalue indicates an unstable mode, while the negative eigenvalue indicates a neutrally stable mode. As always, it only takes one unstable mode to make a fixed point unstable, so $(1,1)^T$ is an unstable fixed point.

Exercises

For Exercises 1–7, linearize the given nonlinear equations around the given fixed points. For each linearization, indicate which modes are stable, which are neutrally stable, and which are unstable.

1. $x_1(n) = -\sin(x_1(n-1) + x_2(n-1))$, $x_2(n) = \sin(x_1(n-1)) - \sin(x_2(n-1))$, fixed point $x_1 = x_2 = 0$.

2. $x_1(n) = x_1(n-1)^3 - x_2(n-1) + 1$, $x_2(n) = x_1(n-1)^2 + x_2(n-1)^2 - 1$, fixed point $x_1 = x_2 = 1$.

3. $x_1(n) = (x_1(n-1)^3 - x_2(n-1))/4 + 1$, $x_2(n) = (x_1(n-1)^2 + x_2(n-1)^2)/4 + 1/2$, fixed point $x_1 = x_2 = 1$. Compare to Exercise 2.

4. $dx_1/dt = e^{x_1} - e^{x_2}$, $dx_2/dt = e^{x_1} + e^{x_2} - 2$, fixed point $x_1 = x_2 = 0$.

5. $dx_1/dt = e^{(x_1^2)} - e^{(x_2^2)}$, $dx_2/dt = 2\sin(x_1 - x_2)$, fixed point $x_1 = x_2 = 1$.

6. $d^2\mathbf{x}/dt^2 = -\nabla V$, where $\mathbf{x} \in \mathbb{R}^3$ and $V(\mathbf{x}) = x_1^2 + 2x_2^2 + 3x_3^2 + (x_1 + x_2 + x_3)^3$, fixed point $(0,0,0)^T$.

7. $d^2\mathbf{x}/dt^2 = -\nabla V$, where $\mathbf{x} \in \mathbb{R}^3$ and $V(\mathbf{x}) = x_1^3 + x_1^2 - 2x_1 - x_2^3 + 2x_2^2 + 4x_2 - x_3^3 + 3x_3^2 - 12x_3 - 3x_1^2x_2 - 3x_1^2x_3 + 3x_2^2x_1 - 3x_2^2x_3 + 3x_3^2x_1 - 3x_3^2x_2 + 6x_1x_2x_3$, and fixed point $(1,-1,2)^T$.

8. Consider the differential equation $\dfrac{d}{dt}\begin{pmatrix}x_1\\x_2\end{pmatrix} = a(x_1^2+x_2^2)\begin{pmatrix}x_1\\x_2\end{pmatrix}$, where a is a real scalar. Show that the linearization at the fixed point $(0,0)^T$ is neutrally stable, but that the long-time behavior of the nonlinear system depends on the sign of a. Show that if $a < 0$ then all solutions converge to $(0,0)$ as $t \to \infty$, while if $a > 0$ then solutions typically go to infinity.

9. Consider the differential equation $\dfrac{d^2}{dt^2}\begin{pmatrix}x_1\\x_2\end{pmatrix} = (x_1^2+x_2^2)\begin{pmatrix}-x_2\\x_1\end{pmatrix}$.

Show that the linearization at the fixed point $(0,0)^T$ is neutrally stable, but that some solutions to the nonlinear system actually grow with time.

5.7. Linear Analysis near Fixed Points of Nonlinear Problems

10. Euler's equations govern the rotation of a rigid body. If the principal moments of inertia are $I_1 > I_2 > I_3 > 0$, and the angular velocity about the k-th principal axis is denoted ω_k, then Euler's equations are

$$\frac{d\omega_1}{dt} = \frac{I_2 - I_3}{I_1}\omega_2\omega_3,$$
$$\frac{d\omega_2}{dt} = \frac{I_3 - I_1}{I_2}\omega_3\omega_1,$$
$$\frac{d\omega_3}{dt} = \frac{I_1 - I_2}{I_3}\omega_1\omega_2. \qquad (5.141)$$

Show that rotation about the first axis or about the third axis is (neutrally) stable, but that rotation about the second axis is unstable. [This is sometimes called the Tennis Racket Theorem, as it explains why a tennis racket can easily be spun about its handle, and can easily be spun in the plane of its head, but is difficult to flip about its other axis. (Try this yourself!) Juggling pins are always made symmetrical, with $I_1 = I_2$, to avoid this instability.]

Chapter 6

Inner Products

We saw in Chapter 2 that not every vector space comes equipped with a notion of length, and so far we have been considering properties of vector spaces that do not depend on length. However, length can be a very useful concept, especially when defined in terms of an inner product. This chapter is an exploration of that concept, beginning with familiar spaces such as $\mathbb{R}^n$ and working up to less familiar spaces of functions. In Chapter 7 we will consider operators (e.g., rotations) that respect the inner product in some way, and the special properties of those operators.

6.1 Real Inner Products; Definitions and Examples

We begin with $\mathbb{R}^n$, where the inner product is probably familiar to you. If $\mathbf{x}$ and $\mathbf{y}$ are elements of $\mathbb{R}^n$, we define

$$\langle \mathbf{x} | \mathbf{y} \rangle = \mathbf{x}^T \mathbf{y} = x_1 y_1 + x_2 y_2 + \cdots + x_n y_n. \quad (6.1)$$

This is called the *standard inner product* or *dot product* on $\mathbb{R}^n$, and is often denoted $\mathbf{x} \cdot \mathbf{y}$, but we will use Dirac's[1] "bracket" notation $\langle \mathbf{x} | \mathbf{y} \rangle$, which generalizes more naturally to other spaces, particularly complex vector spaces. $\mathbb{R}^n$, equipped with the standard inner product, is called Euclidean $\mathbb{R}^n$, and is sometimes denoted E^n.

---

[1] P.A.M. Dirac was one of the greatest theoretical physicists of the 20th century. In addition to discovering the equation that governs relativistic electrons (the Dirac equation), he laid the mathematical groundwork for much of quantum mechanics.

The inner product $\langle \mathbf{x}|\mathbf{y}\rangle$ is a real-valued function of the two vectors $\mathbf{x}$ and $\mathbf{y}$. This function has some important properties, which are easy to verify:

1. Linearity in the first factor:

$$\langle c_1\mathbf{x} + c_2\mathbf{y}|\mathbf{z}\rangle = c_1\langle \mathbf{x}|\mathbf{z}\rangle + c_2\langle \mathbf{y}|\mathbf{z}\rangle, \tag{6.2}$$

2. Linearity in the second factor:

$$\langle \mathbf{x}|c_1\mathbf{y} + c_2\mathbf{z}\rangle = c_1\langle \mathbf{x}|\mathbf{y}\rangle + c_2\langle \mathbf{x}|\mathbf{z}\rangle, \tag{6.3}$$

3. Symmetry:
$$\langle \mathbf{x}|\mathbf{y}\rangle = \langle \mathbf{y}|\mathbf{x}\rangle, \tag{6.4}$$

4. Positivity. If $\mathbf{x} \neq 0$, then

$$\langle \mathbf{x}|\mathbf{x}\rangle > 0. \tag{6.5}$$

This last property, in particular, allows us to define the length, or *norm* of a vector to be

$$|\mathbf{x}| = \sqrt{\langle \mathbf{x}|\mathbf{x}\rangle}. \tag{6.6}$$

Definition *If V is a real vector space, then an* inner product *on V is any real-valued function $\langle \mathbf{x}|\mathbf{y}\rangle$ of pairs of vectors $(\mathbf{x},\mathbf{y})$ in V, that satisfies properties 1–4. A real vector space on which an inner product is defined is called a* real inner product space.

A function of two vectors that is linear in each one is said to be a *bilinear form*, or *bilinear*. A bilinear form is said to be *symmetric* if it satisfies (6.4) and *positive* if it satisfies (6.5). Thus an inner product is, by definition, the same thing as a positive, symmetric, real-valued bilinear form.

Examples of real vector spaces and inner products include

1. $V = \mathbb{R}^n$ and $\langle \mathbf{x}|\mathbf{y}\rangle$ is the standard inner product (6.1).

2. $V = \mathbb{R}^2$ and $\langle \mathbf{x}|\mathbf{y}\rangle = x_1 y_1 + 2 x_2 y_2$.

3. $V = \mathbb{R}_2[t]$ and $\langle a_0+a_1 t+a_2 t^2 | b_0+b_1 t+b_2 t^2\rangle = a_0 b_0 + a_1 b_1 + a_2 b_2$.

4. $V = \mathbb{R}[t]$ and $\langle f|g\rangle = \int_0^1 f(t)g(t)dt$.

6.1. Real Inner Products: Definitions and Examples

A fundamental property of all inner product spaces is

Theorem 6.1 (Schwarz inequality) *Let V be a real inner product space and let $\mathbf{x}$ and $\mathbf{y}$ be vectors in V. Then*

$$|\langle \mathbf{x}|\mathbf{y}\rangle| \leq |\mathbf{x}|\,|\mathbf{y}|, \tag{6.7}$$

with equality if and only if $\mathbf{x}$ and $\mathbf{y}$ are linearly dependent.

Proof: If $\mathbf{x}$ and $\mathbf{y}$ are linearly dependent, then one is a multiple of the other, and the equality follows from linearity. So suppose $\mathbf{x}$ and $\mathbf{y}$ are linearly independent. Define the family of vectors $\mathbf{v}_t = \mathbf{x} + t\mathbf{y}$, where t ranges from $-\infty$ to ∞. Since $\mathbf{x}$ and $\mathbf{y}$ are linearly independent, $\mathbf{v}_t$ is never zero, so

$$\begin{aligned}0 < |\mathbf{v}_t|^2 &= \langle \mathbf{v}_t|\mathbf{v}_t\rangle = \langle \mathbf{x}+t\mathbf{y}|\mathbf{x}+t\mathbf{y}\rangle \\ &= t^2\langle \mathbf{y}|\mathbf{y}\rangle + 2t\langle \mathbf{x}|\mathbf{y}\rangle + \langle \mathbf{x}|\mathbf{x}\rangle.\end{aligned} \tag{6.8}$$

Since this holds for all t, it certainly holds for $t = -\langle \mathbf{x}|\mathbf{y}\rangle/\langle \mathbf{y}|\mathbf{y}\rangle$, which is the value of t that minimizes the right-hand side of (6.8). Plugging this value of t into (6.8) we get

$$0 < \frac{\langle \mathbf{x}|\mathbf{y}\rangle^2}{\langle \mathbf{y}|\mathbf{y}\rangle} - 2\frac{\langle \mathbf{x}|\mathbf{y}\rangle^2}{\langle \mathbf{y}|\mathbf{y}\rangle} + \langle \mathbf{x}|\mathbf{x}\rangle = \frac{\langle \mathbf{x}|\mathbf{x}\rangle\langle \mathbf{y}|\mathbf{y}\rangle - \langle \mathbf{x}|\mathbf{y}\rangle^2}{\langle \mathbf{y}|\mathbf{y}\rangle}, \tag{6.9}$$

and so

$$|\mathbf{x}|\,|\mathbf{y}| = \sqrt{\langle \mathbf{x}|\mathbf{x}\rangle\langle \mathbf{y}|\mathbf{y}\rangle} > |\langle \mathbf{x}|\mathbf{y}\rangle|. \ \blacksquare \tag{6.10}$$

Definition *Let $\mathbf{x}$ and $\mathbf{y}$ be any two nonzero vectors in an inner product space V. The angle between $\mathbf{x}$ and $\mathbf{y}$ is the unique angle $\theta \in [0, \pi]$ such that*

$$\cos(\theta) = \frac{\langle \mathbf{x}|\mathbf{y}\rangle}{|\mathbf{x}|\,|\mathbf{y}|}. \tag{6.11}$$

Thanks to the Schwarz inequality, the right-hand side of (6.11) is always between -1 and 1, so the angle is well-defined. Moreover, the angle equals 0 precisely when the two vectors are positive multiples of one another, and equals π precisely when the two vectors are negative multiples of one another. With this definition, much of our intuition about the geometry of Euclidean $\mathbb{R}^2$ and $\mathbb{R}^3$ carries over to arbitrary inner product spaces.

Generalized inner products in relativity and mechanics[2]

There exist some bilinear forms which, while not satisfying all the properties of an inner product, satisfy enough to be called "generalized inner products". These forms have great importance in relativity and in classical mechanics.

In special relativity, we use the 4-vector $(x_1, x_2, x_3, x_4)^T$, where x_1, x_2, and x_3 denote spatial position and x_4 denotes time. The bilinear form that keeps appearing, which we call a "Lorentzian inner product", is

$$\langle \mathbf{x} | \mathbf{y} \rangle_L = x_1 y_1 + x_2 y_2 + x_3 y_3 - c^2 x_4 y_4, \qquad (6.12)$$

where c is the speed of light. The subscript L stands for Lorentz. This form is linear in each factor and is symmetric, but is not positive. Indeed, we find that some vectors, such as $(0,0,0,1)^T$, have $\langle \mathbf{x} | \mathbf{x} \rangle_L < 0$. These vectors are called *timelike*. Other vectors have $\langle \mathbf{x} | \mathbf{x} \rangle_L > 0$ and are called *spacelike*, while still others, such as $(c,0,0,1)$, have $\langle \mathbf{x} | \mathbf{x} \rangle_L = 0$ and are called *lightlike*. The history of a particle traces out a curve in space-time, called a *world line*. We can parametrize the world line any way we wish. If $\mathbf{x}(s)$ is such a parametrized world line, then one of the axioms of special relativity is that, if our particle is massless, then $\langle d\mathbf{x}/ds | d\mathbf{x}/ds \rangle = 0$. If our particle is massive, then $\langle d\mathbf{x}/ds | d\mathbf{x}/ds \rangle < 0$. Under no circumstances can $\langle d\mathbf{x}/ds | d\mathbf{x}/ds \rangle$ be positive, as that would correspond to traveling faster than light.

Although the bilinear form $\langle \cdot | \cdot \rangle_L$ is not positive, it has the property that, for any nonzero $\mathbf{x}$, there exists a $\mathbf{y}$ such that $\langle \mathbf{y} | \mathbf{x} \rangle_L \neq 0$. Bilinear forms with this property are said to be *nondegenerate*. A positive form is always nondegenerate (just take $\mathbf{y} = \mathbf{x}$), but, as we have just seen, a form can be nondegenerate without being positive.

We next consider a bilinear form on $\mathbb{R}^2$. Let

$$\langle \mathbf{x} | \mathbf{y} \rangle_A = x_1 y_2 - x_2 y_1. \qquad (6.13)$$

The subscript A stands for "area", since this form gives the area of the parallelogram spanned by $\mathbf{x}$ and $\mathbf{y}$ if $\mathbf{y}$ is counterclockwise of $\mathbf{x}$, and minus the area otherwise (see Exercise 9). This form is bilinear and nondegenerate, but it is not symmetric. In fact,

---

[2] The bilinear forms discussed here are not directly used in the remainder of this book. A reader who is not interested in relativity or Hamiltonian mechanics can skip to the beginning of Section 6.2 without loss of continuity.

6.1. Real Inner Products: Definitions and Examples 149

$\langle \mathbf{x} | \mathbf{y} \rangle_A = -\langle \mathbf{y} | \mathbf{x} \rangle_A$. A form that switches sign when you switch the two arguments is called *anti-symmetric* or *skew-symmetric*.

There is an extension of $\langle \cdot | \cdot \rangle_A$ to higher dimensions that is of great importance in classical mechanics. Let $(x_1, \ldots, x_n, p_1, \ldots, p_n)$ be the coordinates of $\mathbb{R}^{2n}$. In mechanics, the x's denote generalized positions, while the p's denote generalized momenta. We define the *standard symplectic form*

$$\langle \mathbf{x}', \mathbf{p}' | \mathbf{x}, \mathbf{p} \rangle_S = \sum_i (x_i' p_i - x_i p_i'). \tag{6.14}$$

This form, like $\langle \cdot | \cdot \rangle_A$, is skew-symmetric.

The symplectic form appears in classical mechanics as follows. If f and g are any two functions of position and momentum, then the *Poisson bracket* of f and g, denoted $\{f, g\}$, is defined to be

$$\{f, g\} = \langle \nabla f | \nabla g \rangle_S = \sum_i \left(\frac{\partial f}{\partial x_i} \frac{\partial g}{\partial p_i} - \frac{\partial f}{\partial p_i} \frac{\partial g}{\partial x_i} \right). \tag{6.15}$$

In particular, $\{x_i, g\} = \partial g / \partial p_i$ and $\{p_i, g\} = -\partial g / \partial x_i$. There is a function $H(p, q)$, called the Hamiltonian, and the Hamiltonian determines the evolution of all functions of $\mathbf{x}$ and $\mathbf{p}$ via the equation

$$df/dt = \{f, H\}. \tag{6.16}$$

In particular, taking $f = x_i$ or $f = p_i$ we have *Hamilton's equations of motion*:

$$dx_i/dt = \partial H / \partial p_i; \qquad dp_i/dt = -\partial H / \partial x_i. \tag{6.17}$$

Exercises

1–3. Show that the inner products of examples 2–4, respectively, are indeed bilinear, symmetric, and positive.

4. Compute $\langle (1, 2)^T | (4, 5)^T \rangle$ using the inner product of Example 2.

5. Compute $\langle t^2 + t - 3 | 3t^2 - 2t + 5 \rangle$ using the inner product of Example 3.

6. Compute $\langle t^2 + 3t + 1 | 4t^2 - 5t + 7 \rangle$ using the inner product of Example 4.

7. On $\mathbb{R}^2$, is the bilinear form $\langle (x_1, x_2)^T | (y_1, y_2)^T \rangle = x_1 y_2 + x_2 y_1$ an inner product? Why or why not?

8. In an inner product space V, consider a triangle with vertices A, B, and C. Let a, b, and c be the lengths of the sides opposite A, B, and C, respectively, and let θ be the angle at C. Prove the law of cosines: $c^2 = a^2 + b^2 - 2ab\cos(\theta)$.

9. Let $\mathbf{x}$ and $\mathbf{y}$ be vectors in Euclidean $\mathbb{R}^2$, and consider the parallelogram with vertices at the origin, $\mathbf{x}$, $\mathbf{x}+\mathbf{y}$, and $\mathbf{y}$. Show that the area of this parallelogram is $\pm\langle\mathbf{x}|\mathbf{y}\rangle_A$, the sign depending on whether $\mathbf{x}$ is clockwise or counterclockwise of $\mathbf{y}$.

10. Given a Hamiltonian $H(\mathbf{x}, \mathbf{p}) = \sum_i p_i^2/2m + V(\mathbf{x})$, show that Hamilton's equations (6.17) reduce to Newton's equations $\mathbf{F} = m\mathbf{a}$, where $\mathbf{F} = -\nabla V$. The function $V(x)$ is called the potential energy.

6.2 Complex Inner Products

In this section we consider inner products for complex vector spaces, beginning with $\mathbb{C}^n$. The simple formula (6.1), which works so well for $\mathbb{R}^n$, fails badly for $\mathbb{C}^n$. To get the squared norm of a complex number you must multiply the complex number by its conjugate, not by itself. Similarly, to get the squared length of a complex vector, you must take

$$|\mathbf{x}|^2 = \sum_i |x_i|^2 = \sum_i \bar{x}_i x_i, \qquad (6.18)$$

not $\sum_i x_i^2$ ($\bar{x}_i$ denotes the complex conjugate of x_i). We extend the formula (6.18) for length to a definition of the standard complex inner product

$$\langle\mathbf{x}|\mathbf{y}\rangle = \bar{\mathbf{x}}^T\mathbf{y} = \bar{x}_1 y_1 + \bar{x}_2 y_2 + \cdots + \bar{x}_n y_n. \qquad (6.19)$$

With this definition,[3] the relation (6.6) between inner product and length continues to hold.

The properties of the standard complex inner product $\langle\mathbf{x}|\mathbf{y}\rangle$ are slightly different from the real case, and may be easily verified:

1. Conjugate-linearity in the first factor:

$$\langle c_1\mathbf{x} + c_2\mathbf{y}|\mathbf{z}\rangle = \bar{c}_1\langle\mathbf{x}|\mathbf{z}\rangle + \bar{c}_2\langle\mathbf{y}|\mathbf{z}\rangle, \qquad (6.20)$$

---

[3] Warning: It is *extremely* common for people to forget about the conjugation and use (6.1) instead of (6.19). This error can lead to absurd results, such as the vector $(1, 2i)^T$ having imaginary length!

6.2. Complex Inner Products

2. Linearity in the second factor:

$$\langle \mathbf{x} | c_1 \mathbf{y} + c_2 \mathbf{z} \rangle = c_1 \langle \mathbf{x} | \mathbf{y} \rangle + c_2 \langle \mathbf{x} | \mathbf{z} \rangle, \qquad (6.21)$$

3. Conjugate-symmetry:

$$\langle \mathbf{x} | \mathbf{y} \rangle = \overline{\langle \mathbf{y} | \mathbf{x} \rangle}, \qquad (6.22)$$

4. Positivity. If $\mathbf{x} \neq 0$, then

$$\langle \mathbf{x} | \mathbf{x} \rangle > 0. \qquad (6.23)$$

Notice the difference between conjugate-linearity and linearity. Multiplying $\mathbf{x}$ by $a + bi$ results in multiplying $\langle \mathbf{x} | \mathbf{y} \rangle$ by $a - bi$, not by $a + bi$. However, multiplying $\mathbf{y}$ by $a + bi$ results in multiplying $\langle \mathbf{x} | \mathbf{y} \rangle$ by $a + bi$.

Definition *A function of two vectors that is conjugate-linear (also called* antilinear*) in the first factor, and linear in the second, is called a* sesquilinear form*. A sesquilinear form that is conjugate-symmetric is called* Hermitian, *or a* Hermitian form.

The standard inner product on $\mathbb{C}^n$ is a positive Hermitian form. As in the real case, we extend this to a definition of an inner product on a general complex vector space.[4]

Definition *If V is a complex vector space, then an* inner product *on V is any complex-valued function $\langle \mathbf{x} | \mathbf{y} \rangle$ of pairs of vectors $(\mathbf{x}, \mathbf{y})$ in V that satisfies properties (6.20–6.23). That is, an inner product is a positive Hermitian form. A complex vector space on which an inner product is defined is called a* complex inner product space.

To get examples of complex inner product spaces we need only make small changes to the real examples:

1. $V = \mathbb{C}^n$ and $\langle \mathbf{x} | \mathbf{y} \rangle$ is the standard inner product (6.19).

2. $V = \mathbb{C}^2$ and $\langle \mathbf{x} | \mathbf{y} \rangle = \bar{x}_1 y_1 + 2\bar{x}_2 y_2$.

---

[4] Many mathematics texts define inner products to be linear in the first factor and conjugate-linear in the second. Almost all physics texts have inner products linear in the second factor and conjugate-linear in the first. We are following the physics convention.

3. $V = \mathbb{C}_2[t]$ and $\langle a_0+a_1t+a_2t^2|b_0+b_1t+b_2t^2\rangle = \bar{a}_0b_0+\bar{a}_1b_1+\bar{a}_2b_2$.

4. $V = \mathbb{C}[t]$ and $\langle f|g\rangle = \int_0^1 \overline{f(t)}g(t)dt$.

The real Schwarz inequality generalizes to

Theorem 6.2 *Let V be a complex inner product space and let $\mathbf{x}$ and $\mathbf{y}$ be vectors in V. Then*

$$|\langle \mathbf{x}|\mathbf{y}\rangle| \leq |\mathbf{x}|\,|\mathbf{y}|, \tag{6.24}$$

with equality if and only if $\mathbf{x}$ and $\mathbf{y}$ are (complex) linearly dependent.

Exercises

1. Check that the examples of complex inner products given above do indeed satisfy properties (6.20–6.23).

2. Prove Theorem 6.2. Follow the proof of Theorem 6.1, making adjustments for the fact that $\langle \mathbf{x}|\mathbf{y}\rangle$ and $\langle \mathbf{y}|\mathbf{x}\rangle$ are conjugate rather than equal. Two possible approaches are: 1) allowing t to be complex, or 2) multiplying $\mathbf{x}$ by a unit complex number chosen so that $\langle \mathbf{y}|\mathbf{x}\rangle$ becomes real.

3. In $\mathbb{C}^2$ with the standard inner product, compute $|\mathbf{x}|$ and $\langle \mathbf{x}|\mathbf{x}\rangle$, where $\mathbf{x} = \begin{pmatrix} 1 \\ i \end{pmatrix}$.

4. In $\mathbb{C}^4$ with the standard inner product, compute $|\mathbf{x}|$, $|\mathbf{y}|$, $\langle \mathbf{x}|\mathbf{y}\rangle$, and $\langle \mathbf{y}|\mathbf{x}\rangle$, where $\mathbf{x} = (1, i, 1+i, 2)^T$ and $\mathbf{y} = (2, -i, 1+i, 3)^T$.

5. In Example 2, compute $|\mathbf{x}|$, $|\mathbf{y}|$, $\langle \mathbf{x}|\mathbf{y}\rangle$, and $\langle \mathbf{y}|\mathbf{x}\rangle$, where $\mathbf{x} = (1, i)^T$ and $\mathbf{y} = (3i, 5)^T$.

6. In Example 3, compute $|\mathbf{x}|$, $|\mathbf{y}|$, $\langle \mathbf{x}|\mathbf{y}\rangle$, and $\langle \mathbf{y}|\mathbf{x}\rangle$, where $\mathbf{x} = 2t^2 + (1+i)t - 3i$ and $\mathbf{y} = t - 2i$.

7. In Example 4, compute $|\mathbf{x}|$, $|\mathbf{y}|$, $\langle \mathbf{x}|\mathbf{y}\rangle$ and $\langle \mathbf{y}|\mathbf{x}\rangle$, where $\mathbf{x} = 2t^2 + (1+i)t - 3i$ and $\mathbf{y} = t - 2i$.

8. Let V be a real inner product space and let V_c be its complexification. If $\langle \cdot|\cdot\rangle$ is an inner product on V, we define a form $\langle \cdot|\cdot\rangle_c$ on V_c by the formula

$$\langle \mathbf{x}|\mathbf{y}\rangle_c = \langle \mathbf{x}_R|\mathbf{y}_R\rangle + \langle \mathbf{x}_I|\mathbf{y}_I\rangle + i(\langle \mathbf{x}_R|\mathbf{y}_I\rangle - \langle \mathbf{x}_I|\mathbf{y}_R\rangle), \tag{6.25}$$

where $\mathbf{x}_R$ and $\mathbf{x}_I$ are the real and imaginary parts of $\mathbf{x}$, and similarly with $\mathbf{y}$. Show that V_c is a complex inner product.

6.3. Bras, Kets, and Duality

9. Let V be Euclidean $\mathbb{R}^n$. Show that the procedure of Exercise 2.8 yields $\mathbb{C}^n$ with $\langle \cdot | \cdot \rangle_c$ being the standard complex inner product (6.19).

6.3 Bras, Kets, and Duality

The notation we have been using suggests that the bracket $\langle \mathbf{x} | \mathbf{y} \rangle$ is somehow the product of a "bra" $\langle \mathbf{x} |$ and a "ket" $| \mathbf{y} \rangle$.[5] In this section we make sense of these separate expressions, beginning as usual with the case of Euclidean $\mathbb{R}^n$.

In $\mathbb{R}^n$, with the standard inner product, we have already defined $\mathbf{x}$ and $\mathbf{y}$ to be column vectors. Let

$$|\mathbf{y}\rangle = \mathbf{y} = \begin{pmatrix} y_1 \\ \vdots \\ y_n \end{pmatrix}. \qquad (6.26)$$

The inner product $\langle \mathbf{x} | \mathbf{y} \rangle$ is then the matrix product of the row

$$\langle \mathbf{x} | = \mathbf{x}^T = (x_1, \ldots, x_n) \qquad (6.27)$$

with the column $|\mathbf{y}\rangle$.

In $\mathbb{C}^n$, with the standard inner product (6.19), things are almost as easy. We again take $|\mathbf{y}\rangle = \mathbf{y} = \begin{pmatrix} y_1 \\ \vdots \\ y_n \end{pmatrix}$, but now we must take

$$\langle \mathbf{x} | = \bar{\mathbf{x}}^T = (\bar{x}_1, \ldots, \bar{x}_n). \qquad (6.28)$$

(On $\mathbb{R}^n$, the formula (6.28) is the same as (6.26), since the complex conjugate of a real number equals the number itself.)

Notice that $\langle \mathbf{x} |$ is not an element of $\mathbb{C}^n$ (or $\mathbb{R}^n$). Elements of $\mathbb{C}^n$ (or $\mathbb{R}^n$) are columns, and $\langle \mathbf{x} |$ is a row, not a column. Rows can be added and multiplied by scalars, so the space of all possible rows is a vector space in its own right, which we denote $\mathbb{C}^{n*}$ (or $\mathbb{R}^{n*}$). $\mathbb{C}^{n*}$ is called the *dual space* of $\mathbb{C}^n$. We will discuss dual spaces at greater length later in this section.

Next we consider $\mathbb{R}^n$ with an arbitrary inner product, not necessarily the one given by (6.1). Let $\mathbf{G}$ be a matrix whose matrix elements are given by

$$\mathbf{G}_{ij} = \langle \mathbf{e}_i | \mathbf{e}_j \rangle. \qquad (6.29)$$

---

[5] This whimsical terminology is due to Dirac, and has become standard in quantum mechanics.

Then we can expand

$$\langle \mathbf{x}|\mathbf{y}\rangle = \langle \sum_i x_i \mathbf{e}_i | \sum_j y_j \mathbf{e}_j \rangle = \sum_{i,j} x_i g_{ij} y_j$$

$$= (x_1, \ldots, x_n) \begin{pmatrix} & & \\ & \mathbf{G} & \\ & & \end{pmatrix} \begin{pmatrix} y_1 \\ \vdots \\ y_n \end{pmatrix} = \mathbf{x}^T \mathbf{G} \mathbf{y}. \quad (6.30)$$

The matrix $\mathbf{G}$ is called the *metric*. It has the property that $\mathbf{G}^T = \mathbf{G}$, since $\langle \mathbf{e}_i | \mathbf{e}_j \rangle = \langle \mathbf{e}_j | \mathbf{e}_i \rangle$. The expression (6.30) is useful in its own right, but we are looking for an expression for the inner product as the matrix product of a row vector and a column vector. To do this we must define the bra $\langle \mathbf{x}|$ to be

$$\langle \mathbf{x}| = \mathbf{x}^T \mathbf{G} = (x_1, \ldots, x_n) \begin{pmatrix} & & \\ & \mathbf{G} & \\ & & \end{pmatrix}, \quad (6.31)$$

rather than $\mathbf{x}^T = (x_1, \ldots, x_n)$. The point of this example is that the vector space $\mathbb{R}^n$ is the same as in the standard case, and so the kets are the same as always, as is the *space* of rows. However, the inner product is nonstandard, so the choice of *which* row corresponds to a given column $\mathbf{x}$ is different. Since the bra $\langle \mathbf{x}|$ means "take the inner product with $\mathbf{x}$", the form of $\langle \mathbf{x}|$ depends not only on the vector $\mathbf{x}$, but also on the metric $\mathbf{G}$.

If we are working with $\mathbb{C}^n$ with an arbitrary inner product, we need both to involve a metric and to keep track of complex conjugates. We define a metric by (6.29), only now $\mathbf{G}$ is *not* symmetric in general. Rather, $g_{ji} = \bar{g}_{ij}$. As before, we compute

$$\langle \mathbf{x}|\mathbf{y}\rangle = \langle \sum_i x_i \mathbf{e}_i | \sum_j y_j \mathbf{e}_j \rangle = \sum_{i,j} \bar{x}_i g_{ij} y_j$$

$$= (\bar{x}_1, \ldots, \bar{x}_n) \begin{pmatrix} & & \\ & \mathbf{G} & \\ & & \end{pmatrix} \begin{pmatrix} y_1 \\ \vdots \\ y_n \end{pmatrix} = \bar{\mathbf{x}}^T \mathbf{G} \mathbf{y}. \quad (6.32)$$

To rewrite this as a product of a bra and a ket we must take

$$\langle \mathbf{x}| = \bar{\mathbf{x}}^T \mathbf{G} = (\bar{x}_1, \ldots, \bar{x}_n) \begin{pmatrix} & & \\ & \mathbf{G} & \\ & & \end{pmatrix}. \quad (6.33)$$

Next, suppose V is a (real or complex) inner product space with the basis $\mathcal{B} = \{\mathbf{b}_1, \ldots, \mathbf{b}_n\}$. We have seen that a basis makes V look

6.3. Bras, Kets, and Duality

like $\mathbb{R}^n$ or $\mathbb{C}^n$, depending on whether V is real or complex. However, the basis generally does not make the inner product on V look like the standard inner product on $\mathbb{R}^n$ (or $\mathbb{C}^n$). Instead, the basis makes V look like the examples considered in the last two paragraphs.

If $\mathbf{y} = \sum_i a_i \mathbf{b}_i$, we associate the ket vector $|\mathbf{x}\rangle = \mathbf{x}$ with the column vector $(a_1, \ldots, a_n)^T = [\mathbf{y}]_\mathcal{B}$. When emphasising bras and kets we will write

$$|\mathbf{y}\rangle_\mathcal{B} = [\mathbf{y}]_\mathcal{B} = \begin{pmatrix} a_1 \\ \vdots \\ a_n \end{pmatrix}. \tag{6.34}$$

To represent $\langle \mathbf{x}|$ in the $\mathcal{B}$ basis we need a row vector which, when multiplied by $|\mathbf{y}\rangle_\mathcal{B}$, gives $\langle \mathbf{x}|\mathbf{y}\rangle$. If $\mathbf{x} = \sum c_i \mathbf{b}_i$, we have

$$\langle \mathbf{x}|\mathbf{y}\rangle = \langle \sum_i c_i \mathbf{b}_i | \sum_j a_j \mathbf{b}_j \rangle = \sum_{i,j} \bar{c}_i a_j \langle \mathbf{b}_i|\mathbf{b}_j\rangle. \tag{6.35}$$

(If V is a real vector space we can dispense with complex conjugation.) Defining the metric matrix $\mathbf{G}_\mathcal{B}$ by

$$(\mathbf{G}_\mathcal{B})_{ij} = \langle \mathbf{b}_i|\mathbf{b}_j\rangle, \tag{6.36}$$

we have

$$\langle \mathbf{x}|\mathbf{y}\rangle = \sum_{i,j} \bar{c}_i g_{ij} a_j = (\bar{c}_1, \ldots, \bar{c}_n) \begin{pmatrix} & & \\ & \mathbf{G}_\mathcal{B} & \\ & & \end{pmatrix} \begin{pmatrix} a_1 \\ \vdots \\ a_n \end{pmatrix}. \tag{6.37}$$

So we must take

$$_\mathcal{B}\langle \mathbf{x}| = (\bar{c}_1, \ldots, \bar{c}_n) \begin{pmatrix} & & \\ & \mathbf{G}_\mathcal{B} & \\ & & \end{pmatrix}. \tag{6.38}$$

For example, in $\mathbb{R}^2$ with the usual inner product, take $\mathcal{B} = \{(1,0)^T, (1,1)^T\}$. See Figure 6.1. Then

$$\mathbf{G}_\mathcal{B} = \begin{pmatrix} 2 & 1 \\ 1 & 1 \end{pmatrix} \tag{6.39}$$

and the inner product of $\mathbf{v} = a_1 \mathbf{b}_1 + a_2 \mathbf{b}_2$ and $\mathbf{w} = c_1 \mathbf{b}_1 + c_2 \mathbf{b}_2$ is

$$\begin{pmatrix} a_1 & a_2 \end{pmatrix} \begin{pmatrix} 2 & 1 \\ 1 & 1 \end{pmatrix} \begin{pmatrix} c_1 \\ c_2 \end{pmatrix}. \tag{6.40}$$

In this case

$$_\mathcal{B}\langle \mathbf{v}| = (2a_1 + a_2, a_1 + a_2) \quad \text{and} \quad |\mathbf{w}\rangle_\mathcal{B} = \begin{pmatrix} c_1 \\ c_2 \end{pmatrix}. \tag{6.41}$$

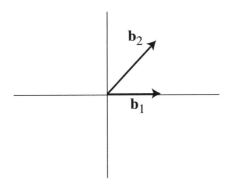

Figure 6.1: A nonstandard basis for $\mathbb{R}^2$ forces us to use $\mathbf{G}_\mathcal{B}$

To sum up, in all cases one can take the ket $|\mathbf{y}\rangle$ to be the same thing as $\mathbf{y}$. If our space is $\mathbb{R}^n$ or $\mathbb{C}^n$, then this is already a column vector. If our space is anything else, we need a basis to make the vector space look like $\mathbb{R}^n$ or $\mathbb{C}^n$. In those cases we take $|\mathbf{y}\rangle_\mathcal{B}$ to be $[\mathbf{y}]_\mathcal{B}$. *Henceforth we will use the symbols $\mathbf{y}$ and $|\mathbf{y}\rangle$ interchangably.*

Defining bras is much trickier, since a bra is not an element of the original vector space. $\langle \mathbf{x}|$ always means "take the inner product with $\mathbf{x}$", so the form of $\langle \mathbf{x}|$ depends not only on the vector space, but also on the inner product. We summarize as follows:

On $\mathbb{R}^n$ with the standard inner product, $\langle \mathbf{x}| = \mathbf{x}^T$. On $\mathbb{C}^n$ with the standard inner product, $\langle \mathbf{x}| = \bar{\mathbf{x}}^T$. On $\mathbb{R}^n$ with nonstandard inner product, $\langle \mathbf{x}| = \mathbf{x}^T \mathbf{G}$. On $\mathbb{C}^n$ with nonstandard inner product, $\langle \mathbf{x}| = \bar{\mathbf{x}}^T \mathbf{G}$. On a general vector space V, we need a basis before we can write bras as row vectors. Given a basis $\mathcal{B}$, the bra $_\mathcal{B}\langle \mathbf{x}|$ is the conjugate of the transpose of $|\mathbf{x}\rangle_\mathcal{B}$ times the metric matrix $\mathbf{G}_\mathcal{B}$. If V happens to be a real vector space, then the conjugation is unnecessary.

Dual spaces[6]

Since it makes sense to add bras and to multiply them by scalars, the set of all bras is a vector space in its own right. It is said to be *dual* to the space of kets. Dual spaces are defined in general:

Definition *Given a (real or complex) vector space V, the* dual space V^* *is the space of linear transformations from V to the scalars.*

---

[6] This subsection can be skipped without loss of continuity.

6.3. Bras, Kets, and Duality

Elements of V^* are sometimes called covectors.

If V has an inner product, then "take the inner product with $\mathbf{x}$" is certainly a map from V to the scalars. However, dual spaces exist even in the absence of an inner product. On $\mathbb{R}^n$ (with or without an inner product), the projection map ϕ_k, defined by $\phi_k(\mathbf{x}) = x_k$ is an element of $\mathbb{R}^{n*}$. In terms of rows,

$$\phi_k = (0, 0, \ldots, 0, 1, 0, \ldots, 0), \tag{6.42}$$

where the 1 appears in the k-th slot. It should be clear that the ϕ_k's form a basis for $\mathbb{R}^{n*}$.

In general, if V is any n-dimensional vector space, then V^* is an n-dimensional vector space. (The proof of this fact is an exercise.) If $\mathcal{B} = \{\mathbf{b}_1, \ldots, \mathbf{b}_n\}$ is a basis for V, then we can define a set of covectors $\{\phi_1, \ldots, \phi_n\}$ by

$$\phi_k(\mathbf{x}) = ([\mathbf{x}]_\mathcal{B})_k. \tag{6.43}$$

The set of ϕ's is called the dual basis to $\mathcal{B}$.

If V is an inner product space with basis $\mathcal{B}$, then there are two natural choices of basis for V^*. One choice is the dual basis $\{\phi_k\}$. The other choice is the set of bras $\{\langle \mathbf{b}_k |\}$. These are typically *not* the same. For example, if $|\mathbf{b}_1| = 2$, then $\langle \mathbf{b}_1 | \mathbf{b}_1 \rangle = 4$, but $\phi_1(\mathbf{b}_1) = 1$, so ϕ_1 and $\langle \mathbf{b}_1 |$ must be different. In the next section we will see that the two bases $\{\phi_k\}$ and $\{\langle \mathbf{b}_k |\}$ agree precisely when the basis $\mathcal{B}$ is orthonormal.

We now consider some properties of the metric matrix $\mathbf{G}_\mathcal{B}$. As with $\mathbb{C}^n$, we always have $(\mathbf{G}_\mathcal{B})_{ji} = \overline{(\mathbf{G}_\mathcal{B})_{ij}}$. If V is a real vector space, then the conjugation is unnecessary, and $(\mathbf{G}_\mathcal{B})_{ji} = (\mathbf{G}_\mathcal{B})_{ij}$. That is, for real vector spaces the metric matrix is always symmetric.

Although $\mathbf{G}_\mathcal{B}$ is a matrix, it should not be viewed as a linear operator on V. The inner product does not really send a vector space to itself. Instead, it sends kets (elements of V) to bras (elements of V^*). As a result, $\mathbf{G}$ does not transform like an operator under a change of basis. If $\mathcal{D}$ is another basis for the real vector space V, then

$$\mathbf{G}_\mathcal{D} = P_{\mathcal{BD}}^T \mathbf{G}_\mathcal{B} P_{\mathcal{BD}}. \tag{6.44}$$

If $\mathbf{G}$ had been an operator, the formula for $\mathbf{G}_\mathcal{D}$ would have involved $P_{\mathcal{BD}}^{-1}$ rather than $P_{\mathcal{BD}}^T$.

If V is complex, then the metric is an *anti*linear map from V to V^*, not a linear map, as converting kets to bras involves complex conjugation:
$$\langle c_1\mathbf{x} + c_2\mathbf{y}| = \bar{c}_1\langle\mathbf{x}| + \bar{c}_2\langle\mathbf{y}|. \tag{6.45}$$

Exercises

1. On $\mathbb{R}^3$ with the standard inner product, compute $\langle\mathbf{x}|$ and $\langle\mathbf{y}|$, where $\mathbf{x} = (1,2,3)^T$ and $\mathbf{y} = (-1,2,5)^T$. Is the inner product $\langle\mathbf{x}|\mathbf{y}\rangle$ the same thing as the product of the row $\langle\mathbf{x}|$ and the column $|\mathbf{y}\rangle$?

2. On $\mathbb{R}^3$ with the inner product $\langle\mathbf{x}|\mathbf{y}\rangle = x_1y_1 + 2x_2y_2 + 3x_3y_3$, compute $\langle\mathbf{x}|$ and $\langle\mathbf{y}|$, where $\mathbf{x} = (1,2,3)^T$ and $\mathbf{y} = (-1,2,5)^T$. Is the inner product $\langle\mathbf{x}|\mathbf{y}\rangle$ the same thing as the product of the row $\langle\mathbf{x}|$ and the column $|\mathbf{y}\rangle$?

3. On $\mathbb{C}^3$ with the standard inner product, compute $\langle\mathbf{x}|$ and $\langle\mathbf{y}|$, where $\mathbf{x} = (1,2,3)^T$ and $\mathbf{y} = (-1,2,5)^T$. Is the inner product $\langle\mathbf{x}|\mathbf{y}\rangle$ the same thing as the product of the row $\langle\mathbf{x}|$ and the column $|\mathbf{y}\rangle$?

4. On $\mathbb{C}^3$ with the standard inner product, compute $\langle\mathbf{x}|$ and $\langle\mathbf{y}|$, where $\mathbf{x} = (1,2i,3+i)^T$ and $\mathbf{y} = (-i,2,5-i)^T$. Is the inner product $\langle\mathbf{x}|\mathbf{y}\rangle$ the same thing as the product of the row $\langle\mathbf{x}|$ and the column $|\mathbf{y}\rangle$? Compute the product of the row $\langle\mathbf{x}|$ with the column $|\mathbf{x}\rangle$.

5. Show that the space of linear maps from a vector space V to the scalars is a vector space in its own right, real if V is real, and complex if V is complex.

6. On $\mathbb{R}_2[t]$ with the inner product $\langle\mathbf{f}|\mathbf{p}\rangle = \int_0^1 f(t)p(t)dt$, and with the standard basis $\mathcal{E} = \{1, t, t^2\}$, compute the metric matrix $\mathbf{G}_\mathcal{E}$. What are $_\mathcal{E}\langle 1+t+t^2|$ and $|1+t+t^2\rangle_\mathcal{E}$? Show that the product of the row $_\mathcal{E}\langle 1+t+t^2|$ and the column $|1+t+t^2\rangle_\mathcal{E}$ is indeed the norm squared of the vector $1+t+t^2$.

7. On $\mathbb{R}_2[t]$ with the inner product $\langle\mathbf{f}|\mathbf{p}\rangle = \int_0^1 f(t)p(t)dt$, consider the basis $\mathcal{B} = \{2, t+t^2, -t+t^2\}$. Compute $\mathbf{G}_\mathcal{B}$ two ways — from the definition (6.36) and from the formula (6.44) — and show that the results agree.

8. Prove formula (6.44) in general.

9. Let $\mathcal{B}$ be a basis for an n-dimensional vector space V, and let ϕ_k be defined as in (6.43). Let α be an arbitrary linear map from V to the scalars, and let $a_k = \alpha(\mathbf{b}_k)$. Show that $\alpha = \sum_{k=1}^n a_k \phi_k$.

6.4. Expansion in Orthonormal Bases: Finding Coefficients

10. Show that the maps ϕ_k of (6.43) are linearly independent (i.e., if $\sum_k a_k \phi_k$ maps every element of V to zero, then $a_k = 0$ for all k).

Exercises 9 and 10 show that V^* is an n-dimensional vector space with basis $\{\phi_k\}$. The α_k's are then the coefficients of α in this (dual) basis, and we write $[\alpha]_\phi = (a_1, \ldots, a_n)$.

11. Let V, $\mathcal{B}$, $\{\phi_k\}$, and α be as above. If $\mathbf{v}$ is an arbitrary element of V, show that $\alpha(\mathbf{v})$ is the matrix product of the row $[\alpha]_\phi$ and the column $[\mathbf{v}]_\mathcal{B}$.

12. Let $\mathcal{B}$ and $\mathcal{D}$ be bases for a real vector space V. If the bases are related by $\mathbf{d}_i = \sum_j A_{ij} \mathbf{b}_j$, show that $P_{\mathcal{BD}} = A^T$.

13. In Exercise 12, let $\{\phi_k\}$ and $\{\psi_k\}$ be bases for V^*, dual to $\mathcal{B}$ and $\mathcal{D}$, respectively. Show that $\psi_i = \sum_j B_{ij} \phi_j$, where $B = (A^{-1})^T$.

14. In the setup of Exercise 13, let $\mathbf{v}$ be an element of V, and let α be an element of V^*. Show that $[\alpha]_\psi = [\alpha]_\phi P_{\mathcal{BD}}$. Note that $[\alpha]_\psi$ and $[\alpha]_\phi$ are rows, not columns, so we must take the product $[\alpha]_\phi P_{\mathcal{BD}}$, not $P_{\mathcal{BD}} [\alpha]_\phi$.

15. Show that the matrix products $[\alpha]_\phi [\mathbf{v}]_\mathcal{B}$ and $[\alpha]_\psi [\mathbf{v}]_\mathcal{D}$ are the same, and equal $\alpha(\mathbf{v})$.

16. If $\mathcal{B}$ and $\mathcal{D}$ are bases for a complex vector space, derive a formula, analagous to (6.44), for $\mathbf{G}_\mathcal{D}$ in terms of $\mathbf{G}_\mathcal{B}$ and $P_{\mathcal{BD}}$.

6.4 Expansion in Orthonormal Bases: Finding Coefficients

As we have seen, computing in a general inner product space, or even $\mathbb{R}^n$ with a nonstandard inner product, can be messy. The matrix $\mathbf{G}$ gets in the way of easy computation. In this section we consider bases in which the matrix $\mathbf{G}$ is simple, allowing for practical applications of the inner product.

Definition *Two vectors, $\mathbf{x}$ and $\mathbf{y}$, are said to be* orthogonal *if $\langle \mathbf{x} | \mathbf{y} \rangle = 0$. A collection of vectors $\{\mathbf{b}_1, \ldots, \mathbf{b}_n\}$ is said to be orthogonal if, whenever $i \neq j$, $\langle \mathbf{b}_i | \mathbf{b}_j \rangle = 0$.*

For the rest of this section, our bases will always be orthogonal. In such cases, the metric $\mathbf{G}_\mathcal{B}$ is diagonal, and multiplying by $\mathbf{G}_\mathcal{B}$ or $\mathbf{G}_\mathcal{B}^{-1}$ becomes quite easy.

Suppose $\mathcal{B}$ is an orthogonal basis for the inner product space V. Then, given a vector $\mathbf{x}$, we can use the inner product to find $[\mathbf{x}]_\mathcal{B}$. If

$$\mathbf{x} = a_1\mathbf{b}_1 + \cdots + a_n\mathbf{b}_n, \tag{6.46}$$

then

$$\langle \mathbf{b}_i | \mathbf{x} \rangle = \sum_j a_j \langle \mathbf{b}_i | \mathbf{b}_j \rangle = a_i \langle \mathbf{b}_i | \mathbf{b}_i \rangle. \tag{6.47}$$

So

$$a_i = \frac{\langle \mathbf{b}_i | \mathbf{x} \rangle}{\langle \mathbf{b}_i | \mathbf{b}_i \rangle}. \tag{6.48}$$

To put it another way,

$$\mathbf{x} = \sum_i \frac{|\mathbf{b}_i\rangle \langle \mathbf{b}_i | \mathbf{x} \rangle}{\langle \mathbf{b}_i | \mathbf{b}_i \rangle}. \tag{6.49}$$

The first term is the part of $\mathbf{x}$ in the $\mathbf{b}_1$ direction, the second term is the part of $\mathbf{x}$ in the $\mathbf{b}_2$ direction, and so on.

Definition *A vector $\mathbf{x}$ is* normalized, *or is a* unit vector, *if $|\mathbf{x}| = 1$. If a set of vectors is both orthogonal and normalized, the set is said to be* orthonormal.

Orthonormal bases are even better than orthogonal bases. If $\mathcal{B}$ is an orthonormal basis, then the matrix $\mathbf{G}_\mathcal{B}$ is the identity, and can be left out of formulas altogether. A bra is then just the conjugate of the transpose of a ket, and the inner product of $\mathbf{x}$ and $\mathbf{y}$ is just the *standard* inner product of $[\mathbf{x}]_\mathcal{B}$ and $[\mathbf{y}]_\mathcal{B}$. In other words, while *any* basis can make a vector space V look like $\mathbb{R}^n$ or $\mathbb{C}^n$, an *orthonormal* basis makes V look like Euclidean $\mathbb{R}^n$ or like $\mathbb{C}^n$ with its standard inner product.

We return to the decomposition of vectors by an orthonormal basis $\mathcal{B}$. If $\mathcal{B}$ is orthonormal, then the denominators in (6.49) are all equal to 1, so we have the simpler expression

$$\mathbf{x} = \sum_i |\mathbf{b}_i\rangle \langle \mathbf{b}_i | \mathbf{x} \rangle. \tag{6.50}$$

This may be abbreviated as saying that

$$I = \sum_i |\mathbf{b}_i\rangle \langle \mathbf{b}_i|, \tag{6.51}$$

where I is the identity operation: $I\mathbf{x} = \mathbf{x}$.

6.4. Expansion in Orthonormal Bases: Finding Coefficients

How are we to understand the right-hand side of (6.51)? The bra $\langle \mathbf{b}_i |$ means "take the inner product with $\mathbf{b}_i$", so the expression $|\mathbf{b}_i\rangle\langle \mathbf{b}_i|$ means "take the inner product with $\mathbf{b}_i$ and then multiply the resulting number by the vector $\mathbf{b}_i$." Applying $|\mathbf{b}_i\rangle\langle \mathbf{b}_i|$ to a vector $\mathbf{x}$ gives the portion of $\mathbf{x}$ along the $\mathbf{b}_i$ direction. Summing over i, we then get all of $\mathbf{x}$.

Another way of viewing the same result is to say that $|\mathbf{b}_i\rangle\langle \mathbf{b}_i|$ is the matrix product of the column $|\mathbf{b}_i\rangle$ and the row $\langle \mathbf{b}_i |$. The product of a $n \times 1$ column vector and a $1 \times n$ row vector is an $n \times n$ matrix, in other words an operator. Summing over i, we must get the identity matrix.

Inner products and the identity (6.51) allow us not only to decompose vectors into standard components, but also to decompose operators. We begin as usual in Euclidean space. On $\mathbb{R}^n$, as we have seen, every operator is just multiplication by a matrix, and $A\mathbf{e}_j$ is the j-th column of the matrix A, so

$$A\mathbf{e}_j = \sum_k A_{kj}\mathbf{e}_k. \quad (6.52)$$

Taking the inner product with $\mathbf{e}_i$ we get the matrix elements of A:

$$A_{ij} = \langle \mathbf{e}_i | A\mathbf{e}_j \rangle. \quad (6.53)$$

Furthermore, $|\mathbf{e}_i\rangle\langle \mathbf{e}_j|$ is a matrix with a one in the i,j slot, and zeroes everywhere else. These matrices form a basis for $M_{n,n}$, the space of $n \times n$ matrices, and

$$A = \sum_{i,j} A_{ij} |\mathbf{e}_i\rangle\langle \mathbf{e}_j|. \quad (6.54)$$

These results depend crucially on the fact that the metric is standard, so $\langle \mathbf{e}_i |$ is just the transpose of $\mathbf{e}_i$. Since an orthonormal basis makes any space look like Euclidean $\mathbb{R}^n$ (or $\mathbb{C}^n$), an analogous result holds for an arbitrary inner product space with an orthonormal basis.

Theorem 6.3 *Let V be an inner product space with an orthonormal basis $\mathcal{B}$. Let L be an operator on V, and let $A = [L]_\mathcal{B}$. Then the matrix elements of A are*

$$A_{ij} = \langle \mathbf{b}_i | L\mathbf{b}_j \rangle. \quad (6.55)$$

Furthermore,

$$L = \sum_{i,j=1}^n A_{ij} |\mathbf{b}_i\rangle\langle \mathbf{b}_j|. \quad (6.56)$$

Proof: By Theorem 2.3, the j-th column of A is $[L\mathbf{b}_j]_\mathcal{B}$, so A_{ij} is the i-th component of $[L\mathbf{b}_j]_\mathcal{B}$. But by (6.48), the i-th component of any vector is obtained by taking the inner product of $\mathbf{b}_i$ with that vector. Applying this to $[A\mathbf{b}_j]_\mathcal{B}$ gives (6.55).

We now give two proofs that the two sides of (6.56) are equal. The first proof is a formal computation using (6.51):

$$\begin{aligned} L = ILI &= (\sum_i |\mathbf{b}_i\rangle\langle\mathbf{b}_i|)L(\sum_j |\mathbf{b}_j\rangle\langle\mathbf{b}_j|) \\ &= \sum_{i,j} |\mathbf{b}_i\rangle(\langle\mathbf{b}_i|L\mathbf{b}_j\rangle)\langle\mathbf{b}_j| \\ &= \sum_{i,j} A_{ij}|\mathbf{b}_i\rangle\langle\mathbf{b}_j|. \end{aligned} \quad (6.57)$$

This proof is slick, and demonstrates the power of the identity (6.51). However, the very ease of this computation obscures what we have done, so we supply a second proof.

Apply the right-hand side of (6.56) to an arbitrary vector $\mathbf{x}$. Now $\langle\mathbf{b}_j|\mathbf{x}\rangle = ([\mathbf{x}]_\mathcal{B})_j$, so $\sum_j A_{ij}\langle\mathbf{b}_j|\mathbf{x}\rangle = \sum_j A_{ij}([\mathbf{x}]_\mathcal{B})_j = (A[\mathbf{x}]_\mathcal{B})_i$. But $A = [L]_\mathcal{B}$, so $A[\mathbf{x}]_\mathcal{B} = [L\mathbf{x}]_\mathcal{B}$, and in particular $(A[\mathbf{x}]_\mathcal{B})_i = ([L\mathbf{x}]_\mathcal{B})_i$. Multiplying by $|\mathbf{b}_i\rangle$ and summing over i then gives $L\mathbf{x}$. ■

If the second proof makes sense to you, you understand what the matrix of a linear transformation really means. If you have trouble following the second proof, now would be a good time to review Section 3.2.

The point of Theorem 6.3 is that the operators $|\mathbf{b}_i\rangle\langle\mathbf{b}_j|$ form a basis for the n^2 dimensional space of operators on V, and that it is easy to compute the coefficients of any operator relative to this basis. Indeed, just as the basis $|\mathbf{b}_i\rangle$ makes V look like $\mathbb{R}^n$ (or $\mathbb{C}^n$), the basis $|\mathbf{b}_i\rangle\langle\mathbf{b}_j|$ makes the space of operators on V look like M_{nn}, the space of $n \times n$ matrices. If the basis is orthonormal, then V looks like Euclidean $\mathbb{R}^n$, $\langle\mathbf{b}_i|$ corresponds to the row $\langle\mathbf{e}_i| = \mathbf{e}_i^T$, and $|\mathbf{b}_i\rangle\langle\mathbf{b}_j|$ corresponds to the simple matrix $|\mathbf{e}_i\rangle\langle\mathbf{e}_j|$.

Example: Let $V = \mathbb{R}_1[t]$ with the inner product $\langle a_0+a_1t|b_0+b_1t\rangle = a_0b_0 + a_1b_1$. With this inner product, the standard basis $\{1,t\}$ is orthonormal. Let $L\mathbf{p}(t) = \mathbf{p}(t) + \mathbf{p}'(t)$. Then

$$\begin{array}{ll} \langle 1|L1\rangle = \langle 1|1\rangle = 1 & \langle 1|Lt\rangle = \langle 1|1+t\rangle = 1 \\ \langle t|L1\rangle = \langle t|1\rangle = 0 & \langle t|Lt\rangle = \langle t|1+t\rangle = 1, \end{array} \quad (6.58)$$

so

$$[L]_\mathcal{B} = \begin{pmatrix} 1 & 1 \\ 0 & 1 \end{pmatrix} \quad \text{and} \quad L = |1\rangle\langle 1| + |1\rangle\langle t| + |t\rangle\langle t|. \quad (6.59)$$

Exercises

1. In Euclidean $\mathbb{R}^3$, the vectors $\mathbf{b}_1 = (1,1,1)^T$, $\mathbf{b}_2 = (-2,1,1)^T$, and $\mathbf{b}_3 = (0,1,-1)^T$ are orthogonal. Use the inner product to write $\mathbf{x} = (-3,7,2)^T$ as a linear combination of $\mathbf{b}_1$, $\mathbf{b}_2$, and $\mathbf{b}_3$.

2. As in Exercise 1, let $\mathbf{b}_1 = (1,1,1)^T$, $\mathbf{b}_2 = (-2,1,1)^T$, and $\mathbf{b}_3 = (0,1,-1)^T$. Let $\mathbf{d}_i = \mathbf{b}_i/|\mathbf{b}_i|$. For each i, compute the matrix $|\mathbf{d}_i\rangle\langle\mathbf{d}_i|$ and confirm that $\sum_i |\mathbf{d}_i\rangle\langle\mathbf{d}_i|$ is the identity.

3. With $\mathbf{d}_i$ as in Exercise 2, decompose the vectors $\mathbf{x} = (-3,7,2)^T$, $\mathbf{y} = (1,3,4)^T$, and $\mathbf{z} = (1,-1,0)^T$ in the $\mathcal{D}$ basis. How do the inner products of $\mathbf{x}$ with $\mathbf{y}$, etc., compare to the inner products of $[\mathbf{x}]_\mathcal{D}$, with $[\mathbf{y}]_\mathcal{D}$, etc?

4. With $\mathbf{d}_i$ as in Exercise 2, decompose the matrix $\begin{pmatrix} 1 & 2 & 3 \\ 4 & 5 & 6 \\ 7 & 8 & 9 \end{pmatrix}$ as a linear combination of $|\mathbf{d}_i\rangle\langle\mathbf{d}_j|$.

5. Let L be reflection about the $x_1 - x_2$ plane (that is, taking $(x_1, x_2, x_3)^T$ to $(x_1, x_2, -x_3)^T$). Find the matrix of L in the $\mathcal{D}$ basis.

6. Let $\mathcal{B} = \{\mathbf{b}_1, \ldots, \mathbf{b}_n\}$ be an orthonormal basis for a vector space V. Show that the dual basis is precisely $\{\langle\mathbf{b}_k|\}$.

7. Let $\mathcal{B} = \{\mathbf{b}_1, \ldots, \mathbf{b}_n\}$ be a basis for a vector space V such that the dual basis is $\{\langle\mathbf{b}_k|\}$. Show that $\mathcal{B}$ is orthonormal.

8. Let $\mathcal{B}$ be an orthogonal (but not necessarily orthonormal) basis. What is the dual basis to $\mathcal{B}$?

6.5 Projections and the Gram-Schmidt Process

In the last section we discussed the value of orthonormal bases, but we did not say how to obtain them. In this section we will show how to convert an arbitrary basis into an orthonormal basis.

The idea is straightforward. Given a collection $\{\mathbf{x}_1, \ldots, \mathbf{x}_n\}$ of linearly independent vectors, we convert the vectors into an orthogonal basis one at a time. At the k-th stage, we have a collection $\{\mathbf{y}_1, \ldots, \mathbf{y}_k\}$ of orthogonal vectors that span the same space as $\{\mathbf{x}_1, \ldots, \mathbf{x}_k\}$. The vector $\mathbf{y}_1$ is simply $\mathbf{x}_1$. The vector $\mathbf{y}_2$ is the part of $\mathbf{x}_2$ that is orthogonal to $\mathbf{y}_1$. Similarly, $\mathbf{y}_3$ is the part of $\mathbf{x}_3$ that is orthogonal to $\mathbf{y}_1$ and $\mathbf{y}_2$. In general, $\mathbf{y}_k$ is the part of $\mathbf{x}_k$ that is orthogonal to the subspace spanned by $\mathbf{y}_1, \ldots, \mathbf{y}_{k-1}$ (which is the

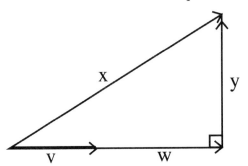

Figure 6.2: Decomposing **x** parallel and perpendicular to **v**

same as the subspace spanned by $\mathbf{x}_1, \ldots, \mathbf{x}_{k-1}$). Once we have generated an orthogonal basis, rescaling the vectors gives an orthonormal basis.

Geometrically, this is a simple idea. Algebraically, the challenge is to devise a way to extract the part of a vector that is orthogonal to a subspace spanned by some orthogonal vectors. To do this we need projection operators. We begin with projections onto lines.

Theorem 6.4 *Let* **x** *and* **v** *be vectors in an inner product space* V, *with* $\mathbf{v} \neq 0$. *Then, there is a unique way to write* **x** *as a sum*

$$\mathbf{x} = \mathbf{w} + \mathbf{y}, \tag{6.60}$$

with **w** *a multiple of* **v** *and with* **y** *orthogonal to* **v**. *Furthermore,*

$$\mathbf{w} = \frac{\langle \mathbf{v} | \mathbf{x} \rangle}{\langle \mathbf{v} | \mathbf{v} \rangle} \mathbf{v}. \tag{6.61}$$

Proof: See Figure 6.2. We first show that the decomposition (6.60) implies the formula (6.61). We then show that this formula actually works.

If $\mathbf{w} = c\mathbf{v}$ for some scalar c, then

$$\mathbf{x} = c\mathbf{v} + \mathbf{y}. \tag{6.62}$$

Taking the inner product of **v** with each side, we get

$$\langle \mathbf{v} | \mathbf{x} \rangle = c \langle \mathbf{v} | \mathbf{v} \rangle + 0, \tag{6.63}$$

from which (6.61) immediately follows. What remains is to check that

$$\mathbf{y} = \mathbf{x} - \frac{\langle \mathbf{v} | \mathbf{x} \rangle}{\langle \mathbf{v} | \mathbf{v} \rangle} \mathbf{v} \tag{6.64}$$

6.5. Projections and the Gram-Schmidt Process

is indeed orthogonal to $|\mathbf{v}\rangle$. Multiplying $\langle\mathbf{v}|$ by each side of (6.64) we get

$$\langle\mathbf{v}|\mathbf{y}\rangle = \langle\mathbf{v}|\mathbf{x}\rangle - \frac{\langle\mathbf{v}|\mathbf{x}\rangle}{\langle\mathbf{v}|\mathbf{v}\rangle}\langle\mathbf{v}|\mathbf{v}\rangle = \langle\mathbf{v}|\mathbf{x}\rangle - \langle\mathbf{v}|\mathbf{x}\rangle = 0. \blacksquare \qquad (6.65)$$

Definition *The vector* $\mathbf{w} = \frac{\langle\mathbf{v}|\mathbf{x}\rangle}{\langle\mathbf{v}|\mathbf{v}\rangle}\mathbf{v}$ *is called the orthogonal projection of* $\mathbf{x}$ *onto* $\mathbf{v}$.

Consider the operator

$$P_{\mathbf{v}} = \frac{|\mathbf{v}\rangle\langle\mathbf{v}|}{\langle\mathbf{v}|\mathbf{v}\rangle}. \qquad (6.66)$$

The gist of (6.61) is that, for any vector $\mathbf{x}$, $P_{\mathbf{v}}|\mathbf{x}\rangle$ is the orthogonal projection of $\mathbf{x}$ onto $\mathbf{v}$.

To build intuition, we consider the simplest possible example. In Euclidean $\mathbb{R}^n$, the projection $P_{\mathbf{e}_1}$ sends a vector $(x_1, x_2, \ldots, x_n)^T$ to $(x_1, 0, \ldots, 0)^T$. You should write down the matrix of $P_{\mathbf{e}_1}$ according to (6.66) and check that multiplying a vector by this matrix has exactly this effect. Exercise 1 shows that this example, while simple, is really quite general.

Now that we understand projecting onto a single vector, we extend this to projecting onto subspaces spanned by several orthogonal vectors. The following is a generalization of Theorem 6.4.

Theorem 6.5 *Let* $\{\mathbf{b}_1, \ldots, \mathbf{b}_m\}$ *be a collection of orthogonal nonzero vectors in an inner product space* V. *Let* $\mathbf{x}$ *be any element of space* V. *Then, there is a unique way to write* $\mathbf{x}$ *as a sum*

$$\mathbf{x} = c_1\mathbf{b}_1 + \cdots c_m\mathbf{b}_m + \mathbf{y}, \qquad (6.67)$$

with $\mathbf{y}$ *orthogonal to every* $\mathbf{b}_i$. *Furthermore, the i-th term is precisely*

$$c_i\mathbf{b}_i = P_{\mathbf{b}_i}|\mathbf{x}\rangle. \qquad (6.68)$$

Proof: Exercise 4.

The Gram-Schmidt process

We are now capable of converting an arbitrary basis to an orthonormal basis. This is called the Gram-Schmidt process.

Let $\mathbf{x}_1, \ldots, \mathbf{x}_m$ be a linearly independent set of vectors in an n-dimensional inner product space V. The $\mathbf{x}_i$'s are a basis of some

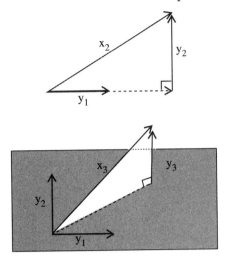

Figure 6.3: Two steps in the Gram-Schmidt process

subspace W of V. If $m = n$, then W is all of V, while if $m < n$ then W is a proper subspace of V. We will find an orthogonal basis $\mathbf{y}_1, \ldots, \mathbf{y}_m$ of W, using Theorem 6.5 to convert the $\mathbf{x}_i$'s to $\mathbf{y}_i$'s, one at a time. See Figure 6.3. Let

$$\mathbf{y}_1 = \mathbf{x}_1. \tag{6.69}$$

For $k > 1$, let

$$\begin{aligned}\mathbf{y}_k &= \mathbf{x}_k - \sum_{i=1}^{k-1} P_{\mathbf{y}_i}|\mathbf{x}_k\rangle \\ &= \mathbf{x}_k - \sum_{i=1}^{k-1} \frac{\langle \mathbf{y}_i|\mathbf{x}_k\rangle}{\langle \mathbf{y}_i|\mathbf{y}_i\rangle}\mathbf{y}_i.\end{aligned} \tag{6.70}$$

In other words, $\mathbf{y}_1$ is all of $\mathbf{x}_1$, $\mathbf{y}_2$ is the part of $\mathbf{x}_2$ that is orthogonal to $\mathbf{y}_1$ (and hence orthogonal to $\mathbf{x}_1$), $\mathbf{y}_3$ is the part of $\mathbf{x}_3$ that is orthogonal to $\mathbf{y}_1$ and $\mathbf{y}_2$ (and hence orthogonal to $\mathbf{x}_1$ and $\mathbf{x}_2$), and so on. Since each $\mathbf{y}_k$ is orthogonal to all the previous $\mathbf{y}_i$'s, the $\mathbf{y}_i$'s form an orthogonal set.

Notice that each $\mathbf{y}_k$ is a linear combination of $\mathbf{x}_1, \ldots, \mathbf{x}_k$, with the last coefficient equal to 1. Since the $\mathbf{x}_i$'s are linearly independent, $\mathbf{y}_i$ is nonzero. Since the $\mathbf{y}_i$'s are nonzero and orthogonal, they must be linearly independent. Since there are m of them, all in the space W (since all are linear combinations of the $\mathbf{x}_i$'s), they form a basis.

6.5. Projections and the Gram-Schmidt Process

Once we have an orthogonal basis for W, finding an orthonormal basis is easy. Just let

$$\mathbf{z}_i = \frac{\mathbf{y}_i}{|\mathbf{y}_i|} = \frac{\mathbf{y}_i}{\sqrt{\langle \mathbf{y}_i | \mathbf{y}_i \rangle}}. \tag{6.71}$$

Example: We work in $\mathbb{R}^3$ with the usual inner product, and let $\mathbf{x}_1 = (1,1,0)^T$, $\mathbf{x}_2 = (3,1,1)^T$, and $\mathbf{x}_3 = (1,1,3)^T$. We apply the Gram-Schmidt process to convert this to an orthonormal basis.

$$\mathbf{y}_1 = \mathbf{x}_1 = \begin{pmatrix} 1 \\ 1 \\ 0 \end{pmatrix},$$

$$\mathbf{y}_2 = \mathbf{x}_2 - P_{\mathbf{y}_1}|\mathbf{x}_2\rangle = \mathbf{x}_2 - \frac{\langle \mathbf{y}_1|\mathbf{x}_2\rangle}{\langle \mathbf{y}_1|\mathbf{y}_1\rangle}\mathbf{y}_1$$

$$= \begin{pmatrix} 3 \\ 1 \\ 1 \end{pmatrix} - \frac{4}{2}\begin{pmatrix} 1 \\ 1 \\ 0 \end{pmatrix} = \begin{pmatrix} 1 \\ -1 \\ 1 \end{pmatrix},$$

$$\mathbf{y}_3 = \mathbf{x}_3 - P_{\mathbf{y}_1}|\mathbf{x}_3\rangle - P_{\mathbf{y}_2}|\mathbf{x}_3\rangle = \mathbf{x}_3 - \frac{\langle \mathbf{y}_1|\mathbf{x}_3\rangle}{\langle \mathbf{y}_1|\mathbf{y}_1\rangle}\mathbf{y}_1 - \frac{\langle \mathbf{y}_2|\mathbf{x}_3\rangle}{\langle \mathbf{y}_2|\mathbf{y}_2\rangle}\mathbf{y}_2$$

$$= \begin{pmatrix} 1 \\ 1 \\ 3 \end{pmatrix} - \frac{2}{2}\begin{pmatrix} 1 \\ 1 \\ 0 \end{pmatrix} - \frac{3}{3}\begin{pmatrix} 1 \\ -1 \\ 1 \end{pmatrix} = \begin{pmatrix} -1 \\ 1 \\ 2 \end{pmatrix}. \tag{6.72}$$

Notice that $\mathbf{y}_1$ and $\mathbf{y}_2$ span the same subspace of $\mathbb{R}^3$ as $\mathbf{x}_1$ and $\mathbf{x}_2$. Just as we can write $\mathbf{y}_1$ and $\mathbf{y}_2$ as linear combinations of $\mathbf{x}_1$ and $\mathbf{x}_2$, we can write $\mathbf{x}_1$ and $\mathbf{x}_2$ as linear combinations of $\mathbf{y}_1$ and $\mathbf{y}_2$. This is a general feature. The first k $\mathbf{y}_i$'s are always an invertible linear combination of the first k $\mathbf{x}_i$'s, and so span the same space.

Finally we compute the $\mathbf{z}_i$'s:

$$\mathbf{z}_1 = \mathbf{y}_1/\sqrt{2} = \frac{1}{\sqrt{2}}\begin{pmatrix} 1 \\ 1 \\ 0 \end{pmatrix},$$

$$\mathbf{z}_2 = \mathbf{y}_2/\sqrt{3} = \frac{1}{\sqrt{3}}\begin{pmatrix} 1 \\ -1 \\ 1 \end{pmatrix},$$

$$\mathbf{z}_3 = \mathbf{y}_3/\sqrt{6} = \frac{1}{\sqrt{6}}\begin{pmatrix} 1 \\ 1 \\ 2 \end{pmatrix}. \tag{6.73}$$

You should check explicitly that these vectors are indeed orthonormal.

There are two *very* common mistakes you should do your best to avoid. First, $\mathbf{y}_k$ is equal to $\mathbf{x}_k - \sum_{i=1}^{k-1} P_{\mathbf{y}_i}|\mathbf{x}_k\rangle$, not to $\mathbf{x}_k - \sum_{i=1}^{k-1} P_{\mathbf{x}_i}|\mathbf{x}_k\rangle$. Theorem 6.5 only applies to orthogonal sets of vectors $\{\mathbf{b}_i\}$. Since the $\mathbf{x}_i$'s are typically not orthogonal, the vector $\mathbf{x}_k - \sum_{i=1}^{k-1} P_{\mathbf{x}_i}|\mathbf{x}_k\rangle$ is typically not orthogonal to the previous $\mathbf{x}_i$'s, or to the previous $\mathbf{y}_i$'s. Second, when taking complex inner products, do not forget to conjugate the coefficients of the first vector. The standard inner product of $\begin{pmatrix} 1 \\ i \end{pmatrix}$ with $\begin{pmatrix} 2 \\ i \end{pmatrix}$ is $1 \times 2 + (-i) \times i = 3$, not $1 \times 2 + i \times i = 1$.

Exercises

1. Let $\mathcal{B} = \{\mathbf{b}_1, \ldots, \mathbf{b}_n\}$ be an orthogonal basis for the inner product space V. Find the matrix of $P_{\mathbf{b}_i}$ with respect to the $\mathcal{B}$ basis.

2. In Euclidean $\mathbb{R}^2$, find the matrices of $P_{\mathbf{v}_1}$ and $P_{\mathbf{v}_2}$, where $\mathbf{v} = (1,1)^T$ and $\mathbf{v}_2 = (1,-1)$. Use your results to find $P_{\mathbf{v}_1}\mathbf{x}$, where $\mathbf{x} = (3,1)^T$.

3. In $\mathbb{C}^2$ with the standard inner product, find the matrices of $P_{\mathbf{v}_1}$ and $P_{\mathbf{v}_2}$, where $\mathbf{v}_1 = \begin{pmatrix} 1 \\ i \end{pmatrix}$ and $\mathbf{v}_2 = \begin{pmatrix} 1 \\ -i \end{pmatrix}$. What is $P_{\mathbf{v}_1} + P_{\mathbf{v}_2}$? Explain.

4. Prove Theorem 6.5 by mimicking the proof of Theorem 6.4. First show that any expansion (6.67) must satisfy (6.68). Then show that $\mathbf{x} - \sum_i P_{\mathbf{b}_i}|\mathbf{x}\rangle$ is indeed orthogonal to all the $\mathbf{b}_i$'s.

5. The results of the Gram-Schmidt process depend on the order of the vectors considered. In $\mathbb{R}^3$, apply the Gram-Schmidt process to the vectors $\mathbf{x}_1 = (3,1,1)^T$, $\mathbf{x}_2 = (1,1,3)^T$, and $\mathbf{x}_3 = (1,1,0)^T$. Your results should look quite different from (6.73).

6. Find an orthonormal basis for the span of the three vectors $(1,1,1,1)^T$, $(1,2,3,4)^T$, $(4,0,2,0)^T$ in Euclidean $\mathbb{R}^4$.

7. Find an orthonormal basis for the span of the three vectors $(1,1,1,1)^T$, $(1,1,1,-1)^T$, $(1,1,-1,1)^T$ in Euclidean $\mathbb{R}^4$.

8. Find an orthonormal basis for the span of the complex vectors $(1+i, 1-i, 2)^T$ and $(1-i, 2+2i, 4)^T$ with the standard complex inner product.

9. In $\mathbb{C}^4$ with the standard inner product, find an orthonormal basis for the span of the vectors $(1, i, -1, -i)^T$, $(1, 2i, -3, -4i)^T$, and $(4, 0, -2, 0)^T$.

10. In the space of smooth function on $[0, \pi]$ with inner product $\langle f|g\rangle = \int_0^\pi f(t)g(t)dt$, find an orthonormal basis for the span of $1, \sin(t)$, and $\sin^2(t)$.

11. Let $V = \mathbb{R}[t]$, with the inner product $\langle f|g\rangle = \int_{-1}^1 f(t)g(t)dt$. Apply the Gram-Schmidt process to the vectors $\mathbf{x}_1(t) = 1$, $\mathbf{x}_2(t) = t$, $\mathbf{x}_3(t) = t^2$, and $\mathbf{x}_4(t) = t^3$. The vector $\mathbf{y}_i(t)$ is called the i-th Legendre polynomial.

12. Let $V = \mathbb{R}[t]$, with the inner product given by $\langle f|g\rangle = (\pi)^{-1/2} \int_{-\infty}^\infty f(t)g(t)e^{-t^2}dt$. Apply the Gram-Schmidt process to the vectors $\mathbf{x}_1(t) = 1$, $\mathbf{x}_2(t) = t$, $\mathbf{x}_3(t) = t^2$, and $\mathbf{x}_4(t) = t^3$. The vector $\mathbf{y}_i(t)$ is called the i-th Hermite polynomial.

6.6 Orthogonal Complements and Projections onto Subspaces

Theorem 6.4 tells us that, given a nonzero vector $\mathbf{v}$ in an inner product space V, we can write V as the (internal) direct sum of two subspaces. One subspace, which we call K, consists of all multiples of $\mathbf{v}$. The other subspace, which we call $K^\perp$, consists of all vectors orthogonal to $\mathbf{v}$. In the terminology of Section 2.5, $\mathbf{w}$ is the projection of $\mathbf{x}$ onto K along $K^\perp$. This is why we called $P_\mathbf{v}$ the *orthogonal projection onto* $\mathbf{v}$.

The essential difference between the situation of Theorem 6.4 and the situation of Section 2.5 is that, in Section 2.5 we needed two subspaces whose direct sum was V, while in Theorem 6.4 we only needed a single vector. That vector defined a 1-dimensional subspace K, and the inner product then defined the complementary subspace $K^\perp$.

In this section we consider an extension of Theorem 6.4 to higher dimensions.

Definition *Let W be any subspace of an inner product space V. The* orthogonal complement *of W, denoted $W^\perp$, is the set of vectors that are orthogonal to every vector in W.*

Here are some examples of subspaces and orthogonal complements:

1. In $\mathbb{R}^{n+m}$ with the standard inner product, let W be the set of

all vectors of the form $(x_1, \ldots, x_n, 0, \ldots, 0)^T$. Then $W^\perp$ is all vectors of the form $(0, \ldots, 0, x_{n+1}, \ldots, x_{n+m})^T$.

2. In any vector space V, let $W = V$. Then $W^\perp = \{0\}$, since the only vector orthogonal to *everything* is zero. Similarly, if $W = \{0\}$, then $W^\perp = V$.

3. If $V = \mathbb{R}^2$ and W is all multiples of $(1,1)^T$, then $W^\perp$ is all multiples of $(1,-1)^T$.

4. In $V = \mathbb{R}^3$, let W be the span of $(1,1,0)^T$ and $(1,2,3)^T$. The vector $(1,-1,1)^T$ is orthogonal to $(1,1,0)^T$, but is not an element of $W^\perp$, since it is not orthogonal to *every* element of W. Indeed, $(1,-1,1)^T$ is not orthogonal to $(1,2,3)^T$. The vector $(3,-3,1)^T$, however, is orthogonal to both $(1,1,0)^T$ and $(1,2,3)^T$, and is in $W^\perp$. In this example $W^\perp$ is all multiples of $(3,-3,1)^T$.

5. Let $\mathcal{D} = \{\mathbf{d}_1, \ldots, \mathbf{d}_m\}$ be a collection of vectors in V, and let W be the span of $\mathcal{D}$. Then $W^\perp$ is the set of vectors perpendicular to all the elements of $\mathcal{D}$. For if $\mathbf{x}$ is orthogonal to all the elements of $\mathcal{D}$, then it must be orthogonal to all linear combinations of the elements of $\mathcal{D}$, that is to every element of W. Conversely, since each element of $\mathcal{D}$ is in W, anything in $W^\perp$ must be orthogonal to all of $\mathcal{D}$.

Theorem 6.6 *Let W be a finite dimensional subspace of a inner product space V. Then V is the internal direct sum of W and $W^\perp$.*

Proof: We must show that any vector $\mathbf{x} \in V$ can be uniquely decomposed as a vector $\mathbf{y} \in W$ plus a vector $\mathbf{z} \in W^\perp$. Since W is finite dimensional, W has a finite basis. By the Gram-Schmidt process this can be converted to an orthonormal basis $|\mathbf{b}_1\rangle, \ldots, |\mathbf{b}_m\rangle$, where m is the dimension of W. By Theorem 6.5, any vector in V can be uniquely decomposed as a linear combination of the $|\mathbf{b}_i\rangle$'s plus a remainder that is orthogonal to all the $|\mathbf{b}_i\rangle$'s. However, a linear combination of the $|\mathbf{b}_i\rangle$'s is nothing more or less than an element of W, while, by example 5, a vector orthogonal to all the $|\mathbf{b}_i\rangle$'s is nothing more or less than an element of $W^\perp$. ∎

Since V is the direct sum of W and $W^\perp$, there exist projections onto W along $W^\perp$ and onto $W^\perp$ along W. Theorem 6.5 essentially constructs these projections for us. Pick an orthogonal basis

6.6. Orthogonal Complements and Projections onto Subspaces

$\mathbf{b}_1, \ldots, \mathbf{b}_m$ of W. Define

$$P_W = \sum_{i=1}^m P_{\mathbf{b}_i} = \sum_{i=1}^m \frac{|\mathbf{b}_i\rangle\langle\mathbf{b}_i|}{\langle\mathbf{b}_i|\mathbf{b}_i\rangle}. \tag{6.74}$$

If the $\mathbf{b}_i$'s are orthonormal, then the denominators in (6.74) are all 1, and (6.74) simplifies to

$$P_W = \sum_{i=1}^m P_{\mathbf{b}_i} = \sum_{i=1}^m |\mathbf{b}_i\rangle\langle\mathbf{b}_i|. \tag{6.75}$$

By Theorem 6.5, if $\mathbf{v}$ is any vector, then $P_W|\mathbf{v}\rangle$ is in W and $\mathbf{v} - P_W|\mathbf{v}\rangle$ is orthogonal to any vector in W. In other words, P_W is a projection onto W along $W^\perp$ and $I - P_W$ is a projection onto $W^\perp$ along W, a projection we then denote $P_{W^\perp}$.

Definition *A projection onto a subspace W along its orthogonal complement $W^\perp$ is called the* orthogonal projection *onto W.*

Example: Consider $V = \mathbb{R}^3$ with W being the span of $(1, 1, 0)^T$ and $(1, 2, 3)^T$, as in example 4 above. Applying the Gram-Schmidt process to $(1, 1, 0)^T$ and $(1, 2, 3)^T$, we get that $\mathbf{b}_1 = (1, 1, 0)^T$ and $\mathbf{b}_2 = (-1/2, 1/2, 3)^T$ form an orthogonal basis for W. The orthogonal projection onto W is then given by the matrix

$$\begin{aligned} P_W &= \sum_i \frac{1}{\langle\mathbf{b}_i|\mathbf{b}_i\rangle}|\mathbf{b}_i\rangle\langle\mathbf{b}_i| \\ &= \frac{1}{2}\begin{pmatrix} 1 \\ 1 \\ 0 \end{pmatrix}(1,1,0) + \frac{2}{19}\begin{pmatrix} -1/2 \\ 1/2 \\ 3 \end{pmatrix}(-1/2, 1/2, 3) \\ &= \frac{1}{2}\begin{pmatrix} 1 & 1 & 0 \\ 1 & 1 & 0 \\ 0 & 0 & 0 \end{pmatrix} + \frac{1}{38}\begin{pmatrix} 1 & -1 & -6 \\ -1 & 1 & 6 \\ -6 & 6 & 36 \end{pmatrix} \\ &= \frac{1}{38}\begin{pmatrix} 20 & 18 & -6 \\ 18 & 20 & 6 \\ -6 & 6 & 36 \end{pmatrix}. \end{aligned} \tag{6.76}$$

The orthogonal projection onto $W^\perp$ is given by the matrix $P_{W^\perp} = I - P_W$. We can also compute $P_{W^\perp}$ directly. $W^\perp$ is 1-dimensional, namely all multiples of $\mathbf{b}_3 = (3, -3, 1)^T$, so $P_{W^\perp} = P_{\mathbf{b}_3}$. By either method, we obtain

$$P_{W^\perp} = \frac{1}{19} \begin{pmatrix} 9 & -9 & 3 \\ -9 & 9 & -3 \\ 3 & -3 & -1 \end{pmatrix}. \qquad (6.77)$$

Exercises

1. On Euclidean $\mathbb{R}^3$, find P_W, where W is the subspace spanned by $(0, 1, -1)^T$ and $(2, -1, -1)^T$. Find $P_W \mathbf{x}$, where $\mathbf{x} = (1, 3, -1)^T$.

2. On Euclidean $\mathbb{R}^4$, find P_W, where W is spanned by $(1, 1, 1, 1)^T$ and $(1, 2, 3, 4)^T$. Write $(1, 1, -1, -1)$ as the sum of an element of W and an element of $W^\perp$. (Note: $(1, 1, 1, 1)^T$ and $(1, 2, 3, 4)^T$ are not orthogonal.)

3. On $\mathbb{R}_3[t]$ with inner product $\langle \mathbf{f} | \mathbf{g} \rangle = \int_0^1 f(t)g(t)dt$, let W be the subspace spanned by 1 and t. Compute $P_W(t^2)$.

4. Let $\mathbf{x}_1, \ldots, \mathbf{x}_m$ be a basis for a subspace W of an n-dimensional inner product space V. Let $\mathbf{x}_{m+1}, \ldots, \mathbf{x}_n$ be chosen so that $\mathbf{x}_1, \ldots, \mathbf{x}_n$ is a basis for V. Apply Gram-Schmidt to $\{\mathbf{x}_i\}$ to get an orthogonal basis $\{\mathbf{y}_i\}$. Show that $\{\mathbf{y}_{m+1}, \ldots, \mathbf{y}_n\}$ is an orthogonal basis for $W^\perp$.

5. Find a basis for $W^\perp$, where W is as in Exercise 2.

6. Find a basis for $W^\perp$, where W is as in Exercise 3.

6.7 Least Squares Solutions

In this section we consider linear equations that have no solution, and try to find a vector that comes as close to solving the equations as possible. This has many applications, the most common of which is fitting a straight line to some data. Typically the data points are not colinear, so there is no line that goes exactly through the points, but there is a unique line that comes closest.

A system of n linear equations in m unknowns is written in matrix form as

$$A\mathbf{x} = \mathbf{b}, \qquad (6.78)$$

where A is an $n \times m$ matrix, $\mathbf{x} \in \mathbb{R}^m$, and $\mathbf{b} \in \mathbb{R}^n$. If we denote the i-th column of A by $\mathbf{a}_i$, then the system of equations (6.78) can be rewritten as a vector equation

$$x_1 \mathbf{a}_1 + \cdots + x_n \mathbf{a}_n = \mathbf{b}. \qquad (6.79)$$

6.7. Least Squares Solutions

This has an exact solution if and only if $\mathbf{b}$ lies in the column space of A (i.e., the span of the $\mathbf{a}_i$'s). In this case the solution is unique if (and only if) the $\mathbf{a}_i$'s are linearly independent.

Let W be the column space of A. If $\mathbf{b}$ is not in W, then there are no solutions to (6.78) (or equivalently (6.79)). However, we can look for the vector $\mathbf{x}$ that makes the error $|\mathbf{b} - A\mathbf{x}|$ as small as possible. Of course, to do that we need a way to measure length. If $\mathbb{R}^n$ has the standard inner product, then the squared error is

$$|\mathbf{b} - A\mathbf{x}|^2 = \langle \mathbf{b} - A\mathbf{x} | \mathbf{b} - A\mathbf{x} \rangle = \sum_{i=1}^{m}(b_i - (A\mathbf{x})_i)^2. \qquad (6.80)$$

This explains the term "least squares".

Definition *A least squares solution to the system of equations* $A\mathbf{x} = \mathbf{b}$ *is a vector* $\mathbf{x}$ *that minimizes* $|\mathbf{b} - A\mathbf{x}|$, *using the standard inner product on* $\mathbb{R}^n$.

Let $\mathbf{b}_{\parallel}$ and $\mathbf{b}_{\perp}$ be the projections of $\mathbf{b}$ onto W and $W^{\perp}$, respectively. See Figure 6.4. That is,

$$\mathbf{b}_{\parallel} = P_W \mathbf{b}, \qquad \mathbf{b}_{\perp} = P_{W^{\perp}} \mathbf{b} = \mathbf{b} - \mathbf{b}_{\parallel}. \qquad (6.81)$$

Since $\mathbf{b}_{\perp}$ is in $W^{\perp}$, it is orthogonal to both $\mathbf{b}_{\parallel}$ and to $A\mathbf{x}$, for every $\mathbf{x}$, so

$$|\mathbf{b} - A\mathbf{x}|^2 = |\mathbf{b}_{\perp} + \mathbf{b}_{\parallel} - A\mathbf{x}|^2 = |\mathbf{b}_{\perp}|^2 + |\mathbf{b}_{\parallel} - A\mathbf{x}|^2. \qquad (6.82)$$

The way to minimize $|\mathbf{b} - A\mathbf{x}|$, then, is to set

$$A\mathbf{x} = \mathbf{b}_{\parallel}. \qquad (6.83)$$

Since $\mathbf{b}_{\parallel}$ is in the column space of A, a solution to (6.83) exists. If the columns of A are linearly independent, this solution is unique.

Occasionally it is possible to find $\mathbf{b}_{\parallel}$, and thereby to solve (6.83) directly. This is especially true if the columns of A are orthogonal, in which case the projection operator P_W is easy to compute. However, most of the time it is difficult to compute $\mathbf{b}_{\parallel}$. Instead, we use the following theorem.

Theorem 6.7 *Every least squares solution to* $A\mathbf{x} = \mathbf{b}$ *is an actual solution to*

$$A^T A\mathbf{x} = A^T \mathbf{b}. \qquad (6.84)$$

Conversely, every solution to (6.84) is a least squares solution to $A\mathbf{x} = \mathbf{b}$.

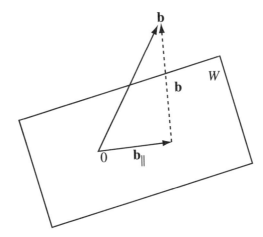

Figure 6.4: Decomposing **b** along W and $W^\perp$

Proof: Since $\mathbf{b}_\perp$ is orthogonal to every column of A, we must have

$$A^T \mathbf{b}_\perp = 0. \tag{6.85}$$

Now suppose **x** is a least squares solution. Then **x** satisfies (6.83), so

$$A^T A\mathbf{x} = A^T \mathbf{b}_\parallel = A^T(\mathbf{b} - \mathbf{b}_\perp) = A^T \mathbf{b}. \tag{6.86}$$

This proves the first half of the theorem.

Furthermore, since $\mathbf{b}_\parallel$ is in the range of A, there exists a least squares solution **v**, which we have shown solves (6.84). What remains is to show that any other solution to (6.84) also solves (6.83). Let **y** be another solution to (6.84), so $A^T A\mathbf{y} = A^T \mathbf{b} = A^T A\mathbf{v}$. Then $A^T A(\mathbf{v} - \mathbf{y}) = 0$. By Exercise 1, below, this implies that $A(\mathbf{v} - \mathbf{y}) = 0$, and hence that $A\mathbf{y} = A\mathbf{v} = \mathbf{b}_\parallel$. Thus **y** is also a least squares solution to $A\mathbf{x} = \mathbf{b}$. ∎

If the columns of A are linearly independent, then $A^T A$ is invertible, there is exactly one solution to (6.84), and there is exactly one least squares solution to (6.78).

Example: Suppose we want to find the best (i.e., least squares) solution to

$$\begin{pmatrix} 1 & 1 \\ 1 & 3 \\ 2 & 4 \end{pmatrix} \begin{pmatrix} x_1 \\ x_2 \end{pmatrix} = \begin{pmatrix} 5 \\ 1 \\ 4 \end{pmatrix}. \tag{6.87}$$

This has no exact solutions, as the third row of A is the sum of the first two rows, while the third entry of **b** is not the sum of the

6.7. Least Squares Solutions

first two entries. However, there is a least squares solution. In this example, equation (6.84) becomes

$$\begin{pmatrix} 6 & 12 \\ 12 & 26 \end{pmatrix} \begin{pmatrix} x_1 \\ x_2 \end{pmatrix} = \begin{pmatrix} 14 \\ 24 \end{pmatrix}, \qquad (6.88)$$

whose solution is $x_1 = \frac{19}{3}$, $x_2 = -2$. $\mathbf{b} - A\mathbf{x}$ equals $(\frac{2}{3}, \frac{2}{3}, -\frac{2}{3})^T$ which is indeed orthogonal to the columns of A.

It sometimes happens that $\mathbf{b}$ is in the column space of A. In that case, the original equation (6.78) has a solution, and this exact solution certainly minimizes the squared error. In short, when a true solution exists, it is also a least squares solution.

Best lines and curves through data

The most common use of least squares is to find the "best" line through a set of points in the plane. Finding a line $y = c_0 + c_1 x$ through a set of n points $\{(x_1, y_1), \ldots, (x_n, y_n)\}$ in the x-y plane is the same thing as solving the equations

$$\begin{aligned} c_0 + c_1 x_1 &= y_1, \\ c_0 + c_1 x_2 &= y_2, \\ &\vdots \\ c_0 + c_1 x_n &= y_n. \end{aligned} \qquad (6.89)$$

In matrix form, this is $A\mathbf{x} = \mathbf{b}$, with

$$A = \begin{pmatrix} 1 & x_1 \\ 1 & x_2 \\ \vdots & \vdots \\ 1 & x_n \end{pmatrix}; \quad \mathbf{x} = \begin{pmatrix} c_0 \\ c_1 \end{pmatrix}; \quad \mathbf{b} = \begin{pmatrix} y_1 \\ y_2 \\ \vdots \\ y_n \end{pmatrix}. \qquad (6.90)$$

By Theorem 6.7, we are looking for a true solution to (6.84), which takes the form

$$\begin{pmatrix} n & \sum_{i=1}^{n} x_i \\ \sum_{i=1}^{n} x_i & \sum_{i=1}^{n} x_i^2 \end{pmatrix} \begin{pmatrix} c_0 \\ c_1 \end{pmatrix} = \begin{pmatrix} \sum_{i=1}^{n} y_i \\ \sum_{i=1}^{n} x_i y_i \end{pmatrix}. \qquad (6.91)$$

This 2×2 system of equations is easy to solve. The solution is

$$c_0 = \frac{(\sum_{i=1}^{n} x_i^2)(\sum_{i=1}^{n} y_i) - (\sum_{i=1}^{n} x_i)(\sum_{i=1}^{n} x_i y_i)}{n(\sum_{i=1}^{n} x_i^2) - (\sum_{i=1}^{n} x_i)^2};$$

$$c_1 = \frac{n(\sum_{i=1}^n x_i y_i) - (\sum_{i=1}^n x_i)(\sum_{i=1}^n y_i)}{n(\sum_{i=1}^n x_i^2) - (\sum_{i=1}^n x_i)^2}. \tag{6.92}$$

Sometimes you expect your quantity y to depend quadratically on x, rather than linearly. In such cases you want to find the best parabola $y = c_0 + c_1 x + x_2 x^2$ through the data points. This is similar to, and a little more difficult than, solving for the best line. The equations for passing exactly through the data points are now

$$\begin{pmatrix} 1 & x_1 & x_1^2 \\ \vdots & \vdots & \vdots \\ 1 & x_n & x_n^2 \end{pmatrix} \begin{pmatrix} c_0 \\ c_1 \\ c_2 \end{pmatrix} = \begin{pmatrix} y_1 \\ \vdots \\ y_n \end{pmatrix}. \tag{6.93}$$

Typically these equations cannot be solved exactly, but they have a least squares solution, namely the exact solution to

$$\begin{pmatrix} n & \sum_{i=1}^n x_i & \sum_{i=1}^n x_i^2 \\ \sum_{i=1}^n x_i & \sum_{i=1}^n x_i^2 & \sum_{i=1}^n x_i^3 \\ \sum_{i=1}^n x_i^2 & \sum_{i=1}^n x_i^3 & \sum_{i=1}^n x_i^4 \end{pmatrix} \begin{pmatrix} c_0 \\ c_1 \\ c_2 \end{pmatrix} = \begin{pmatrix} \sum_{i=1}^n y_i \\ \sum_{i=1}^n x_i y_i \\ \sum_{i=1}^n x_i^2 y_i \end{pmatrix}. \tag{6.94}$$

Unfortunately, the formula for the inverse of a general 3×3 matrix is much more complicated than the formula for the inverse of a 2×2 matrix, so the formula for the least squares solution to (6.93) is much more complicated than (6.92), and is not useful. Rather, the way to do a parabolic fit is to plug the data into (6.94), get a 3×3 system of equations, then solve the 3×3 system by Gaussian elimination.

There are many other data-fitting problems that can be attacked with least squares. Any time you can represent the exact passage of a curve through the data as a set $A\mathbf{x} = \mathbf{b}$ of linear equations, one can find the least squares solution by solving $A^T A \mathbf{x} = A^T \mathbf{b}$. Some of these additional applications are explored in the exercises.

Weighted least squares

In doing ordinary least squares, the quantity we are trying to minimize, $\sum (b_i - (A\mathbf{x})_i)^2$, gives equal weight to each component of $\mathbf{b}$. Put another way, we are computing $|\mathbf{b} - A\mathbf{x}|^2$ using the standard inner product on $\mathbb{R}^n$. There are occasions, however, where equal weighting is not appropriate. For example, in fitting a straight line it may happen that some data points are more reliable than others. It is then preferable to give the more reliable points greater weight than the less reliable points.

6.7. Least Squares Solutions

We therefore need a way to minimize $|\mathbf{b} - A\mathbf{x}|^2$, where the norm is computed using an inner product that is not standard. In this case the derivation of (6.83) proceeds exactly as before, only $\mathbf{b}_\|$ is the projection of $\mathbf{b}$ onto W using the *nonstandard* inner product, and the nonstandard inner product of $\mathbf{b}_\perp$ with each column of A. Since

$$0 = \langle \mathbf{a}_i | \mathbf{b}_\perp \rangle = \mathbf{a}_i^T G \mathbf{b}_\perp, \tag{6.95}$$

where G is the metric, we now have

$$A^T G \mathbf{b}_\perp = 0, \tag{6.96}$$

instead of (6.85), and so

$$A^T G A \mathbf{x} = A^T \mathbf{g}(\mathbf{b} - \mathbf{b}_\perp) = A^T \mathbf{g} \mathbf{b}. \tag{6.97}$$

In short, a weighted least squares solution to (6.78) using the metric G is obtained by solving (6.97), not (6.84).

As an application, consider fitting a straight line to a set of data points, where the i-th data point is given weight w_i. In other words, the inner product we are taking on $\mathbb{R}^n$ is

$$\langle \mathbf{v} | \mathbf{z} \rangle = \sum_{i=1}^{n} w_i v_i z_i, \tag{6.98}$$

instead of $\sum v_i z_i$. The metric G is then a diagonal matrix with diagonal entries $w_1, w_2, \ldots, w_n$. For the system of equations (6.89), equation (6.97) takes the form

$$\begin{pmatrix} \sum_{i=1}^{n} w_i & \sum_{i=1}^{n} w_i x_i \\ \sum_{i=1}^{n} w_i x_i & \sum_{i=1}^{n} w_i x_i^2 \end{pmatrix} \begin{pmatrix} c_0 \\ c_1 \end{pmatrix} = \begin{pmatrix} \sum_{i=1}^{n} w_i y_i \\ \sum_{i=1}^{n} w_i x_i y_i \end{pmatrix}. \tag{6.99}$$

Exercises

1. Show that, if $A^T A \mathbf{x} = 0$, then $A \mathbf{x} = 0$. [Hint: consider the inner product $\langle A\mathbf{x} | A\mathbf{x} \rangle$.] Use this to show that, if the columns of A are linearly independent, then $A^T A$ is invertible.

2. Find a least squares solution to $A\mathbf{x} = \mathbf{b}$, where

$$A = \begin{pmatrix} 1 & 2 \\ 3 & 4 \\ 5 & 6 \end{pmatrix}, \quad \mathbf{b} = \begin{pmatrix} 4 \\ 5 \\ 1 \end{pmatrix}. \tag{6.100}$$

3. Find a least squares solutions to $A\mathbf{x} = \mathbf{b}$, where

$$A = \begin{pmatrix} 1 & 2 & 7 \\ 3 & 4 & 8 \\ 5 & 6 & 9 \end{pmatrix}, \quad \mathbf{b} = \begin{pmatrix} 4 \\ 5 \\ 1 \end{pmatrix}. \quad (6.101)$$

Is this solution unique? Explain.

4. Find the equation of the best line through the points $(-1,-2)$, $(0,0)$, $(1,1)$, and $(2,3)$.

5. Find the equation of the best line through the points $(0,-1)$, $(1,1)$, $(2,4)$, $(3,9)$, and $(4,15)$.

6. Find the equation of the best parabola through the data points of Exercise 5.

7. Consider planes of the form $z = a + bx + cy$. Find the equation of the best plane through the points $(0,0,0)$, $(0,1,0)$, $(1,0,1)$, $(1,1,3)$, and $(1,-1,2)$.

8. Suppose we want to fit a cubic curve $y = c_0 + c_1 x + c_2 x^2 + c_3 x^3$ to a set of data. Set up the $n \times 4$ matrix system whose least squares solution gives $(c_0, c_1, c_2, c_3)^T$. Set up the 4×4 system whose true solution gives $(c_0, c_1, c_2, c_3)^T$.

9. The most general equation for a circle is $x^2 + y^2 + ax + by + c = 0$. Given a set of data points, show how to find the values of a, b, and c that give the best circular fit to the data.

10. Using the results of Exercise 9, find the circle that best fits the data points $(1,1)$, $(2,0)$, $(3,1)$, $(3,2)$, $(2,2)$.

11. The most general equation for an ellipse is $ax^2 + by^2 + cxy + dx + ey + f = 0$, with $ab > c^2$. However, if an ellipse is described by $(a, b, c, d, e, f)^T$, then it is also described by all nonzero multiples of $(a, b, c, d, e, f)^T$, so to get a unique answer we must set one of the variables equal to 1. Setting $a = 1$, show how to fit an ellipse to a set of data points. How do things differ if we instead set $b = 1$?

12. Find the best ellipse through the points $(-3,-3)$, $(-2,-3)$, $(-2,-1)$, $(-1,-2)$, $(-1,0)$, $(0,-1)$, $(0,1)$, $(1,0)$, $(1,2)$, $(2,1)$, $(2,3)$, and $(3,3)$, using both methods of Exercise 11. (You may wish to use MATLAB.) Explain any discrepancies between the two answers.

13. The results of this section are all about real vector spaces. How would things change if the spaces were complex? Specifically, suppose A is a complex matrix and $\mathbf{b}$ is a complex vector. How would you find a complex vector $\mathbf{x}$ that minimizes $|\mathbf{b} - A\mathbf{x}|^2$?

6.8 The Spaces ℓ_2 and $L^2[0,1]$[7]

Our basic model of a finite dimensional real inner product space is (Euclidean) $\mathbb{R}^n$ with the standard inner product (6.1), and our basic model of a finite dimensional complex inner product space is $\mathbb{C}^n$ with the standard inner product (6.19). We have seen how, with the choice of an orthonormal basis, any finite dimensional inner product space can be made to look like one of these two examples. But what if our space is infinite dimensional? What we do not yet have, but need, is a standard model of an infinite dimensional inner product space.

Definition $\ell_2(\mathbb{R})$ *is the space of infinite sequences of real scalars $x_1, x_2, \ldots$, such that the infinite sum $\sum_i |x_i|^2$ converges.*

Theorem 6.8 $\ell_2(\mathbb{R})$ *is a real vector space. Moreover, the form*

$$\langle \mathbf{x} | \mathbf{y} \rangle = \sum_{i=1}^{\infty} x_i y_i \qquad (6.102)$$

is a well-defined inner product on $\ell_2(\mathbb{R})$.

Proof: We must show that, if $\mathbf{x}$ and $\mathbf{y}$ are in $\ell_2(\mathbb{R})$, then so is $c_1 \mathbf{x} + c_2 \mathbf{y}$ for any scalars c_1 and c_2. This is not automatic; we must show that the sum

$$\sum_i (c_1 \mathbf{x} + c_2 \mathbf{y})_i^2 = c_1^2 \sum_i x_i^2 + c_2^2 \sum_i y_i^2 + c_1 c_2 \sum_i x_i y_i \qquad (6.103)$$

converges. The first sum on the right-hand side converges since $\mathbf{x} \in \ell_2(\mathbb{R})$, and the second sum converges since $\mathbf{y} \in \ell_2(\mathbb{R})$. Now we use the fact that, for any two real numbers a and b, $|ab| \leq a^2 + b^2$. Taking $a = c_1 x_i$ and $b = c_2 y_i$, we see that each term in the third sum is bounded by the sum of corresponding terms in the first two. Thus $\sum_{i=1}^{\infty} x_i y_i$ converges absolutely, and so converges. This also shows that the inner product (6.102) is well-defined. Checking that the inner product is linear in $\mathbf{x}$, linear in $\mathbf{y}$, symmetric, and positive is straightforward. ∎

$\ell_2(\mathbb{R})$ is our standard example of an infinite dimensional real inner product space. We can similarly define $\ell_2(\mathbb{C})$ as the space of

---

[7] This section can be skipped by a reader who is content with a heuristic understanding of function spaces.

infinite sequences of complex scalars x_i such that $\sum_i |x_i|^2$ converges. On $\ell_2(\mathbb{C})$ the inner product is

$$\langle \mathbf{x} | \mathbf{y} \rangle = \sum_{i=1}^{\infty} \bar{x}_i y_i. \tag{6.104}$$

$\ell_2(\mathbb{C})$ is the complexification of $\ell_2(\mathbb{R})$, the same way that $\mathbb{C}^n$ is the complexification of $\mathbb{R}^n$. Applying the procedure of Exercise 8 of Section 6.2 to (6.102) yields (6.104).

Finite and infinite linear combinations

There is a key difference between an infinite dimensional inner product space, such as ℓ_2, and a space such as $\mathbb{R}[t]$. In ℓ_2 we have a notion of distance, and so we can take limits of vectors. In particular, we can take infinite sums and infinite linear combinations (as long as they converge), since these are just limits of sequences of finite sums or finite linear combinations.

This possibility allows for two different conventions for what linear independence, span, and basis mean. In $\mathbb{R}[t]$, a linear combination can only have a finite number of terms. After all, $\mathbb{R}[t]$ is a space of polynomials, and an infinite power series is not a polynomial. A basis for $\mathbb{R}[t]$ is a set of linearly independent polynomials such that every polynomial in t is a finite linear combination of the basis elements. The standard basis $\{1, t, t^2, \ldots\}$ clearly qualifies.

In ℓ_2, however, a linear combination can have an infinite number of terms, and the span of a set of vectors includes such infinite linear combinations. Linear independence means that one cannot write a nontrivial linear combination — even an infinite one — that converges to zero. As always, a basis is a set that is linearly independent and spans. With these extended definitions, the collection of vectors $\{\mathbf{e}_1 = (1, 0, 0, \ldots)^T, \mathbf{e}_2 = (0, 1, 0, \ldots)^T, \ldots\}$ is a basis for ℓ_2 since it is linearly independent, and since every element of ℓ_2 can be written as an infinite linear combination of the $\mathbf{e}_i$'s. Equivalently, every element of ℓ_2 can be approximated arbitrarily well by a finite linear combination.

Complete normed vector spaces are called *Banach spaces*. An example is $C^0[0, 1]$, the space of continuous functions on the interval $[0, 1]$, for which the norm of $\mathbf{f}$ is the maximum value of $|f(t)|$. A space such as ℓ_2, where the norm comes from an inner product, is called a *Hilbert space*. When working with Banach spaces, and in particular with Hilbert spaces, we will always allow infinite linear

6.8. The Spaces ℓ_2 and $L^2[0,1]$

combinations. When working with infinite dimensional spaces, such as $\mathbb{R}[t]$, that have no concept of limit, we will only allow finite linear combinations. When working with finite dimensional spaces the issue is moot, as the only linear combinations we ever care to take are finite.

Heuristic definition of L^2

Now that we have a standard model ℓ_2 of an infinite dimensional Hilbert space, we consider a space of functions on $[0,1]$, denoted $L^2([0,1])$, where an orthonormal basis makes the space look like ℓ_2. The orthonormal basis will be sine functions, and the conversion from a function (an element of L^2) to an infinite set of coefficients (an element of ℓ_2) is called Fourier series. Fourier series is discussed at length in the next section.

We begin with a heuristic definition, and then sketch the technical construction needed to make this definition precise.

Let U be a subset of $\mathbb{R}$. Then $L^2(U, \mathbb{R})$ is the space of real-valued functions $f(t)$ on U such that $\int_U |f(t)|^2 dt$ converges. $L^2(U, \mathbb{R})$ is a real inner product space with inner product

$$\langle f|g\rangle = \int_U f(t)g(t)dt. \tag{6.105}$$

$L^2(U, \mathbb{C})$ is the complexification of $L^2(U, \mathbb{R})$. Equivalently, $L^2(U, \mathbb{C})$ is the space of complex-valued functions $f(t)$ on U such that $\int_U |f(t)|^2$ converges. $L^2(U, \mathbb{C})$ is a complex inner product space with inner product

$$\langle f|g\rangle = \int_U \overline{f(t)}g(t)dt. \tag{6.106}$$

When it is clear whether our functions are real or complex, or when it does not matter, we often write $L^2(U)$ or even just L^2 rather than $L^2(U, \mathbb{R})$ or $L^2(U, \mathbb{C})$.

Definition *A function whose square can be integrated is called* square-integrable.

Rigorous construction

The trouble with these definitions is that we have not said which functions are allowed. Are we just considering continuous functions? Differentiable functions? All square-integrable functions? Whichever space of functions we pick, we run into some technical problems. In most applications, these technical problems can be ignored, and the heuristic definition works just fine. For the careful reader, however, we include the standard construction of $L^2(U, \mathbb{R})$. In this construction the problems, while not eliminated, are minimized. The construction of $L^2(U, \mathbb{C})$ is almost identical and is not included.

We begin with $C_0^0(U)$, the space of continuous, bounded functions on U that are zero outside a finite region. (If U is itself a bounded region, this last condition imposes no constraints, but this construction also applies to the case that U is an unbounded region, perhaps even all of $\mathbb{R}$.) It is easy to check that this space is a vector space, on which the bilinear form (6.105) is an inner product (i.e., symmetric and positive).

Unfortunately, however, the limit of a sequence of functions in $C_0^0(U)$ is not necessarily in $C_0^0(U)$. For example, if $U = [-1, 1]$, consider the functions $f_n(t) = e^{nt}/(1+e^{nt})$. As $n \to \infty$, this function approaches the discontinuous function

$$f_\infty(t) = \begin{cases} 0 & \text{if } t < 0; \\ 1/2 & \text{if } t = 0; \\ 1 & \text{if } t > 0. \end{cases} \tag{6.107}$$

This limit occurs pointwise; for every t, $\lim_{n \to \infty} f_n(t) = f_\infty(t)$. It also occurs in the sense of our vector space norm:

$$\lim_{n \to \infty} \int_{-1}^{1} |f_\infty(t) - f_n(t)|^2 dt = 0. \tag{6.108}$$

To do calculus and analysis we need to be able to take limits, so we must expand our space to include functions that are limits, in the sense of (6.108), of elements of C_0^0. We let $V(U)$ be the space of such limits. In the process of taking limits we not only allow functions such as those given by (6.107), but we also allow functions that are unbounded, or that are nonzero on infinite regions. Here are some examples of functions that are in $V(U)$ but not in C_0^0.

- The function $f(t) = e^{-t^2}$ is in $V(\mathbb{R})$. To see it as a limit of functions in $C_0(\mathbb{R})$, let

$$b_n(t) = \begin{cases} 1 & \text{for } t \leq n; \\ n+1-t & \text{for } n < t < n+1; \text{ and} \\ 0 & \text{for } t \geq n+1, \end{cases} \tag{6.109}$$

and let $f_n(t) = b_n(|t|)e^{-t^2}$. Notice that f_n is continuous, and that $f_n(t) = 0$ for $|t| > n+1$, so f_n is in C_0^0. As $n \to \infty$, f_n approaches e^{-t^2}, both pointwise and in vector space norm.

6.8. The Spaces ℓ_2 and $L^2[0,1]$

- The function
$$g(t) = \begin{cases} t^{-1/3} & \text{if } t > 0; \\ 0 & \text{if } t = 0. \end{cases} \quad (6.110)$$
is in $V([0,1])$. It is the limit of functions
$$g_n(t) = \begin{cases} n^{1/3} & \text{if } t \leq 1/n; \\ t^{-1/3} & \text{if } t > 1/n. \end{cases} \quad (6.111)$$
As $n \to \infty$, $g_n(t)$ approaches $g(t)$ everywhere except at $t = 0$. However, one point does not affect an integral, and it is not hard to check that $\lim_{n \to \infty} \int_0^1 |g(t) - g_n(t)|^2 dt = 0$.

- The function
$$h(t) = \begin{cases} 0 & \text{if } t < 1; \\ 1 & \text{if } t = 1. \end{cases} \quad (6.112)$$
is in $V([0,1])$. It is a limit of the functions $h_n(t) = t^n$.

On V, we can certainly take limits. However, limits are not unique! On $V([0,1])$, both $h(t)$ and the zero function are limits of the functions $h_n(t) = t^n$, in that
$$\lim_{n \to \infty} \int_0^1 |h_n(t) - h(t)|^2 = \lim_{n \to \infty} \int_0^1 |h_n(t) - 0|^2 = 0. \quad (6.113)$$

This is related to the fact that our bilinear form (6.105), while symmetric, is no longer positive. There are nonzero vectors, such as $h(t)$, that have zero norm. Indeed, for a function to have zero norm, it does not have to be zero everywhere; it just has to be nonzero only on a set of measure zero.

Let $W(U)$ be the subspace of $V(U)$ of vectors that have zero norm. To get a space on which (6.105) is an inner product, and where limits are uniquely defined, we take the quotient of V by W. (Recall the definition of quotient spaces in Section 2.5.) $L^2(U)$ is, by definition, the quotient space of $V(U)$ by $W(U)$.

To sum up, $L^2(U)$ is not *exactly* the space of square-integrable functions on U. Rather, it is the space of square-integrable functions that can be approximated by functions in C_0^0, with the caveat that two functions in V that differ only on a set of measure zero are considered to be the same.

Exercises

1. Suppose that we have a basis $\{\mathbf{b}_n(t)\}$ for $L^2([0,1])$. Show that $\mathbf{d}_n(t) = a^{-1/2}\mathbf{b}(t/\sqrt{a})$ is a basis for $L^2[0,a]$. Construct a basis for $L^2[a,b]$.

2. Given a basis for $L^2[0,1]$, construct a basis for $L^2(\mathbb{R})$. Make your procedure such that, if the original basis was orthonormal, then the basis for $L^2(\mathbb{R})$ is orthonormal.

3. Let $C_0^n(\mathbb{R})$ denote the space of real n-times differentiable functions that vanish outside a bounded region. Show that the bilinear form

$$\langle \mathbf{f}|\mathbf{g}\rangle = \int_{-\infty}^{\infty} (f(x)g(x) + f'(x)g'(x) + \cdots + f^{(n)}(x)g^{(n)}(x))dx$$

is an inner product on $C_0^n(\mathbb{R})$.

4. Show that ℓ_2 is a complete metric space. That is, every Cauchy sequence in ℓ_2 has a (unique) limit in ℓ_2.

5. Show that $C_0^n(\mathbb{R})$, with the inner product of Exercise 3, is not a complete metric space.

The *Sobolev space* $H_n(\mathbb{R})$ is the completion of $C_0^n(\mathbb{R})$ with the inner product of Exercise 3. That is, one considers limits of elements of $C_0^n(\mathbb{R})$, modulo limits of norm zero.

6. Prove the Rellich lemma: every element of $H_1(\mathbb{R})$ is continuous. [Note: this is not true in higher dimensions. One can find a sequence of smooth functions on $\mathbb{R}^2$ whose limit, in the H_1 norm, is not continuous.]

7. Show that the derivative of an element of $H_k(\mathbb{R})$ is in $H_{k-1}(\mathbb{R})$. Conclude that every element of $H_k(\mathbb{R})$ is $k-1$ times differentiable. [In higher dimensions the situation is described by the Sobolev Embedding Theorem: $H_k(\mathbb{R}^n)$ embeds in $C^\ell(\mathbb{R}^n)$ if and only if $k > \ell + n/2$.]

6.9 Fourier Series on an Interval

In this section we show how to decompose any reasonable function on an interval (specifically, any element of L^2) into an (infinite) sum of sine waves. This decomposition, called Fourier series, is on the one hand a straightforward application of orthogonal bases, and on the other hand an incredibly powerful tool in physics and engineering. In this section we merely introduce the subject. Fourier series will be revisited periodically throughout the remainder of the book.

We work on the interval $U = [0, L]$. Consider the functions

$$\xi_n(t) = \sin(n\pi t/L). \tag{6.114}$$

6.9. Fourier Series on an Interval

We compute the inner products of these functions in $L^2([0, L])$:

$$\begin{aligned}
\langle \boldsymbol{\xi}_n | \boldsymbol{\xi}_m \rangle &= \int_0^L \sin(n\pi t/L)\sin(m\pi t/L)dt \\
&= \frac{1}{2}\int_0^L \cos((n-m)\pi t/L) - \cos((n+m)\pi t/L)dt \\
&= \begin{cases} L/2 & \text{if } n = m; \\ 0 & \text{otherwise.} \end{cases}
\end{aligned} \quad (6.115)$$

In other words, the functions are orthogonal. The fact that these functions span L^2 is somewhat subtler, and will be explained in Chapter 8. Since the functions $\{\boldsymbol{\xi}_n\}$ form an orthogonal basis for L^2, we can easily expand *any* element of L^2 as a linear combination of sine waves. The function $f(t)$ is expanded as

$$f(t) = \sum_{n=1}^{\infty} c_n \boldsymbol{\xi}_n(t) = \sum_{n=1}^{\infty} c_n \sin(n\pi t/L), \quad (6.116)$$

with

$$c_n = \frac{\langle \boldsymbol{\xi}_n | f \rangle}{\langle \boldsymbol{\xi}_n | \boldsymbol{\xi}_n \rangle} = \frac{2}{L}\int_0^L f(t)\sin(n\pi t/L)dt. \quad (6.117)$$

When $f(t)$ is a reasonably smooth function with nice behavior at the endpoints, the series (6.116) converges extremely quickly. One can then get a very good approximation to $f(t)$ by adding up the first few terms and ignoring the rest. This involves keeping track of just a few numbers, namely the first few coefficients c_n. This is a simple job, especially on a computer. In many cases you can get as good an approximation to a function by keeping track of 10 or 20 Fourier coefficients, as by keeping track of the actual value of the function at hundreds or thousands of points.

Since our basis is orthogonal, it is very easy to take inner products of Fourier series. Namely,

$$\langle \sum_n c_n \sin(n\pi t/L) | \sum_n c'_n \sin(n\pi t/L) \rangle = (L/2) \sum_n \bar{c}_n c'_n. \quad (6.118)$$

Fourier decomposition maps a function (an element of L^2) to an infinite set of coefficients (an element of ℓ_2), such that, aside from an overall factor of $L/2$, the inner product is preserved. Instead of computing inner products by doing integrals, we need only do sums.

Fourier analysis also makes taking derivatives very easy, as we already know the derivative of a sine wave. The first two derivatives of $f(t)$ are

$$\frac{df(t)}{dt} = \frac{\pi}{L} \sum_{n=1}^{\infty} nc_n \cos(n\pi t/L),$$

$$\frac{d^2 f(t)}{dt^2} = \frac{-\pi^2}{L^2} \sum_{n=1}^{\infty} n^2 c_n \sin(n\pi t/L). \quad (6.119)$$

This makes Fourier series a powerful tool for any application involving differential equations. In Chapter 8 we will use Fourier series to understand the motion of a vibrating string, and thereby to understand the tones that one hears when plucking a guitar string or striking a note on a piano.

We will now work out the Fourier coefficients of several functions. In preparation, we list several standard integrals:

$$\int \sin(kt)dt = \frac{-\cos(kt)}{k}; \quad \int \cos(kt)dt = \frac{\sin(kt)}{k};$$

$$\int t \sin(kt) = \frac{-t\cos(kt)}{k} + \frac{1}{k}\int \cos(kt)dt = \frac{-t\cos(kt)}{k} + \frac{\sin(kt)}{k^2};$$

$$\int t \cos(kt) = \frac{t\sin(kt)}{k} - \frac{1}{k}\int \sin(kt)dt = \frac{t\sin(kt)}{k} + \frac{\cos(kt)}{k^2};$$

$$\int t^2 \sin(kt) = \frac{-t^2\cos(kt)}{k} + \frac{2}{k}\int t\cos(kt)dt$$

$$= \frac{-t^2\cos(kt)}{k} + \frac{2t\sin(kt)}{k^2} + \frac{2\cos(kt)}{k^3}. \quad (6.120)$$

One can easily extend this list by successively integrating by parts.

In the following examples we take $L = 1$ to simplify the algebra. These expansions are illustrated in Figures 6.9–6.7. In each figure we show the actual function, the first nonzero term of the Fourier series (6.116), the sum of the first two nonzero terms and the sum of the first three nonzero terms.

Example 1: Suppose $f(t)$ is a square wave:

$$f(t) = \begin{cases} 1 & \text{if } t < 1/2; \\ -1 & \text{if } t \geq 1/2. \end{cases} \quad (6.121)$$

Then our coefficients are

$$c_n = 2\left(\int_0^{1/2} \sin(n\pi t)dt - \int_{1/2}^{1} \sin(n\pi t)dt\right)$$

6.9. Fourier Series on an Interval

$$= \frac{2}{n\pi}\left(1 - 2\cos(n\pi/2) + \cos(n\pi)\right). \tag{6.122}$$

This equals zero if n is odd or if n is divisible by 4, and equals $8/n\pi$ if n is even and not divisible by 4. These coefficients decay relatively slowly, proportional to $1/n$. This slow decay reflects the fact that $f(t)$ is discontinuous at $t = 1/2$, and is also nonzero at the endpoints. We are trying to write $f(t)$ as a sum of terms that are all continous and are all zero at the endpoints. It should be no surprise that it takes quite a few terms to do this well. What is amazing is that we can do it at all.

Example 2: Next suppose that $f(t)$ is a triangle wave:

$$f(t) = \begin{cases} t & \text{if } t < 1/2; \\ 1-t & \text{if } t \geq 1/2. \end{cases} \tag{6.123}$$

Then

$$c_n = 2\left(\int_0^{1/2} t \sin(n\pi t)dt + \int_{1/2}^1 (1-t)\sin(n\pi t)dt\right)$$

$$= \frac{4}{n^2\pi^2}\sin(n\pi/2). \tag{6.124}$$

This equals zero if n is even, $4/n^2\pi^2$ if $n - 1$ is divisible by 4, and $-4/n^2\pi^2$ if $n - 3$ is divisible by 4. Notice that the coefficients decay like $1/n^2$ rather than $1/n$. This reflects the fact that $f(t)$ is continuous (albeit not smooth) and equals zero at the endpoints.

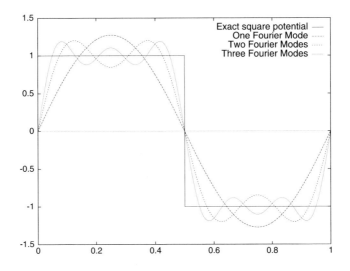

Figure 6.5: Approximations to a square potential

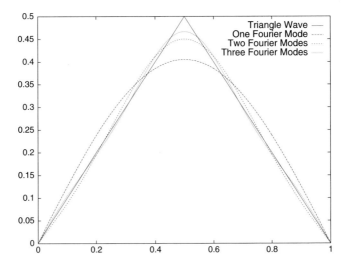

Figure 6.6: Approximations to a triangle potential

6.9. Fourier Series on an Interval

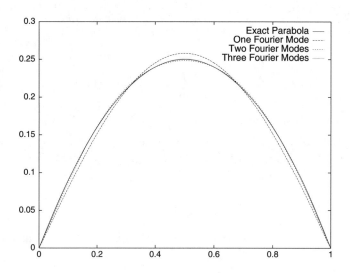

Figure 6.7: Approximations to a parabola

Example 3: Suppose that $f(t) = t - t^2$. Our coefficients are then

$$\begin{aligned} c_n &= 2\int_0^1 (t-t^2)\sin(n\pi t)dt \\ &= 2\left[\frac{(t^2-t)\cos(n\pi t)}{n\pi} + \frac{(2t-1)\sin(n\pi t)}{n^2\pi^2} - \frac{2\cos(n\pi t)}{n^3\pi^3}\right]_0^1 \\ &= \frac{4(1-\cos(n\pi))}{n^3\pi^3}. \end{aligned} \qquad (6.125)$$

This equals 0 for n even and $8/n^3\pi^3$ for n odd. The coefficients decay quite rapidly, like $1/n^3$. Indeed, the first two nonzero terms approximate the function to within 0.002, and the sum of the first three nonzero terms is practically indistinguishable from the function itself (see Figure 6.7). This rapid rate of decay is due to the function being zero on the boundary and smooth on the interior.

Example 4: Suppose the function $f(t)$ is any twice differentiable function with $f(0) = f(1) = 0$. Then c_n equals

$$\begin{aligned} 2\int_0^1 f(t)\sin(n\pi t)dt &= \frac{-2f(t)\cos(n\pi t)}{n\pi}\Big|_0^1 + \frac{2}{n\pi}\int_0^1 f'(t)\cos(n\pi t)dt \\ &= \frac{2f'(t)\sin(n\pi t)}{n^2\pi^2}\Big|_0^1 - \frac{2}{n^2\pi^2}\int_0^1 f''(t)\sin(n\pi t)dt \\ &= \frac{-2}{n^2\pi^2}\int_0^1 f''(t)\sin(n\pi t)dt. \end{aligned} \quad (6.126)$$

That is, the Fourier coefficients of $f(t)$ are $-1/n^2\pi^2$ times the Fourier coefficients of $f''(t)$. Since $f''(t)$ is continuous, the Fourier coefficients of f decay at least as fast as n^{-3}, and if $f''(t)$ is itself fairly smooth, then the Fourier coefficients of $f(t)$ decay even faster.

Exercises

1. On the interval $[0,1]$, decompose the function $f(t) = t$ as a Fourier series.

2. Evaluating the result of Exercise 1 at $t = 1/2$, show that $\pi/4 = 1 - 1/3 + 1/5 - 1/7 + \cdots$.

3. On the interval $[0,1]$, decompose the function $\cos(\pi t)$ as a Fourier series.

4. By considering the expansion of the triangle wave (6.123) applied to the point $t = 1/2$, derive a formula for $\sum_{n=0}^\infty (2n+1)^{-2}$.

5. Use the result of Exercise 4 to show that $\sum_{n=1}^\infty n^{-2} = \pi^2/6$.

6. Using the fact that $|\sin(kx) - \sin(ky)| \le |k(x-y)|$, show that, if $\sum n|c_n|$ converges, then f is continuous.

7. Show that, if $\sum n^k|c_n|$ converges, then f is $k-1$ times differentiable.

The results of Exercises 6 and 7 are not optimal and are improved upon in Chapter 8. We will show that the convergence of $\sum |c_n|$ is actually enough to imply continuity, and that the convergence of $\sum n^k|c_n|$ implies k-fold differentiability.

Chapter 7

Adjoints, Hermitian Operators, and Unitary Operators

In Chapters 4 and 5 we saw how useful a basis of eigenvectors can be. In Chapter 6 we saw how useful a basis of orthogonal vectors can be. For a typical operator L, the eigenvectors are not orthogonal, so we have to choose whether to work with eigenvectors or with orthogonal vectors. For example, the matrix

$$A = \begin{pmatrix} 1 & 2 \\ 1 & 2 \end{pmatrix} \tag{7.1}$$

has eigenvectors $(1,1)^T$ and $(2,-1)^T$, which are decidedly not orthogonal. We could apply Gram-Schmidt to these vectors to get an orthonormal basis, but then the second basis vector would no longer be an eigenvector. We can't have our cake and eat it too.

There are four classes of operators, however, that *do* have orthogonal eigenvectors. These classes are quite common and appear in many applications. Each class is defined by a relation between an operator and its adjoint. In Section 7.1 we define the adjoint of an operator.[1] In subsequent sections we study the four classes in turn: Hermitian operators, real symmetric operators, orthogonal operators, and unitary operators.

---

[1] The word "adjoint" is sometimes used in the setting of Cramer's rule to mean a matrix of minors. This has *nothing* to do with our (more common) usage of the word! Also, some texts denote the adjoint of an operator L by L^* rather than $L^\dagger$. We follow the prevailing convention in physics of denoting adjoints by $\dagger$ and reserving $*$ for other purposes.

7.1 Adjoints and Transposes

Before defining adjoints for operators on a general inner product space, we examine the situation in $\mathbb{R}^n$ with the standard inner product. Suppose $\mathbf{x}$ and $\mathbf{y}$ are vectors and A is an $n \times n$ matrix. The inner product $\langle \mathbf{x} | A\mathbf{y} \rangle$ is then the product of a row vector $\langle \mathbf{x} |$, a matrix A, and a column vector $|\mathbf{y}\rangle$:

$$\langle \mathbf{x} | A\mathbf{y} \rangle = \mathbf{x}^T A \mathbf{y} = \overbrace{(x_1, \ldots, x_n) \begin{pmatrix} A \end{pmatrix}}^{\mathbf{x}^T A} \underbrace{\begin{pmatrix} y_1 \\ \vdots \\ y_n \end{pmatrix}}_{A\mathbf{y}} \quad (7.2)$$

This can be computed in two ways. We can multiply A by $\mathbf{y}$ and then multiply $\mathbf{x}^T$ by the result. That is what the notation $\langle \mathbf{x} | A\mathbf{y} \rangle$ suggests. Alternatively, we can first multiply $\mathbf{x}^T$ by A and then multiply the product by $\mathbf{y}$. That is, there is a linear operation on $\mathbf{x}$, which we denote $A^\dagger$, such that $\langle A^\dagger \mathbf{x} | \mathbf{y} \rangle = \langle \mathbf{x} | A\mathbf{y} \rangle$ for all $\mathbf{x}$ and $\mathbf{y}$. In this example, $A^\dagger$ is just multiplication by A^T, since

$$\langle A^T \mathbf{x} | \mathbf{y} \rangle = (A^T \mathbf{x})^T \mathbf{y} = (\mathbf{x}^T A)\mathbf{y} = \mathbf{x}^T (A\mathbf{y}) = \langle \mathbf{x} | A\mathbf{y} \rangle. \quad (7.3)$$

Definition *Suppose L is an operator on an inner product space V. The* adjoint *of L is an operator $L^\dagger$ such that, for any vectors $\mathbf{x}, \mathbf{y} \in V$, $\langle L^\dagger \mathbf{x} | \mathbf{y} \rangle = \langle \mathbf{x} | L\mathbf{y} \rangle$.*

We sometimes write $\langle \mathbf{x} | L | \mathbf{y} \rangle$ instead of $\langle \mathbf{x} | L\mathbf{y} \rangle$. We have seen that we can evaluate this expression by having L act on $\mathbf{y}$ or by having $L^\dagger$ act on $\mathbf{x}$. Neither way is intrinsically better than the other, and placing L in the middle emphasizes this duality.

It is not immediately obvious that $L^\dagger$ exists for every L, or that $L^\dagger$ is unique. We will show that $L^\dagger$ exists (and is unique) whenever V is finite dimensional, and we will derive a formula for $L^\dagger$ in terms of L. This formula also works in infinite dimensions as long as the relevant sums converge. The basic properties of adjoints are

Theorem 7.1 *If A and B are operators on an inner product space V, and c is a scalar, then*

- $(A + B)^\dagger = A^\dagger + B^\dagger$
- $(AB)^\dagger = B^\dagger A^\dagger$

7.1. Adjoints and Transposes

- $(cA)^\dagger = \bar{c}A^\dagger$
- $(A^\dagger)^\dagger = A$.

Proof:

$$\begin{aligned}
\langle \mathbf{x}|(A+B)\mathbf{y}\rangle &= \langle \mathbf{x}|A\mathbf{y}\rangle + \langle \mathbf{x}|B\mathbf{y}\rangle = \langle A^\dagger\mathbf{x}|\mathbf{y}\rangle + \langle B^\dagger\mathbf{x}|\mathbf{y}\rangle \\
&= \langle (A^\dagger\mathbf{x}+B^\dagger\mathbf{x})|\mathbf{y}\rangle = \langle (A^\dagger+B^\dagger)\mathbf{x}|\mathbf{y}\rangle.
\end{aligned}$$

$$\begin{aligned}
\langle \mathbf{x}|AB\mathbf{y}\rangle &= \langle \mathbf{x}|A(B\mathbf{y})\rangle = \langle A^\dagger\mathbf{x}|B\mathbf{y}\rangle \\
&= \langle B^\dagger(A^\dagger\mathbf{x})|\mathbf{y}\rangle = \langle (B^\dagger A^\dagger)\mathbf{x}|\mathbf{y}\rangle.
\end{aligned}$$

$$\langle \mathbf{x}|cA\mathbf{y}\rangle = c\langle \mathbf{x}|A\mathbf{y}\rangle = c\langle A^\dagger\mathbf{x}|\mathbf{y}\rangle = \langle \bar{c}A^\dagger\mathbf{x}|\mathbf{y}\rangle.$$

$$\langle \mathbf{x}|A^\dagger\mathbf{y}\rangle = \overline{\langle A^\dagger\mathbf{y}|\mathbf{x}\rangle} = \overline{\langle \mathbf{y}|A\mathbf{x}\rangle} = \langle A\mathbf{x}|\mathbf{y}\rangle. \blacksquare \qquad (7.4)$$

Of course, if V is a real vector space, then c is real, and $\bar{c} = c$. In the case that V is Euclidean $\mathbb{R}^n$ and A is a matrix, it is not hard to check that the properties of Theorem 7.1 are satisfied by the transpose operation.

Now let $V = \mathbb{C}^n$ with the standard inner product, and let our operator L be multiplication by a matrix A. Then the matrix elements of $L^\dagger$ must be

$$(L^\dagger)_{ij} = \langle \mathbf{e}_i|L^\dagger\mathbf{e}_j\rangle = \overline{\langle L^\dagger\mathbf{e}_j|\mathbf{e}_i\rangle} = \overline{\langle \mathbf{e}_j|L\mathbf{e}_i\rangle} = \overline{A_{ji}}. \qquad (7.5)$$

In other words, the matrix of $L^\dagger$ is the complex conjugate of the transpose of A. (By the complex conjugate of a matrix we mean the matrix obtained by conjugating each element. This is different from the function of a matrix as defined in Section 4.8.) Of course, if A happens to be a real matrix, then this is the same as the transpose of A.

We have seen the operation of taking the complex conjugate of a transpose before. That is exactly how we turn a ket $|\mathbf{x}\rangle = \begin{pmatrix} x_1 \\ \vdots \\ x_n \end{pmatrix}$ in $\mathbb{C}^n$ into a bra $\langle \mathbf{x}| = (\bar{x}_1, \ldots, \bar{x}_n)$. Indeed, with this in mind we can see that, in $\mathbb{C}^n$ with the standard inner product,

$$\langle \mathbf{x}|A\mathbf{y}\rangle = \begin{pmatrix} \bar{\mathbf{x}}^T \end{pmatrix} \begin{pmatrix} A \end{pmatrix} \begin{pmatrix} \mathbf{y} \end{pmatrix} = (A^T\bar{\mathbf{x}})^T \begin{pmatrix} \mathbf{y} \end{pmatrix}$$

$$= \overline{\left((A^T)\mathbf{x}\right)^T}\begin{pmatrix}\mathbf{y}\end{pmatrix} = \langle \bar{A}^T\mathbf{x}|\mathbf{y}\rangle, \tag{7.6}$$

showing that the transpose of the conjugate of a matrix is indeed an adjoint.

Definition *The* Hermitian conjugate *of a matrix A is the transpose of its complex conjugate, and is denoted $A^\dagger$.*

We have just shown that, on $\mathbb{R}^n$ or $\mathbb{C}^n$ with the standard inner product, the adjoint of a matrix A is the same as its Hermitian conjugate, which justifies our denoting both by $A^\dagger$. In general, however, we should distinguish between Hermitian conjugation, which is something we do to matrices, and taking the adjoint, which is something we do to operators. This distinction is explored in the exercises.

Next we consider the adjoint in a general inner product space. Since an orthonormal basis makes any inner product space look like $\mathbb{R}^n$ or $\mathbb{C}^n$ (with the standard inner product), it should be no surprise that the results for $\mathbb{R}^n$ and $\mathbb{C}^n$ carry over.

Theorem 7.2 *Let L be an operator on a finite dimensional inner product space V with orthonormal basis $\mathcal{B}$. Then $L^\dagger$ is the operator whose matrix in the $\mathcal{B}$ basis is the Hermitian conjugate of $[L]_\mathcal{B}$. In other words,*

$$[L^\dagger]_\mathcal{B} = ([L]_\mathcal{B})^\dagger = \overline{([L]_\mathcal{B})^T}. \tag{7.7}$$

In terms of bras and kets,

$$L^\dagger = \sum_{i,j=1}^{n} |\mathbf{b}_i\rangle\langle L\mathbf{b}_i|\mathbf{b}_j\rangle\langle \mathbf{b}_j|. \tag{7.8}$$

Proof: See Figure 7.1. First we assume that $L^\dagger$ exists, and show it must take the form (7.8), and that its matrix in the $\mathcal{B}$ basis must be given by (7.7). Then we show that the operator defined in (7.8) actually is an adjoint to L.

If $L^\dagger$ exists, then by Theorem 6.3

$$\begin{aligned}([L^\dagger]_\mathcal{B})_{ij} &= \langle \mathbf{b}_i|L^\dagger|\mathbf{b}_j\rangle = \langle L^{\dagger\dagger}\mathbf{b}_i|\mathbf{b}_j\rangle \\ &= \langle L\mathbf{b}_i|\mathbf{b}_j\rangle = \overline{\langle \mathbf{b}_j|L\mathbf{b}_i\rangle} = \overline{([L]_\mathcal{B})_{ji}} = ([L]_\mathcal{B}^\dagger)_{ij}.\end{aligned} \tag{7.9}$$

7.1. Adjoints and Transposes

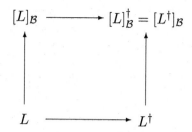

Figure 7.1: The adjoint of an operator and of its matrix

Also by Theorem 6.3,

$$
\begin{aligned}
L^\dagger &= \sum_{i,j=1}^n |\mathbf{b}_i\rangle\langle\mathbf{b}_i|L^\dagger\mathbf{b}_j\rangle\langle\mathbf{b}_j| \\
&= \sum_{i,j=1}^n |\mathbf{b}_i\rangle\langle L^{\dagger\dagger}\mathbf{b}_i|\mathbf{b}_j\rangle\langle\mathbf{b}_j| \\
&= \sum_{i,j=1}^n |\mathbf{b}_i\rangle\langle L\mathbf{b}_i|\mathbf{b}_j\rangle\langle\mathbf{b}_j|. \quad (7.10)
\end{aligned}
$$

To prove existence we show, for $L^\dagger$ given by (7.8) and for arbitrary vectors $\mathbf{x} = \sum x_i \mathbf{b}_i$ and $\mathbf{y} = \sum y_i \mathbf{b}_i$, that $\langle L^\dagger \mathbf{x}|\mathbf{y}\rangle$ equals $\langle \mathbf{x}|L\mathbf{y}\rangle$. Now $\langle \mathbf{x}|\mathbf{b}_i\rangle = \bar{x}_i$ and $\langle \mathbf{b}_j|\mathbf{y}\rangle = y_j$, so

$$
\begin{aligned}
\langle \mathbf{x}|L^\dagger\mathbf{y}\rangle &= \sum_{i,j=1}^n \langle \mathbf{x}|\mathbf{b}_i\rangle\langle L\mathbf{b}_i|\mathbf{b}_j\rangle y_j = \sum_{i,j=1}^n \bar{x}_i\langle L\mathbf{b}_i|\mathbf{b}_j\rangle y_j \\
&= \sum_{i,j=1}^n \langle Lx_i\mathbf{b}_i|y_j\mathbf{b}_j\rangle = \langle L\mathbf{y}|\mathbf{x}\rangle. \quad\blacksquare \quad (7.11)
\end{aligned}
$$

We have already seen what the adjoint means for matrices. Here are some additional examples of adjoints in various vector spaces.

1. Let $V = \mathbb{R}^3$ with the usual inner product, and let L_θ be a rotation by an angle θ about the x_1-axis. Then $L_\theta^\dagger$ is a rotation by an angle $-\theta$ about the x_1-axis. In other words, $L_\theta^\dagger = L_{-\theta} = L_\theta^{-1}$. You can check that by looking at the matrices:

$$
L = \begin{pmatrix} 1 & 0 & 0 \\ 0 & \cos(\theta) & -\sin(\theta) \\ 0 & \sin(\theta) & \cos(\theta) \end{pmatrix}; \; L^\dagger = \begin{pmatrix} 1 & 0 & 0 \\ 0 & \cos(\theta) & \sin(\theta) \\ 0 & -\sin(\theta) & \cos(\theta) \end{pmatrix}. \quad (7.12)
$$

2. Let $V = L^2(\mathbb{R}, \mathbb{C})$, let $h(t)$ be a bounded complex-valued function, and let H be the operation of multiplying by h. In other words, $(Hf)(t) = h(t)f(t)$. Then $H^\dagger$ is multiplication by $\overline{h(t)}$, since $\langle H^\dagger f | g \rangle = \int \overline{h(t)f(t)} g(t) dt = \int \overline{f(t)} h(t) g(t) = \langle f | Hg \rangle$. In particular, if $h(t)$ is a real-valued function, then $H^\dagger = H$.

3. Let $V = L^2(\mathbb{R})$ and let D be the operation d/dt. Then $D^\dagger$ is the operation $-D$, since

$$\begin{aligned}\langle -Df | g \rangle &= -\int_{-\infty}^{\infty} \frac{df(t)}{dt} g(t) dt \\ &= -f(t)g(t)\Big|_{-\infty}^{\infty} + \int_{-\infty}^{\infty} f(t) \frac{dg(t)}{dt} dt \\ &= 0 + \langle f | Dg \rangle.\end{aligned} \quad (7.13)$$

Here we have used the fact that f and g are square-integrable, so the product $f(t)g(t)$ must go to zero as $t \to \pm\infty$.

Exercises

1. Check that the properties of Theorem 7.1 are satisfied by the operation of taking the complex conjugate of the transpose of a matrix.

2. On Euclidean $\mathbb{R}^3$, let $L(\mathbf{x}) = (3x_1 + x_2, x_2 - x_3, 5x_1 + x_3)^T$. Compute $L^\dagger(\mathbf{x})$.

3. On $\mathbb{C}^3$ with the standard inner product, let $L(\mathbf{x}) = (3x_1 + ix_2, ix_2 - 2x_3, (1+i)x_1 + 5x_3)^T$. Compute $L^\dagger(\mathbf{x})$.

The form of the adjoint depends on the form of the inner product, as these examples show:

4. Let V be $\mathbb{R}^n$ with a nonstandard inner product. Let L be multiplication by a matrix A. Find the matrix of $L^\dagger$ in terms of the matrix A and the metric matrix $\mathbf{g}$. (The answer is not simply $A^\dagger$.)

5. Repeat Exercise 2, only with $\mathbb{R}^3$ having the nonstandard inner product $\langle \mathbf{x} | \mathbf{y} \rangle = x_1 y_1 + 2x_2 y_2 + 3x_3 y_3$.

6. Let $V = \mathbb{R}_2[t]$ with the inner product $\langle a_0 + a_1 t + a_2 t^2 | b_0 + b_1 t + b_2 t^2 \rangle = a_0 b_0 + a_1 b_1 + a_2 b_2$. Let $D = d/dt$, and let $\mathbf{x} = 1 + 2t - t^2$. Find $D\mathbf{x}$ and $D^\dagger(\mathbf{x})$.

7. Let V be a space of smooth real-valued functions on $\mathbb{R}$ with the inner product $\langle \mathbf{f} | \mathbf{g} \rangle = \int_{-\infty}^{\infty} f(t) g(t) e^{-t^2} dt$. Find the adjoint of $D = d/dt$.

7.2 Hermitian Operators

In this section we define and analyze the first class of operators that have real eigenvalues and orthogonal eigenvectors.

Definition *An operator L on an inner product space is called* self-adjoint *or* Hermitian *if $L^\dagger = L$. An operator L is called* anti-Hermitian *if $L^\dagger = -L$. Similarly, a matrix A is called Hermitian if $A^\dagger = A$, and anti-Hermitian if $A^\dagger = -A$. A Hermitian operator on a real vector space (or a real Hermitian matrix) is called* symmetric. *A real anti-Hermitian operator (or matrix) is called* antisymmetric *or* skew.

In quantum mechanics, every physically measurable quantity, such as the position, momentum, or energy of a particle, is described by a Hermitian operator. The possible results of a measurement (e.g., the energy levels) are the eigenvalues of the operator, and the states corresponding to those values are the eigenvectors. As a result, Hermitian operators are central to modern physics.

Real symmetric operators are closely related to quadratic functions, a relation we will explore further in Section 7.3. Since many quantities in physics and engineering are quadratic (kinetic energy is quadratic in velocity, energy dissipation in a circuit is quadratic in currect, etc.), real symmetric matrices occur very frequently.

Here are some examples of Hermitian operators and matrices:

1. Let V be an inner product space with an arbitrary (not necessarily orthonormal) basis $\mathcal{B}$. The metric matrix $\mathbf{g}$ is Hermitian, since
$$g_{ij} = \langle \mathbf{b}_i | \mathbf{b}_j \rangle = \overline{\langle \mathbf{b}_j | \mathbf{b}_i \rangle} = \bar{g}_{ji}. \qquad (7.14)$$

2. Let $f(\mathbf{x})$ be any smooth, real-valued function on $\mathbb{R}^n$. Then at every point the Hessian matrix $H_{ij} = (\partial^2 f / \partial x_i \partial x_j)$ is real symmetric, since mixed partial derivatives are equal.

3. On $L^2(\mathbb{R}, \mathbb{C})$, let T be multiplication by t. That is, $(Tf)(t) = tf(t)$. T is Hermitian by example 2 of Section 7.1. In quantum mechanics one studies functions of a variable x (rather than t) and the operator "multiply the function by x" is associated to the position of the particle being studied.

4. On $L^2(\mathbb{R}, \mathbb{C})$, we have seen that $D = d/dt$ is anti-Hermitian. But $-iD$ is Hermitian, since $(-iD)^\dagger = \overline{-i}D^\dagger = (i)(-D) =$

Chapter 7. Adjoints and Hermitian Operators

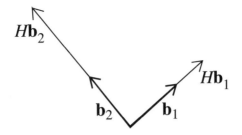

Figure 7.2: Real eigenvalues and orthogonal eigenvectors

$-iD$. In general, if an operator L on a complex inner product space V is Hermitian, then $\pm iL$ is anti-Hermitian, and if L is anti-Hermitian then $\pm iL$ is Hermitian. (We need V to be complex in order for multiplication by i to make sense.) In quantum mechanics, the operator $-id/dx$ is associated with the momentum of a particle.

Hermitian operators, orthonormal bases, and Hermitian matrices are very closely related.

Theorem 7.3 *Let $\mathcal{B}$ be an orthonormal basis for the inner product space V, and let L be an operator on V. The operator L is Hermitian if, and only if, the matrix $[L]_\mathcal{B}$ is Hermitian.*

Proof:

$$\begin{aligned}
L \text{ is Hermitian} &\Leftrightarrow L = L^\dagger & \text{(by definition)} \\
&\Leftrightarrow [L]_\mathcal{B} = [L^\dagger]_\mathcal{B} \\
&\Leftrightarrow [L]_\mathcal{B} = [L]_\mathcal{B}^\dagger & \text{(by Theorem 7.2)} \\
&\Leftrightarrow [L]_\mathcal{B} \text{ is a Hermitian matrix.} \blacksquare
\end{aligned} \quad (7.15)$$

There are three key properties of Hermitian operators. The first two are shown in Figure 7.2 and given in:

Theorem 7.4 *Let H be a Hermitian operator on an inner product space. Then the eigenvalues of H are real and eigenvectors corresponding to distinct eigenvalues are orthogonal.*

Proof: Let $\mathbf{b}_1$ and $\mathbf{b}_2$ be eigenvectors of H with eigenvalues λ_1 and λ_2. Rescale $\mathbf{b}_1$ so that $|\mathbf{b}_1| = 1$. Then

$$\lambda_1 = \lambda_1 \langle \mathbf{b}_1 | \mathbf{b}_1 \rangle = \langle \mathbf{b}_1 | \lambda_1 \mathbf{b}_1 \rangle$$

7.2. Hermitian Operators

$$= \langle \mathbf{b}_1 | H \mathbf{b}_1 \rangle = \langle H \mathbf{b}_1 | \mathbf{b}_1 \rangle = \langle \lambda_1 \mathbf{b}_1 | \mathbf{b}_1 \rangle = \bar{\lambda}_1. \quad (7.16)$$

Since $\lambda_1 = \bar{\lambda}_1$, λ_1 is real. Similarly, λ_2 is real. Next we compute $\langle \mathbf{b}_1 | H | \mathbf{b}_2 \rangle$ in two different ways:

$$\begin{aligned} \lambda_1 \langle \mathbf{b}_1 | \mathbf{b}_2 \rangle &= \langle \lambda_1 \mathbf{b}_1 | \mathbf{b}_2 \rangle = \langle H \mathbf{b}_1 | \mathbf{b}_2 \rangle = \langle \mathbf{b}_1 | H | \mathbf{b}_2 \rangle = \langle \mathbf{b}_1 | H \mathbf{b}_2 \rangle \\ &= \langle \mathbf{b}_1 | \lambda_2 \mathbf{b}_2 \rangle = \lambda_2 \langle \mathbf{b}_1 | \mathbf{b}_2 \rangle. \end{aligned} \quad (7.17)$$

If $\lambda_1 \neq \lambda_2$, we must have $\langle \mathbf{b}_1 | \mathbf{b}_2 \rangle = 0$. ∎

Note that we did not say that two eigenvectors corresponding to the same eigenvalue are orthogonal. That statement would be patently false, since $\mathbf{b}_1$ and $2\mathbf{b}_1$ are both eigenvectors with eigenvalue λ_1, and are certainly not orthogonal.

Example: Consider the Hermitian matrix

$$A = \begin{pmatrix} 1 & 0 & 0 \\ 0 & 0 & 1 \\ 0 & 1 & 0 \end{pmatrix}. \quad (7.18)$$

The vectors $\mathbf{b}_1 = (1, 0, 0)^T$, $\mathbf{b}_2 = (1, 1, 1)^T$, and $\mathbf{b}_3 = (0, 1, -1)^T$ are all eigenvectors, the first two with eigenvalue 1, and the last with eigenvalue -1. Since $1 \neq -1$, $\mathbf{b}_3$ must be orthogonal to $\mathbf{b}_1$, and to $\mathbf{b}_2$. However, since $\mathbf{b}_1$ and $\mathbf{b}_2$ correspond to the same eigenvalue they do not have to be orthogonal, and indeed they are not.

Although Theorem 7.4 does not guarantee that *every* basis of eigenvectors is orthogonal, it does guarantee the existence of such a basis:

Corollary 7.5 *Let H be a Hermitian operator on an inner product space V. If H is diagonalizable, then there exists an orthonormal basis for V consisting of eigenvectors of H.*

Proof: We construct the orthonormal basis as follows. Start with any basis of eigenvectors of H and group these basis vectors by their eigenvalues. For each eigenvalue λ, apply Gram-Schmidt to all the basis vectors in the eigenspace E_λ. Since E_λ is a subspace, linear combinations of elements of E_λ are in E_λ, so the result of the Gram-Schmidt process is an orthonormal basis for E_λ. The vectors that emerge from this process form a basis of eigenvectors. Two vectors with different eigenvalues are orthogonal by Theorem 7.4. Two vectors in the same eigenspace are orthogonal by Gram-Schmidt. ∎

Returning to our example, the vectors $\mathbf{b}_1$ and $\mathbf{b}_2$ lie in the eigenspace E_1. Applying Gram-Schmidt to them we get vectors $\mathbf{d}_1 =

$(1,0,0)^T$ and $\mathbf{d}_2 = (0,1,1)^T/\sqrt{2}$, which are still eigenvectors of A with eigenvalue 1. Applying Gram-Schmidt to E_{-1} just gives us $\mathbf{d}_3 = (0,1,-1)^T/\sqrt{2}$. The triple of vectors $\{\mathbf{d}_1, \mathbf{d}_2, \mathbf{d}_3\}$ is an orthonormal basis for $\mathbb{C}^3$ consisting of eigenvectors of A.

So far we have assumed that our Hermitian operators are diagonalizable. The third essential property of Hermitian operators makes this assumption unnecessary:

Theorem 7.6 *Let H be a Hermitian operator on an finite dimensional complex vector space V. Then H is diagonalizable.*

***Proof*:** Given an orthonormal basis $\mathcal{B}$ of V, diagonalizing the Hermitian matrix $[H]_\mathcal{B}$ is tantamount to diagonalizing H. We therefore need only prove that every $n \times n$ Hermitian matrix is diagonalizable.

We prove this by induction on n. If $n = 1$, the result is trivial, as every 1×1 matrix is diagonal. Now suppose the theorem is true for $n = k$; we must prove it for $n = k+1$.

The characteristic polynomial of H has at least one root λ_1, and hence at least one eigenvalue. Let $\mathbf{b}_1$ be an eigenvector with eigenvalue λ_1. Rescale $\mathbf{b}_1$ to make it a unit vector. Pick vectors $\mathbf{b}_2, \ldots, \mathbf{b}_{k+1}$ so that $\mathbf{b}_1, \ldots, \mathbf{b}_{k+1}$ is an orthonormal basis. (For example, pick any basis that starts with $\mathbf{b}_1$ and apply Gram-Schmidt to it.) Let $A = [H]_\mathcal{B}$. Then A is Hermitian, and, for $i \neq 1$,

$$A_{i1} = \langle \mathbf{b}_i | H | \mathbf{b}_1 \rangle = \lambda_1 \langle \mathbf{b}_i | \mathbf{b}_1 \rangle = 0, \tag{7.19}$$

since the $\mathbf{b}_i$'s are orthonormal. Since A is Hermitian, $A_{1i} = \overline{A_{i1}} = 0$. Thus A must take the general form

$$A = \begin{pmatrix} \lambda_1 & 0 \\ 0 & B \end{pmatrix}, \tag{7.20}$$

where B is a $k \times k$ Hermitian matrix. But, by the inductive hypothesis, every $k \times k$ Hermitian matrix is diagonalizable, so there exist k linearly independent eigenvectors $\boldsymbol{\xi}_1, \ldots, \boldsymbol{\xi}_k$ of B. But then $\mathbf{e}_1$ and $\begin{pmatrix} 0 \\ \boldsymbol{\xi}_i \end{pmatrix}$ are $k+1$ linearly independent eigenvectors of A, and correspond to $k+1$ linearly independent eigenvectors of H. ∎

This proof sheds light on the issue of diagonalizability in general. If H is a general (not necessarily Hermitian) operator, we can still find a single eigenvector $\mathbf{b}_1$; we still can extend that to an orthonormal basis, and it is still true that $A_{i1} = 0$ for $i \neq 1$. However, if H is

7.2. Hermitian Operators

not Hermitian, then A_{1i} is not related to A_{i1}, and does not have to equal zero. As a result, when H is not Hermitian, A takes the form

$$A = \begin{pmatrix} \lambda_1 & C \\ 0 & B \end{pmatrix} \tag{7.21}$$

instead of the form (7.20). The vectors $\begin{pmatrix} 0 \\ \boldsymbol{\xi}_i \end{pmatrix}$ would then no longer be eigenvectors of A. In Section 4.6, trick 4, we saw how to get eigenvectors of a block triangular matrix in terms of the eigenvectors of its blocks. If $\boldsymbol{\xi}_i$ is an eigenvector of B with eigenvalue λ_i, then $\begin{pmatrix} (\lambda_1 - \lambda_i)^{-1} C \boldsymbol{\xi}_i \\ \boldsymbol{\xi}_i \end{pmatrix}$ is an eigenvector of A, as long as $\lambda_i \ne \lambda_i$. However, if $\lambda_i = \lambda_1$, and $C\boldsymbol{\xi}_i \ne 0$, then $(\lambda_1 - \lambda_i)^{-1} C \boldsymbol{\xi}_i$ is undefined, and there is no eigenvector of A corresponding to $\boldsymbol{\xi}_i$. That is how we can get a deficiency. The simplest example of this is where B is a 1×1 matrix and $\lambda_1 = B = C = 1$. In that case $A = \begin{pmatrix} 1 & 1 \\ 0 & 1 \end{pmatrix}$. The vector $\mathbf{e}_1 = (1, 0)^T$ is an eigenvector of A, but there is no second eigenvector of A corresponding to the eigenvector (1) of B.

We have shown that Hermitian operators are diagonalizable, have orthogonal eigenvectors, and have real eigenvalues. The converse is also true:

Theorem 7.7 *Let H be an operator on a finite dimensional space V, whose eigenvalues are all real, and whose eigenvectors form an orthonormal basis for V. Then H is Hermitian.*

Proof: Let $\mathcal{B}$ be an orthonormal basis of eigenvectors of H. The matrix $[H]_\mathcal{B}$ is diagonal, with diagonal entries $\lambda_1, \ldots, \lambda_n$. Since $[H]_\mathcal{B}$ is diagonal and real, it equals its Hermitian conjugate. Since $\mathcal{B}$ is orthonormal and $[H]_\mathcal{B}$ is Hermitian, H is Hermitian by Theorem 7.3.
∎

There is an extension of Theorems 7.4–7.7 to Hermitian operators on infinite dimensional vector spaces. This extension, called the *spectral theorem*, is at the heart of functional analysis. However, it is somewhat difficult to state, much less to prove, because operators on infinite dimensional spaces may have continuous ranges of eigenvalues. Sums over eigenvalues are then replaced by integrals — ordinary Riemann integrals in the simplest cases, and generalized integrals in more exotic examples. We will discuss continuous eigenvalues and state the spectral theorem precisely in Chapter 9. For now we merely

202   Chapter 7. Adjoints and Hermitian Operators

argue, by analogy to finite dimensions, that Hermitian operators on infinite dimensional spaces are very well-behaved.

Exercises

In exercises 1–4, find an orthonormal basis of $\mathbb{C}^2$ or $\mathbb{C}^3$ consisting of eigenvectors of the given matrix.

1. $\begin{pmatrix} 1 & 1+i \\ 1-i & 2 \end{pmatrix}.$

2. $\begin{pmatrix} 2 & 1 & 1 \\ 1 & 2 & 1 \\ 1 & 1 & 2 \end{pmatrix}.$

3. $\begin{pmatrix} 2 & 0 & i \\ 0 & 1 & 0 \\ -i & 0 & 2 \end{pmatrix}.$

4. $\begin{pmatrix} 0 & 3 & 0 \\ 3 & 0 & 4 \\ 0 & 4 & 0 \end{pmatrix}.$

5. Let A be an arbitrary operator. Show that $H = A^\dagger A$ is Hermitian, and that all of the eigenvalues of H are non-negative.

6. Show that the operator $L = -d^2/d\theta^2$ on $L^2(S^1)$ (the space of square-integrable functions on the unit circle) is Hermitian, with all its eigenvalues non-negative. Can you write L in the form $A^\dagger A$ for some operator A?

7.3 Quadratic Forms and Real Symmetric Matrices

The theory of eigenvalues and eigenvectors grew out of the study of quadratic functions on a circle or a sphere. The most general purely quadratic function of two variables, x_1 and x_2, takes the form

$$f(x_1, x_2) = ax_1^2 + 2bx_1 x_2 + cx_2^2, \tag{7.22}$$

where a, b, and c are constants. To maximize this function on the unit circle $g(x_1, x_2) = x_1^2 + x_2^2 - 1 = 0$, we apply the method of Lagrange multipliers. Setting $\nabla f = \lambda \nabla g$ and dividing by two, we obtain

$$ax_1 + bx_2 = \lambda x_1,$$

7.3. Quadratic Forms and Real Symmetric Matrices

$$bx_1 + cx_2 = \lambda x_2, \tag{7.23}$$

which is precisely the eigenvalue[2] equation $A\mathbf{x} = \lambda \mathbf{x}$ for $A = \begin{pmatrix} a & b \\ b & c \end{pmatrix}$.
The maxima and minima of the function $f(x_1, x_2)$ on the unit circle are to be found at the eigenvectors of A, and it is not hard to see that these maximum and minimum values are the eigenvalues themselves.

More generally, if A is a real $n \times n$ matrix, we consider the quadratic function of n variables

$$f_A(\mathbf{x}) = \langle \mathbf{x} | A | \mathbf{x} \rangle = \sum_{i,j} A_{ij} x_i x_j, \tag{7.24}$$

and look for maxima, minima, and saddle points of $f_A(\mathbf{x})$ on the unit sphere $|\mathbf{x}| = 1$. Notice that $f_A(\mathbf{x}) = f_S(\mathbf{x})$, where $S = (A+A^T)/2$, so every quadratic function can be viewed as coming from a symmetric matrix S. Without loss of generality, then, we assume that A was symmetric to begin with.

Example: On the unit circle in $\mathbb{R}^2$, consider the function

$$f_A(\mathbf{x}) = 2x_1^2 + 2x_1 x_2 + 2x_2^2, \tag{7.25}$$

where

$$A = \begin{pmatrix} 2 & 1 \\ 1 & 2 \end{pmatrix}. \tag{7.26}$$

Setting $x_1 = \cos(\theta)$ and $x_2 = \sin(\theta)$, we have

$$f_A(\mathbf{x}) = 2\cos^2(\theta) + 2\sin(\theta)\cos(\theta) + 2\sin^2(\theta) = 2 + \sin(2\theta). \tag{7.27}$$

This has a maximum value of 3 at $\theta = \pi/4$, where $x_1 = x_2 = \sqrt{2}/2$, and a minimum value of 1 at $\theta = 3\pi/4$, where $x_1 = -x_2 = \sqrt{2}/2$. As expected, the maximum and minimum occur at the eigenvectors of A, and the maximum and minimum values are the eigenvalues of A. Figure 7.3 shows a related quantity, the vector $A\mathbf{x}$ as $\mathbf{x}$ varies in the unit circle. $A\mathbf{x}$ is longest at one eigenvector of A and is shortest at the other.

We return to the case of a general symmetric $n \times n$ matrix A. To apply the method of Lagrange multipliers to maximizing f_A while

---

[2] This is why the letter λ is generally used for eigenvalues — it is really a Lagrange multiplier!

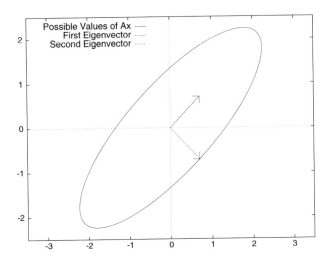

Figure 7.3: Possible values of $A\mathbf{x}$ with $|\mathbf{x}| = 1$

holding $|\mathbf{x}| = 1$, we need the gradient of f_A and the gradient of the constraint function $g(\mathbf{x}) = \sum_i x_i^2 - 1$. We have

$$\frac{\partial f}{\partial x_i} = \sum_j (A_{ij} x_j + A_{ji} x_j) = 2 \sum_j (A_{ij} \mathbf{x}_j), \qquad (7.28)$$

since A is symmetric. Alternatively, in vector form we have

$$\nabla f = 2A\mathbf{x}. \qquad (7.29)$$

Similarly, $\nabla g = 2\mathbf{x}$, so the Lagrange multiplier equation

$$\nabla f = \lambda \nabla g \qquad (7.30)$$

is the same as the eigenvalue equation

$$A\mathbf{x} = \lambda \mathbf{x}. \qquad (7.31)$$

The critical points of f on the unit sphere are then just the eigenvectors of A, normalized to have length one.

The construction of quadratic functions on $\mathbb{R}^n$ extends to a general real inner product space V. For every symmetric operator S we can define the quadratic function

$$f_S(\mathbf{x}) = \langle \mathbf{x} | S | \mathbf{x} \rangle. \qquad (7.32)$$

7.3. Quadratic Forms and Real Symmetric Matrices

$f_S(\mathbf{x})$ is sometimes called the *expectation* of the operator S in the state $\mathbf{x}$. Once again, the critical points of the function f_S, restricted to the unit sphere $\langle \mathbf{x}|\mathbf{x}\rangle = 1$, are the eigenvectors of S.

The study of quadratic functions thus reduces to diagonalizing symmetric operators (or matrices). Since real symmetric operators are Hermitian, we already know that such operators are diagonalizable, with real eigenvalues and orthogonal eigenvectors. All that remains is to check that the eigenvectors can all be chosen real.

Theorem 7.8 *Let V be a finite dimensional real inner product space, and let S be a real symmetric operator on V. Then S is diagonalizable, with real eigenvalues and with real eigenvectors that can be chosen to form an orthonormal basis for V. Furthermore, these properties characterize real symmetric operators; if S' is a diagonalizable operator on V with real eigenvalues and real orthonormal eigenvectors, then S' is real symmetric.*

We give two proofs of this theorem, one relying on the results for Hermitian matrices, and the other using only real numbers.
First Proof: Let S_c be the extension of S to the complexification of V. That is

$$S_c(\mathbf{v}_R + i\mathbf{v}_I) = S\mathbf{v}_R + iS\mathbf{v}_I. \tag{7.33}$$

Since S is real symmetric, it is not hard to check that S_c is Hermitian. Therefore S_c is diagonalizable with real eigenvalues and orthogonal eigenvectors. Now pick an orthonormal basis $\mathcal{B}$ for V. This is also an orthonormal basis for V_c. The matrices of S and S_c relative to this basis are the same, so the eigenvalues and eigenvectors of S and S_c are the same. All we must show is that the eigenvectors of S_c may be chosen real.

Let $A = [S_c]_\mathcal{B} = [S]_\mathcal{B}$, and suppose $\mathbf{v}$ is an eigenvector of A with eigenvalue λ. Since A is a real matrix, the real and imaginary parts of $A\mathbf{v}$ are $A(\mathbf{v}_R)$ and $A(\mathbf{v}_I)$, respectively. Since λ is a real number, the real and imaginary parts of $\lambda\mathbf{v}$ are $\lambda\mathbf{v}_R$ and $\lambda\mathbf{v}_I$, respectively. Thus we must have

$$A\mathbf{v}_R = \lambda\mathbf{v}_R; \qquad A\mathbf{v}_I = \lambda\mathbf{v}_I. \tag{7.34}$$

In other words, the real and imaginary parts of $\mathbf{v}$ are themselves eigenvectors of A. So, in making a list of eigenvectors, we can replace the complex eigenvector $\mathbf{v}$ with the real eigenvectors $\mathbf{v}_R$ and $\mathbf{v}_I$. If the $\mathbf{v}$'s span V_c, then so do the $\mathbf{v}_R$'s and $\mathbf{v}_I$'s. Pruning this list down to a linearly independent set, we get a real basis of eigenvectors of S_c, and hence a real basis of eigenvectors of S. Applying Gram-Schmidt to the vectors in each eigenspace, we then get an orthonormal basis of real eigenvectors of S. ∎

Second Proof: We establish the three properties of S one at a time, using real numbers only. The first property, namely that all the eigenvalues of S are real, is essentially a definition. If **v** is a vector in V, then the right-hand side of the equation $S\mathbf{v} = \lambda\mathbf{v}$ only makes sense if λ is real. But you don't get something for nothing. By excluding complex eigenvalues and eigenvectors we have made the third property (diagonalizability, or finding enough eigenvectors) that much harder.

The second property is established just as in the proof of Theorem 7.4. If $\mathbf{b}_1$ and $\mathbf{b}_2$ are eigenvectors, then

$$\lambda_1\langle\mathbf{b}_1|\mathbf{b}_2\rangle = \langle S\mathbf{b}_1|\mathbf{b}_2\rangle = \langle\mathbf{b}_1|S\mathbf{b}_2\rangle = \lambda_2\langle\mathbf{b}_1|\mathbf{b}_2\rangle, \qquad (7.35)$$

so if $\lambda_1 \neq \lambda_2$ then $\langle\mathbf{b}_1|\mathbf{b}_2\rangle = 0$.

The third property is proven by induction on the dimension of V, as in the proof of Theorem 7.6. Call this dimension n. The property certainly holds when $n = 1$. We assume it holds for $n = k$, and prove it for $n = k + 1$.

Unlike the Hermitian case, we cannot find an eigenvalue simply by finding a root of the characteristic polynomial, since a real polynomial may not have any real roots. Instead, we find our first eigenvector by considering the function f_S and maximizing on the unit sphere. Since spheres are compact sets, there exists a point (i.e., unit vector) $\mathbf{b}_1$, where f_S achieves its maximum. By our previous discussion of Lagrange multipliers, $\mathbf{b}_1$ is an eigenvector of S.

The rest of the proof is exactly as in Theorem 7.6. We extend $\mathbf{b}_1$ to an orthonormal basis $\mathcal{B}$, and see that the matrix $[S]_\mathcal{B}$ is block diagonal, with λ_1 in the upper left corner and a $k \times k$ real symmetric matrix in the lower right. By assumption, this $k \times k$ matrix is diagonalizable, and its eigenvectors are easily converted to eigenvectors of $[S]_\mathcal{B}$, yielding $k + 1$ linearly independent eigenvectors of $[S]_\mathcal{B}$, and hence $k + 1$ linearly independent eigenvectors of S.

Finally, if S' is diagonalizable with real eigenvalues and (real) orthogonal eigenvectors, then we pick a basis $\mathcal{B}$ of orthonormal eigenvectors and look at $[S']_\mathcal{B}$. This is a diagonal matrix with real entries on the diagonal and so is real symmetric. By Theorem 7.3, S is then a Hermitian operator. Since it acts on a real vector space, it is a real symmetric operator. ∎

7.3. Quadratic Forms and Real Symmetric Matrices

Now suppose S is a real symmetric operator on a real inner product space V, and let $\mathbf{b}_1, \ldots, \mathbf{b}_n$ be an orthonormal basis of V consisting of eigenvectors of S with eigenvalues $\lambda_1 \leq \lambda_2 \leq \cdots \leq \lambda_n$. If $\mathbf{x} = a_1 \mathbf{b}_1 + \cdots + a_n \mathbf{b}_n$, then

$$\begin{aligned} f_S(\mathbf{x}) &= \langle \mathbf{x}|S|\mathbf{x}\rangle = \sum_{i,j}\langle a_i \mathbf{b}_i|S|a_j \mathbf{b}_j\rangle \\ &= \sum_{i,j} a_i a_j \langle \mathbf{b}_i|\lambda_j \mathbf{b}_j\rangle = \sum_i a_i^2 \lambda_i. \end{aligned} \qquad (7.36)$$

That is, the function $f_S(\mathbf{x})$ is just a weighted average of the eigenvalues λ_i, with the squared coefficients a_i^2 providing the weights. If $|\mathbf{x}| = 1$, then $\sum a_i^2 = 1$, and the weighted average is somewhere between λ_1 and λ_n. If $\mathbf{x}$ is not a unit vector, we rescale and obtain

Theorem 7.9 *If S is a symmetric operator on a real inner product space, then the quadratic function $f_S(\mathbf{x}) = \langle \mathbf{x}|S|\mathbf{x}\rangle$ satisfies the inequalities*

$$\lambda_1 |\mathbf{x}|^2 \leq f_S(\mathbf{x}) \leq \lambda_n |\mathbf{x}|^2, \qquad (7.37)$$

where λ_1 and λ_n are the smallest and largest eigenvalues of S, respectively.

Quadratic functions occur frequently in math, physics, and engineering. If g is any function on $\mathbb{R}^n$ (say, a potential energy function), and $\mathbf{a}$ is a point where $\nabla g = 0$ (an equilibrium point), then near $\mathbf{x} = \mathbf{a}$ we have the Taylor series expansion

$$g(\mathbf{x}) - g(\mathbf{a}) = (1/2) \sum_{i,j=1}^n H_{ij}(x_i - a_i)(x_j - a_j) + O(|\mathbf{x}-\mathbf{a}|^3), \quad (7.38)$$

where $H_{ij} = \partial^2 g/\partial x_i \partial x_j|_{\mathbf{x}=\mathbf{a}}$ is the matrix of second derivatives (Hessian) of g at $\mathbf{a}$. In other words, near $\mathbf{x} = \mathbf{a}$, $g(\mathbf{x}) - g(\mathbf{a})$ is approximately the quadratic function $\frac{1}{2} f_H(\mathbf{x} - \mathbf{a})$. Quadratic functions thus appear whenever we study functions near their critical points.

Quadratic functions also appear as kinetic energies. Imagine a system with N point particles, with m_i and $\mathbf{x}_i$ denoting the mass and position of the i-th particle. The kinetic energy of the entire system is $(1/2)\sum_i m_i \langle \dot{\mathbf{x}}_i | \dot{\mathbf{x}}_i\rangle$, which is a quadratic function of $3N$ variables (the x, y, and z velocities of each of the N particles).

One can also consider extended objects. If an extended object (such as a chair) is rotating in space, then the angular velocity ω is a vector in $\mathbb{R}^3$, and the kinetic energy is a quadratic function

$(1/2)f_I(\omega)$, where the 3 × 3 matrix I is called the *moment of inertia tensor*.[3] The eigenvectors of I are called the *principal axes* of the rotating body, and the eigenvalues are called the principal moments of inertia.

Competing functions and duality

In many applications, one has to compare two quadratic functions f_A and f_B (e.g., the potential and kinetic energy), and maximize f_A while holding f_B constant. This can be solved in one of three ways.

The first way is to apply Lagrange multipliers directly. This leads to the equations

$$A\mathbf{x} = \lambda B\mathbf{x}, \qquad (7.39)$$

or

$$B^{-1}A\mathbf{x} = \lambda \mathbf{x}. \qquad (7.40)$$

In some cases these equations can be solved directly. However, since $B^{-1}A$ is not symmetric, it is difficult to say anything about the solution, and this approach is frequently a dead end.

If B is a positive operator (i.e., all the eigenvalues of B are positive), then the operator $B^{1/2}$ is well-defined, and we can define a new variable $\mathbf{y} = B^{1/2}\mathbf{x}$. Then $f_B(\mathbf{x}) = |\mathbf{y}|^2$, and

$$\begin{aligned} f_A(\mathbf{x}) &= f_A(B^{-1/2}\mathbf{y}) = \langle B^{-1/2}\mathbf{y}|A|B^{-1/2}\mathbf{y}\rangle \\ &= \langle \mathbf{y}|B^{-1/2}AB^{-1/2}|\mathbf{y}\rangle = f_C(\mathbf{y}), \end{aligned} \qquad (7.41)$$

where

$$C = B^{-1/2}AB^{-1/2}. \qquad (7.42)$$

Notice that C is itself real symmetric (why?), so we have reduced our problem involving two functions, f_A and f_B, into a problem of maximizing a single function, $f_C(\mathbf{y})$, subject to the constraint that $|\mathbf{y}|$ is fixed. This is something we already know how to do. The critical points are where $\mathbf{y}$ is an eigenvector of C, and the critical values of $f_A(\mathbf{x})/f_B(\mathbf{x})$ are the eigenvalues of C. The possible values of $f_A(\mathbf{x})/f_B(\mathbf{x})$ range from the smallest eigenvalue of C to the largest eigenvalue, and the critical points $\mathbf{x} = \mathbf{b}_i$ satisfy the generalized orthogonality relation

$$\langle \mathbf{b}_i|B|\mathbf{b}_j\rangle = 0 \text{ if } \lambda_i \neq \lambda_j. \qquad (7.43)$$

---

[3] Unfortunately, the moment of inertia is traditionally denoted by the same letter, I, as the identity matrix, which can lead to some confusion.

7.3. Quadratic Forms and Real Symmetric Matrices

The third way to solve the problem is to turn it inside-out. Maximizing f_A with f_B constant is the same thing as maximizing f_A/f_B, which is the same thing as minimizing f_B/f_A, which is the same thing as minimizing f_B with f_A held constant. This involves diagonalizing the operator $C' = A^{-1/2}BA^{-1/2}$. This is particularly useful when A is simpler than B, so that it is easier to compute $A^{-1/2}$ than $B^{-1/2}$. This inversion trick shows that, in addition to satisfying the orthogonality relations (7.43), the solutions also satisfy

$$\langle \mathbf{b}_i | A | \mathbf{b}_j \rangle = 0 \text{ if } \lambda_i \neq \lambda_j. \tag{7.44}$$

Exercises

1. Let $f(\mathbf{x}) = 6x_1^2 + 4x_1x_2 + 3x_2^2$. Setting $\mathbf{x} = (\cos(\theta), \sin(\theta))^T$, write $f(\mathbf{x})$ as a function of θ, and find the maxima and minima. How many are there of each?

2. Let $f(\mathbf{x})$ be as in Exercise 1. Show that the curve $f(\mathbf{x}) = 1$ is an ellipse. What are the semi-major and semi-minor axes?

3. Let D be a real diagonal 2×2 matrix. What are the possible shapes of the curve $f_D(\mathbf{x}) = 1$? Under what conditions are each of these curves realized?

4. Let A be an arbitrary real symmetric 2×2 matrix. What are the possible shapes of the curve $f_A(\mathbf{x}) = 1$? Under what conditions are each of these curves realized? Express your answer both in terms of the eigenvalues of A and in terms of the matrix elements.

5. Let A be an arbitrary real symmetric 3×3 matrix. What are the possible shapes of the surface $f_A(\mathbf{x}) = 1$? Under what conditions are each of these surfaces realized?

6. Let $g(\mathbf{x})$ be a function of two variables with a critical point at $\mathbf{x} = \mathbf{x}_0$. Let

$$H = \begin{pmatrix} \partial^2 g/\partial x_1^2 & \partial^2 g/\partial x_1 \partial x_2 \\ \partial^2 g/\partial x_1 \partial x_2 & \partial^2 g/\partial x_2^2 \end{pmatrix}\bigg|_{\mathbf{x}=\mathbf{x}_0} = \begin{pmatrix} a & b \\ b & c \end{pmatrix}.$$

Prove the second derivative test: If $ac > b^2$, then $\mathbf{x}_0$ is either a minimum (if $a > 0$), or a maximum (if $a < 0$), while if $ac < b^2$ then $\mathbf{x}_0$ is a saddle point.

7. Consider the quadratic function $x_1^2 + 2x_2^2 + 3x_3^2$ on $\mathbb{R}^3$. Restrict this function to the unit sphere $|\mathbf{x}|^2 = 1$. Find the maxima, the minima, and the saddle points.

8. Consider the quadratic function $x_1^2 + x_1x_2 + x_2^2 + x_2x_3 + x_3^2$ on $\mathbb{R}^3$. Restrict this function to the unit sphere $|\mathbf{x}|^2 = 1$. Find the maxima, the minima, and the saddle points.

9. On $\mathbb{R}^3$, maximize $|\mathbf{x}|^2$ subject to the constraint $x_1^2 + x_1x_2 + x_2^2 + x_2x_3 + x_3^2 = 1$.

10. Let A and B be real symmetric matrices, with B having all positive eigenvalues. Show that $B^{-1}A$ and $B^{-1/2}AB^{-1/2}$ have the same eigenvalues. If A is invertible, show that the eigenvalues of $A^{-1}B$ are the reciprocals of the eigenvalues of $B^{-1}A$. [Hint: Consider conjugates of matrices.]

7.4 Rotations, Orthogonal Operators, and Unitary Operators

In this section we study rotations in $\mathbb{R}^n$ and their generalizations to real and complex inner product spaces. Remarkably, all of these operators are diagonalizable (if you allow complex eigenvalues and eigenvectors) and have orthogonal eigenvectors.

Rotations in $\mathbb{R}^n$

A key feature of rotations in $\mathbb{R}^n$ is that they preserve length. If you rotate a yardstick, it stays 1 yard long. So if R is a matrix that rotates vectors in $\mathbb{R}^n$, and $\mathbf{x}$ is a vector, then

$$|R\mathbf{x}| = |\mathbf{x}|. \tag{7.45}$$

Moreover, the inner product can be derived from the norm via the identity

$$2\langle \mathbf{x}|\mathbf{y}\rangle = |\mathbf{x}+\mathbf{y}|^2 - |\mathbf{x}|^2 - |\mathbf{y}|^2. \tag{7.46}$$

This implies that R preserves the inner product:

$$\begin{aligned}\langle R\mathbf{x}|R\mathbf{y}\rangle &= (|R\mathbf{x}+R\mathbf{y}|^2 - |R\mathbf{x}|^2 - |R\mathbf{y}|^2)/2 \\ &= (|\mathbf{x}+\mathbf{y}|^2 - |\mathbf{x}|^2 - |\mathbf{y}|^2)/2 \\ &= \langle \mathbf{x}|\mathbf{y}\rangle.\end{aligned} \tag{7.47}$$

From this we deduce several facts about R. Since the i-th column of R is $R\mathbf{e}_i$, and since the $\mathbf{e}_i$'s are orthonormal, the columns of R must be orthonormal. This then implies that R^TR is the identity, since $(R^TR)_{ij}$ is the inner product of the i-th column of R with the j-th

7.4. Rotations, Orthogonal Operators, and Unitary Operators

column. In other words, $R^{-1} = R^T$. But then RR^T is the identity, so the columns of R^T (that is, the transposes of the rows of R) are orthonormal.

Example: The matrix

$$R = \begin{pmatrix} 1/\sqrt{3} & -1/\sqrt{6} & 1/\sqrt{2} \\ 1/\sqrt{3} & -1/\sqrt{6} & -1/\sqrt{2} \\ 1/\sqrt{3} & 2/\sqrt{6} & 0 \end{pmatrix} \qquad (7.48)$$

has columns that are orthonormal. A little computation shows that the columns of R^T are also orthonormal. In fact, R is a rotation by approximately 1.9992 radians (114.55 degrees) about the axis $(0.83750, 0.07132, 0.54177)^T$. We will show how to find the axis and angle of a rotation in $\mathbb{R}^3$ later in this section.

Theorem 7.10 *Let R be a real $n \times n$ matrix. Then the following conditions are equivalent:*

1. *For every $\mathbf{x} \in \mathbb{R}^n$, $|R\mathbf{x}| = |\mathbf{x}|$.*

2. *For every $\mathbf{x}, \mathbf{y} \in \mathbb{R}^n$, $\langle R\mathbf{x} | R\mathbf{y} \rangle = \langle \mathbf{x} | \mathbf{y} \rangle$.*

3. *The columns of R are orthonormal.*

4. *$R^T R = I$.*

5. *$RR^T = I$.*

6. *The columns of R^T are orthonormal.*

Proof: We have already shown that 1 implies 2, which implies 3, which implies 4, which implies 5, which implies 6. To complete the proof we must show that 6 implies 1. The same argument as above, with the roles of R and R^T reversed, shows that 6 implies 5, which implies 4. So we just need to show that 4 implies 1. We compute

$$|R\mathbf{x}|^2 = \langle R\mathbf{x} | R\mathbf{x} \rangle = \langle R^\dagger R\mathbf{x} | \mathbf{x} \rangle = \langle R^T R\mathbf{x} | \mathbf{x} \rangle = \langle I\mathbf{x} | \mathbf{x} \rangle = |\mathbf{x}|^2. \qquad (7.49)$$

Taking the square root of both sides we get $|R\mathbf{x}| = |\mathbf{x}|$. ∎

Definition *A real matrix that has one (and therefore all) of the properties of Theorem 7.10 is called an* **orthogonal** *matrix. (Common sense suggests that such matrices should be called "orthonormal", but they aren't.)*

In addition to representing rotations, orthogonal matrices come up frequently as change-of-basis matrices. In practice, we often want to change from one orthonormal basis to another, say from the standard basis in $\mathbb{R}^n$ to a basis of eigenvectors of a real symmetric operator. If $\mathcal{B}$ and $\mathcal{D}$ are both orthonormal bases for a real vector space V, then the matrix $P_{\mathcal{DB}}$ is orthogonal. The proof of this fact is left as an exercise.

These two roles that orthogonal matrices play are related. In $\mathbb{R}^n$, and even in $\mathbb{R}^3$, there are two ways to make an object seem to rotate. The first, called an *active* rotation, is to actually rotate the object. The second is to tilt your head! Mathematically, that means changing your coordinate axes, i.e., your basis, from one orthonormal set to another. This change of coordinates is called a *passive* rotation. The effect is indistinguishable from an active rotation, and a given orthogonal matrix might describe either one of these operations. When watching a movie, you cannot tell the difference between the world spinning clockwise and the camera being rotated counterclockwise. When you get dizzy you feel that the world is spinning around you, when in fact it is only your head that is turning.

If you rotate an object in physical space once, and then rotate it again, the result is a (possibly more complicated) rotation. This fact has a natural generalization to n dimensions.

Theorem 7.11 *The $n \times n$ orthogonal matrices form a group under matrix multiplication. That is, if A and B are orthogonal $n \times n$ matrices, then so are A^{-1} and AB.*

Proof: If A is orthogonal, then the columns of A^T are orthonormal (property 6), so A^T is an orthogonal matrix (property 3). But, by properties 4 and 5, $A^{-1} = A^T$, so A^{-1} is orthogonal. To show that AB is orthogonal, let $\mathbf{x}$ be any vector. Since A is orthogonal, $|AB\mathbf{x}| = |A(B\mathbf{x})| = |B\mathbf{x}|$. Since B is orthogonal, $|B\mathbf{x}| = |\mathbf{x}|$. Thus $|AB\mathbf{x}| = |\mathbf{x}|$ for every $\mathbf{x}$, so AB is orthogonal (property 1). ∎

There is a natural extension of the notion of orthogonal matrices to operators on a general real inner product space.

Definition *An operator R, on a real inner product space is orthogonal if $R^\dagger R = I = RR^\dagger$.*

Note: In finite dimensions, the equations $R^\dagger R = I$ and $I = RR^\dagger$ are equivalent, but in infinite dimensions they are not. See (3.31) for a counterexample. An operator for which $R^\dagger R = I$ is called an *isometry*, whether or not $RR^\dagger$ equals the identity.

7.4. Rotations, Orthogonal Operators, and Unitary Operators 213

Unitary matrices and operators

We next consider the complex analog of orthogonal matrices and operators.

Theorem 7.12 *Let U be an $n \times n$ matrix. Then the following conditions are equivalent:*

1. *For every $\mathbf{x} \in \mathbb{C}^n$, $|U\mathbf{x}| = |\mathbf{x}|$.*
2. *For every $\mathbf{x}, \mathbf{y} \in \mathbb{C}^n$, $\langle U\mathbf{x} | U\mathbf{y} \rangle = \langle \mathbf{x} | \mathbf{y} \rangle$.*
3. *The columns of U are orthonormal.*
4. *$U^\dagger U = I$.*
5. *$UU^\dagger = I$.*
6. *The columns of $U^\dagger$ are orthonormal.*

Proof: The proof is a variation of the proof of Theorem 7.10. As in that case, we will show that 1 implies 2, 2 implies 3, 3 implies 4, 4 implies 5, 5 implies 6, 6 implies 5, 5 implies 4, and 4 implies 1. This is enough to show that, starting with any one condition, we can derive all the others.

On $\mathbb{C}^n$, the inner product may be recovered from the norm via the identity

$$4\langle \mathbf{x} | \mathbf{y} \rangle = |\mathbf{x}+\mathbf{y}|^2 - i|\mathbf{x}+i\mathbf{y}|^2 - |\mathbf{x}-\mathbf{y}|^2 + i|\mathbf{x}-i\mathbf{y}|^2. \quad (7.50)$$

This identity may be confirmed by expanding the right-hand side. As a result, anything that preserves the norm preserves the inner product, and 1 implies 2. Since the vectors $\{\mathbf{e}_i\}$ are orthonormal, if U preserves the inner product, then the vectors $\{U\mathbf{e}_i\}$ are orthonormal. But $U\mathbf{e}_i$ is the i-th column of U, so 2 implies 3. The ij entry of $U^\dagger U$ is the inner product of the i-th column of U with the j-th column (in that order), so 3 implies 4. 4 and 5 are both equivalent to U^{-1} equalling $U^\dagger$, so they are equivalent to each other. The ij entry of $UU^\dagger$ is the inner product of the i-th column of $U^\dagger$ with the j-th column of $U^\dagger$, so 5 and 6 are equivalent. All that remains is to show that 4 implies 1. But if $U^\dagger U = I$, then

$$|U\mathbf{x}|^2 = \langle U\mathbf{x} | U\mathbf{x} \rangle = \langle U^\dagger U\mathbf{x} | \mathbf{x} \rangle = \langle \mathbf{x} | \mathbf{x} \rangle = |\mathbf{x}|^2, \quad (7.51)$$

so $|U\mathbf{x}| = |\mathbf{x}|$. ∎

Definition *A matrix that has one (and therefore all) of the properties of Theorem 7.12 is called a* unitary matrix. *An operator U on a complex vector space is called a* unitary operator *if $U^\dagger U = UU^\dagger = I$. Of course, a unitary matrix that is also real is an orthogonal matrix.*

As with orthogonal matrices, unitary matrices occur frequently as change-of-basis matrices from one orthonormal basis to another, often from the standard basis in $\mathbb{C}^n$ to a basis of eigenvectors of a Hermitian matrix.

Eigenvalues and eigenvectors

The eigenvectors of unitary and orthogonal matrices (and operators) turn out to always be orthogonal, just as the eigenvectors of Hermitian and real-symmetric operators are. However, the eigenvalues and eigenvectors are almost always complex. As a result, most orthogonal matrices cannot be diagonalized using only real eigenvectors. However, they can always be diagonalized using complex eigenvectors. For example, we have already seen that the 2 dimensional rotation matrix

$$\begin{pmatrix} \cos(\theta) & -\sin(\theta) \\ \sin(\theta) & \cos(\theta) \end{pmatrix} \tag{7.52}$$

has eigenvalues $e^{\pm i\theta}$ and eigenvectors $(\pm i, 1)^T/\sqrt{2}$, which are orthonormal, but complex.

Theorem 7.13 *Let U be a unitary operator on a finite dimensional complex vector space V. Then U is diagonalizable, the eigenvalues λ_i of U all have $|\lambda_i| = 1$, and eigenvectors corresponding to different eigenvalues are orthogonal. Furthermore, this completely characterizes unitary operators; if an operator U' has eigenvectors that form an orthonormal basis for V, and if the eigenvalues of U' all have norm 1, then U' is unitary.*

Proof: If $\mathbf{b}_i$ is an eigenvector of U with eigenvalue λ_i, then

$$|\mathbf{b}_i| = |U\mathbf{b}_i| = |\lambda_i \mathbf{b}_i| = |\lambda_i||\mathbf{b}_i|. \tag{7.53}$$

Since $|\mathbf{b}_i|$ is nonzero, $|\lambda_i|$ must equal 1. Now suppose $\mathbf{b}_1$ and $\mathbf{b}_2$ are eigenvectors with $\lambda_1 \neq \lambda_2$. Then

$$\langle \mathbf{b}_1 | \mathbf{b}_2 \rangle = \langle U\mathbf{b}_1 | U\mathbf{b}_2 \rangle = \langle \lambda_1 \mathbf{b}_1 | \lambda_2 \mathbf{b}_2 \rangle = \bar{\lambda}_1 \lambda_2 \langle \mathbf{b}_1 | \mathbf{b}_2 \rangle. \tag{7.54}$$

7.4. Rotations, Orthogonal Operators, and Unitary Operators

However, $|\lambda_1| = 1$, so $\bar{\lambda}_1 = \lambda_1^{-1}$. Since $\lambda_1 \neq \lambda_2$, $\bar{\lambda}_1 \lambda_2 = \lambda_2/\lambda_1$ cannot equal 1, so we must have $\langle \mathbf{b}_1 | \mathbf{b}_2 \rangle = 0$.

The proof of diagonalizability proceeds almost exactly as for Theorem 7.6. We proceed by induction on the dimension n of V. The result is trivial for $n = 1$, so we need only prove the inductive step, assuming the result for $n = k$ and proving it for $n = k+1$. There is one eigenvector (call it $\mathbf{b}_1$) because the characteristic polynomial has to have a root. We pick vectors $\mathbf{b}_2, \ldots, \mathbf{b}_{k+1}$ so that $\mathcal{B} = \{\mathbf{b}_1, \ldots, \mathbf{b}_{k+1}\}$ is an orthonormal basis, and we look at the matrix $[U]_\mathcal{B}$. Note that, for $i \neq 1$,

$$\langle \mathbf{b}_i | U \mathbf{b}_1 \rangle = \langle \mathbf{b}_i | \lambda_1 \mathbf{b}_1 \rangle = \lambda_1 \langle \mathbf{b}_i | \mathbf{b}_1 \rangle = 0, \tag{7.55}$$

while

$$\langle \mathbf{b}_1 | U \mathbf{b}_i \rangle = \langle U^\dagger \mathbf{b}_1 | \mathbf{b}_i \rangle = \langle U^{-1} \mathbf{b}_1 | \mathbf{b}_i \rangle = \langle \lambda_1^{-1} \mathbf{b}_1 | \mathbf{b}_i \rangle = 0. \tag{7.56}$$

Thus the matrix $M = [U]_\mathcal{B}$ takes the form

$$M = \begin{pmatrix} \lambda_1 & 0 \\ 0 & A \end{pmatrix}. \tag{7.57}$$

Since M is unitary, so is A. Since A is a $k \times k$ unitary matrix it can be diagonalized, by assumption, so M can be diagonalized, so U can be diagonalized.

Now suppose U' can be diagonalized with an orthonormal basis $\mathcal{B}'$ of eigenvectors and with eigenvalues of norm 1. In the $\mathcal{B}'$ basis, $[U']_{\mathcal{B}'}$ is a diagonal matrix with entries of norm 1. This is easily checked to be a unitary matrix. By Exercise 3, U' is then a unitary operator. ∎

Since real orthogonal matrices are special cases of unitary matrices, we can specialize the results to orthogonal matrices.

Corollary 7.14 *If R is a real orthogonal matrix, then R is diagonalizable if we allow complex eigenvalues and eigenvectors. The eigenvalues all have norm 1; the complex eigenvalues occur in conjugate pairs $e^{\pm i\theta}$. The eigenvectors may be chosen (complex) orthonormal.*

Proof: Most of this corollary is immediate. The only new ingredient is that the complex eigenvalues come in $e^{\pm i\theta}$ pairs. Since R is a real matrix, the characteristic polynomial is real, and so its complex roots come in pairs. Since R is orthogonal, hence unitary, the eigenvalues have norm one. A pair of conjugate numbers of norm one always takes the form $e^{\pm i\theta}$. ∎

Chapter 7. Adjoints and Hermitian Operators

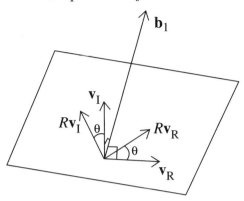

Figure 7.4: R is a rotation by θ about the $\mathbf{b}_1$ axis

Since $R^T R = I$, $\det(R)^2 = \det(R^T)\det(R) = \det(R^T R) = 1$, so $\det(R) = \pm 1$. True rotations are described by orthogonal matrices of determinant $+1$. Orthogonal matrices of determinant -1 describe rotations followed by a reflection.

Rotations in $\mathbb{R}^3$

These observations allow us to characterize rotations in $\mathbb{R}^3$. We are used to describing rotations by axes and angles. The following theorem relates these to eigenvalues and eigenvectors.

Theorem 7.15 *If R is an 3×3 orthogonal matrix with determinant $+1$, then $+1$ is an eigenvalue of R. The other two eigenvalues are $e^{\pm i\theta}$ for some angle θ. Let $\mathbf{b}_1$ be a normalized eigenvector with eigenvalue $+1$. R is a rotation by an angle $\pm\theta$ about the $\mathbf{b}_1$ axis.*

Proof: The determinant is the product of the eigenvalues. If R has three real eigenvalues, they must all be $+1 = e^{i0}$, or two of them are $-1 = e^{\pm i\pi}$, and one is $+1$. If R has a complex pair of eigenvalues, their product is $e^{i\theta}e^{-i\theta} = 1$, so the third eigenvalue must be $+1$. In any case, $+1$ is an eigenvalue, and we can find a corresponding eigenvector $\mathbf{b}_1$.

If the eigenvalues are all $+1$, then R is the identity, which may be viewed as a rotation by 0 radians about the $\mathbf{b}_1$ axis. If the eigenvalues are $1, -1$, and -1, then R preserves $\mathbf{b}_1$ but flips the sign of any vector orthogonal to $\mathbf{b}_1$. In that case, R is a rotation by π radians about the $\mathbf{b}_1$ axis.

7.4. Rotations, Orthogonal Operators, and Unitary Operators

Finally, we consider the case where the eigenvalues are not all real. Let $\mathbf{b}_2$ be the eigenvector of eigenvalue $\lambda_2 = e^{-i\theta}$. We decompose this vector into real and imaginary parts:

$$\mathbf{b}_2 = (\mathbf{v}_R + i\mathbf{v}_I)/\sqrt{2}. \tag{7.58}$$

See Figure 7.4. Since real matrices have eigenvectors that come in complex conjugate pairs, the third eigenvector must be $\mathbf{b}_3 = \mathbf{v}_R - i\mathbf{v}_I$, with $\lambda_3 = e^{i\theta}$. I claim that $\mathbf{b}_1$, $\mathbf{v}_R$, and $\mathbf{v}_I$ form an orthonormal basis for $\mathbb{R}^3$. Orthogonality comes from

$$\begin{aligned} 0 &= \langle \mathbf{b}_1 | \mathbf{b}_2 \rangle = (\langle \mathbf{b}_1 | \mathbf{v}_R \rangle + i \langle \mathbf{b}_1 | \mathbf{v}_I \rangle)/\sqrt{2} \\ 0 &= Im \langle \mathbf{b}_2 | \mathbf{b}_3 \rangle = -\langle \mathbf{v}_R | \mathbf{v}_I \rangle. \end{aligned} \tag{7.59}$$

Normalization comes from

$$\begin{aligned} 1 &= \langle \mathbf{b}_2 | \mathbf{b}_2 \rangle = (\langle \mathbf{v}_R | \mathbf{v}_R \rangle + \langle \mathbf{v}_I | \mathbf{v}_I \rangle)/2; \\ 0 &= Re \langle \mathbf{b}_2 | \mathbf{b}_3 \rangle = (\langle \mathbf{v}_R | \mathbf{v}_R \rangle - \langle \mathbf{v}_I | \mathbf{v}_I \rangle)/2. \end{aligned} \tag{7.60}$$

Thus $\mathcal{D} = \{\mathbf{b}_1, \mathbf{v}_R, \mathbf{v}_I\}$ is an orthonormal real basis for $\mathbb{R}^3$. Either it is a right-handed basis (that is, $\mathbf{v}_I$ is the cross product of $\mathbf{b}_1$ and $\mathbf{v}_R$), or we can make it one by flipping the sign of $\mathbf{b}_1$. Thus, we can imagine drawing our coordinate axes in the $\mathbf{b}_1$, $\mathbf{v}_R$, and $\mathbf{v}_I$ directions. Now, we have already seen (Section 4.4) how a real operator with a complex eigenvalue acts on the plane spanned by the real and imaginary parts of the eigenvector. Applying that to our case we find that R, relative to the $\mathcal{D} = \{\mathbf{b}_1, \mathbf{v}_R, \mathbf{v}_I\}$ basis, takes the form

$$[R]_\mathcal{D} = \begin{pmatrix} 1 & 0 & 0 \\ 0 & \cos(\theta) & -\sin(\theta) \\ 0 & \sin(\theta) & \cos(\theta) \end{pmatrix}. \tag{7.61}$$

In other words, R is a rotation by an angle $+\theta$ about our $\mathbf{b}_1$ axis. Since this axis is plus or minus our original $\mathbf{b}_1$ axis, R is a rotation by $\pm\theta$ about our original $+1$ eigenvector of R. ∎

To sum up, the eigen*values* of a rotation matrix give the angle of rotation, up to sign. The eigen*vector* with eigenvalue $+1$ gives the axis of rotation. However, a rotation by θ about the $\mathbf{b}_1$ axis is the same as a rotation by $-\theta$ about the $-\mathbf{b}_1$ axis, and both $\pm\mathbf{b}_1$ are eigenvectors, so knowing the $+1$ eigenspace and the eigenvalues is not enough to determine whether the rotation is clockwise or counterclockwise. To do that you have to examine the complex eigenvector with eigenvalue $e^{-i\theta}$. The axis about which R is a counterclockwise rotation by θ is the cross product of $\mathbf{v}_R$ and $\mathbf{v}_I$ (in that order).

Exercises

1. Let $\mathcal{B}$ be an orthonormal basis for a finite dimensional real inner product space V. Show that the operator R on V is orthogonal if and only if the matrix $[R]_\mathcal{B}$ is orthogonal.

2. Let R be an orthogonal operator on a (possibly infinite dimensional) real inner product space V. Show that, for any $\mathbf{x}, \mathbf{y} \in V$, $|R\mathbf{x}| = |\mathbf{x}|$ and $\langle R\mathbf{x}|R\mathbf{y}\rangle = \langle \mathbf{x}|\mathbf{y}\rangle$.

3. Let $\mathcal{B}$ be an orthonormal basis for a finite dimensional complex vector space V. Let U be an operator on that space. Show that U is a unitary operator if and only if $[U]_\mathcal{B}$ is a unitary matrix.

4. Let $\mathcal{B}$ and $\mathcal{D}$ be orthonormal bases for a finite dimensional complex inner product space V. Show that the change-of-basis matrix $P_{\mathcal{D}\mathcal{B}}$ is unitary.

5. Not all changes of basis are unitary. Suppose $\mathcal{B}$ is an orthonormal basis and $\mathcal{D}$ is not. Show that $P_{\mathcal{D}\mathcal{B}}$ is not unitary.

In Exercises 6–8, diagonalize the unitary matrices:

6. $\frac{1}{2}\begin{pmatrix} 1+i & 1+i \\ -1+i & 1-i \end{pmatrix}$.

7. $\frac{1}{\sqrt{5}}\begin{pmatrix} 1 & 2 \\ 2i & -i \end{pmatrix}$.

8. $\begin{pmatrix} 1/\sqrt{6} & (1+i)/2 & 1/\sqrt{3} \\ 2i/\sqrt{6} & 0 & -i/\sqrt{3} \\ 1/\sqrt{6} & -(1+i)/2 & 1/\sqrt{3} \end{pmatrix}$.

Each of these matrices in Exercises 9–11 represents a rotation in $\mathbb{R}^3$. Find the axis and the angle.

9. $\begin{pmatrix} 0 & 1 & 0 \\ \sqrt{2}/2 & 0 & \sqrt{2}/2 \\ \sqrt{2}/2 & 0 & -\sqrt{2}/2 \end{pmatrix}$.

10. $\begin{pmatrix} 0 & \sqrt{2}/2 & \sqrt{2}/2 \\ 1 & 0 & 0 \\ 0 & \sqrt{2}/2 & -\sqrt{2}/2 \end{pmatrix}$.

11. $\begin{pmatrix} 0 & 1 & 0 \\ 0 & 0 & 1 \\ 1 & 0 & 0 \end{pmatrix}$.

7.5 How the Four Classes are Related

We have seen that Hermitian, real symmetric, real orthogonal, and unitary operators all have orthonormal eigenvectors. In particular, we can write them in a fairly standard form:

If H is a Hermitian matrix (or a Hermitian operator expressed in an orthonormal basis), then we can find orthonormal eigenvectors $\mathbf{b}_1, \ldots, \mathbf{b}_n$ of H and package them into a matrix

$$U_H = \begin{pmatrix} \mathbf{b}_1 & \cdots & \mathbf{b}_n \end{pmatrix}. \tag{7.62}$$

Note that U_H is unitary, since its columns are orthonormal. We package the eigenvalues $\lambda_1, \ldots, \lambda_n$ into a real diagonal matrix

$$D_H = \begin{pmatrix} \lambda_1 & \cdots & 0 \\ 0 & \ddots & 0 \\ 0 & 0 & \lambda_n \end{pmatrix}, \tag{7.63}$$

and the matrix H then takes the form

$$H = U_H D_H U_H^{-1} = U_H D_H U_H^\dagger. \tag{7.64}$$

If S is a real symmetric matrix, then we can construct an orthogonal matrix R_S, whose columns are the eigenvectors of S, and a real diagonal matrix D_S, whose entries are the corresponding eigenvalues of S. Then

$$S = R_S D_S R_S^{-1} = R_S D_S R_S^T. \tag{7.65}$$

If T is a unitary matrix, then we can package the (complex) eigenvectors into a unitary matrix U_T, and form a diagonal matrix D_T, whose entries are the eigenvalues of T. Note that D_T is typically not a real matrix, as its entries are numbers of norm 1. The only time D_T is real is when all the only eigenvalues are $+1$ and -1. We have

$$T = U_T D_T U_T^{-1} = U_T D_T U_T^\dagger. \tag{7.66}$$

Finally, if R is an orthogonal matrix, then we can package the eigenvectors into a matrix U_R and the eigenvalues into a diagonal matrix D_R. As in the unitary case, the matrix U_R is unitary, and usually not orthogonal, while the matrix D_R is generally not real. We have

$$R = U_R D_R U_R^{-1} = U_R D_R U_R^\dagger, \tag{7.67}$$

again as in the unitary case. The only distinction between (7.66) and (7.67) is that the non-real eigenvalues of an orthogonal matrix come in complex conjugate pairs, with the corresponding eigenvectors also being complex conjugates of each other, while there need not be any relation between the different eigenvalues and eigenvectors of a general unitary operator.

Expressions (7.64) and (7.66) are quite similar. The unitary matrices U_H and U_T that describe the two sets of eigenvectors might well be the same. The only difference is in the eigenvalues. Hermitian matrices have real eigenvalues, while unitary matrices have eigenvalues of norm 1. However, there is an important connection between real numbers and complex numbers of norm 1. If you take the exponential of a pure imaginary number you get a complex number of norm 1. Thus if we exponentiate i times the real diagonal matrix D_H, we would get a diagonal matrix whose diagonal entries are all of norm 1. Since $\exp(i(UDU^{-1}) = U\exp(iD)U^{-1}$, we have proven the following:

Theorem 7.16 *Let H be a Hermitian matrix. Then $\exp(iH)$ is a unitary matrix.*

Corollary 7.17 *Let H be a Hermitian operator on a finite dimensional complex vector space V. Then $\exp(iH)$ is a unitary operator.*

Thus to every Hermitian operator we can associate a unitary operator. This process is not 1-1, since $\exp(i(\theta + 2\pi)) = \exp(i\theta)$. Adding a multiple of 2π to any of the eigenvalues of a Hermitian operator H does not change the exponential of iH. One can also apply the process in reverse, taking the log of a unitary operator to get i times a Hermitian operator. (See Section 4.8 for a discussion of taking functions of a matrix or an operator.) However, since the exponential is not 1-1, the log function is multivalued; both $i\theta$ and $i(\theta + 2\pi)$ are logs of $e^{i\theta}$, and there are quite a few different logs of any given unitary operator.

Corollary 7.18 *Let A be a real anti-symmetric matrix (or operator). Then $\exp(A)$ is a real orthogonal matrix (or operator).*

Proof: Since A is real and anti-symmetric, A is anti-Hermitian, so $-iA$ is Hermitian. Thus $\exp(A) = \exp(i(-iA))$ is unitary. However, A is real, so $\exp(A) = \sum_n A^n/n!$ is a real matrix. Since $\exp(A)$ is unitary and real, it is orthogonal. ∎

7.5. How the Four Classes are Related

The factor of i in Theorem 7.16 has caused confusion between physicists and mathematicians. Physicists like to work with Hermitian operators, especially when doing quantum mechanics, so they tend to speak of a unitary operator being the exponential of $\pm i$ times a Hermitian operator. Mathematicians, on the other hand, generally prefer to speak of a unitary operator being the exponential of an anti-Hermitian operator. Since i (or $-i$) times a Hermitian operator is an anti-Hermitian operator, and vice-versa, this difference is merely a matter of terminology, not of content. However, it can lead to confusion, especially when discussing Lie groups. (Lie groups are continuous groups, such as the set of all rotations in $\mathbb{R}^3$.) When a physicist speaks of a Lie algebra, he means the set of Hermitian matrices which, when multiplied by $-i$ and exponentiated, generate a Lie group. When a mathematician speaks of a Lie algebra, he means a set of anti-Hermitian matrices which, when exponentiated (without an additional factor of $-i$), generate the Lie group. The advantage of the physics convention is that you rarely have to deal with anti-Hermitian objects. The advantage of the math approach is that you don't have to introduce factors of i into problems on real vector spaces.

Example: Consider the Hermitian 2×2 matrix

$$H = \begin{pmatrix} 0 & -i\theta \\ i\theta & 0 \end{pmatrix}, \tag{7.68}$$

which has eigenvalues $\mp\theta$ and eigenvectors $(\pm i, 1)^T$. Multiplying H by $-i$ gives the real anti-symmetric (and hence anti-Hermitian) matrix

$$A = -iH = \begin{pmatrix} 0 & -\theta \\ \theta & 0 \end{pmatrix}, \tag{7.69}$$

which has eigenvalues $\pm i\theta$ and eigenvectors $(\pm i, 1)^T$. Exponentiating A gives the real orthogonal matrix

$$R = e^A = e^{-iH} = \begin{pmatrix} \cos(\theta) & -\sin(\theta) \\ \sin(\theta) & \cos(\theta) \end{pmatrix}, \tag{7.70}$$

with eigenvalues $e^{\pm i\theta} = \cos(\theta) \pm i\sin(\theta)$ and eigenvectors $(\pm i, 1)^T$. See Figure 7.5.

Rotations in $\mathbb{R}^3$ are generated similarly. Consider the Hermitian matrices

$$L_1 = \begin{pmatrix} 0 & 0 & 0 \\ 0 & 0 & -i \\ 0 & i & 0 \end{pmatrix}; \; L_2 = \begin{pmatrix} 0 & 0 & i \\ 0 & 0 & 0 \\ -i & 0 & 0 \end{pmatrix}; \; L_3 = \begin{pmatrix} 0 & -i & 0 \\ i & 0 & 0 \\ 0 & 0 & 0 \end{pmatrix}; \tag{7.71}$$

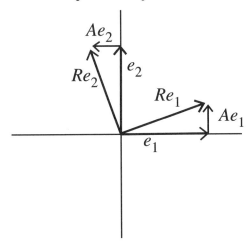

Figure 7.5: A is an infinitesimal rotation in $\mathbb{R}^2$

and the real anti-symmetric matrices $K_j = -iL_j$

$$K_1 = \begin{pmatrix} 0 & 0 & 0 \\ 0 & 0 & -1 \\ 0 & 1 & 0 \end{pmatrix}; K_2 = \begin{pmatrix} 0 & 0 & 1 \\ 0 & 0 & 0 \\ -1 & 0 & 0 \end{pmatrix}; K_3 = \begin{pmatrix} 0 & -1 & 0 \\ 1 & 0 & 0 \\ 0 & 0 & 0 \end{pmatrix}. \quad (7.72)$$

In the exercises you are asked to construct rotations by exponentiating combinations of the K_j's.

Obtaining unitary operators from Hermitian operators is an important operation in infinite dimensional spaces as well. Consider the Hermitian operator $H = -id/dt$ on $L^2(\mathbb{R})$. Let $T_a = \exp(-iaH) = \exp(-a\frac{d}{dt})$ and let f be an analytic function. Then, by Taylor's Theorem,

$$(T_a f)(t) = \sum_{n=0}^{\infty} \frac{\left(-a\frac{d}{dt}\right)^n f(t)}{n!} = \sum_{n=0}^{\infty} \frac{(-a)^n}{n!} \frac{d^n f}{dt^n}(t) = f(t-a). \quad (7.73)$$

In other words, T_a just translates functions to the right by a. (If $f(t)$ is peaked at $t = 0$, then $f(t-a)$ is peaked at $t - a = 0$, or at $t = a$.) It is easy to check that T_a is unitary. In fact, $T_a^{-1} = T_a^\dagger = T_{-a}$.

7.5. How the Four Classes are Related

Exercises

1. For $j = 1, 2, 3$, compute $R_j = \exp(-i\theta L_j) = \exp(\theta K_j)$, either by diagonalization or by summing the power series. The matrix R_j represents a rotation by the angle θ about the j-th coordinate axis in $\mathbb{R}^3$.

2. Let $\mathbf{v} = (v_1, v_2, v_3)$ be any unit vector in $\mathbb{R}^3$. Let $R_{\mathbf{v},\theta} = \exp(-i\theta(v_1 L_1 + v_2 L_2 + v_3 L_3)) = \exp(\theta(v_1 K_1 + v_2 K_2 + v_3 K_3))$. Show that $R_{\mathbf{v},\theta}$ is a rotation by an angle θ about the $\mathbf{v}$ axis. [Hint: Diagonalize $v_1 K_1 + v_2 K_2 + v_3 K_3$.]

3. Find a basis $\mathcal{B}$ of eigenvectors of L_3. Find $[L_1]_\mathcal{B}$, $[L_2]_\mathcal{B}$, and $[L_3]_\mathcal{B}$. The results should be familiar to students of quantum mechanics.

Let $\sigma_1 = \begin{pmatrix} 0 & 1 \\ 1 & 0 \end{pmatrix}$, $\sigma_2 = \begin{pmatrix} 0 & -i \\ i & 0 \end{pmatrix}$, $\sigma_3 = \begin{pmatrix} 1 & 0 \\ 0 & -1 \end{pmatrix}$. These are called the *Pauli matrices*.

4. Show that any 2×2 Hermitian matrix can be written as a real linear combination of the identity matrix, σ_1, σ_2, and σ_3.

5. Suppose that $\mathbf{x} \in \mathbb{R}^4$ and that $|\mathbf{x}| = 1$. Show that $ix_1\sigma_1 + ix_2\sigma_2 + ix_3\sigma_3 + x_4 I$ is unitary, where I is the identity matrix.

6. Show that $e^{-ix_1\sigma_1}$ is unitary and equals $\cos(x_1)I - i\sin(x_1)\sigma_1$.

7. More generally, show that, for $k = 1, 2, 3$, $e^{-ix_k\sigma_k}$ is unitary and equals $\cos(x_k)I - i\sin(x_k)\sigma_k$.

8. Still more generally, let $\mathbf{v}$ be a unit vector in $\mathbb{R}^3$. Show that $\exp(-i\theta(v_1\sigma_1 + v_2\sigma_2 + v_3\sigma_3))$ is unitary, and equals $\cos(\theta)I - i\sin(\theta)M$, where $M = v_1\sigma_1 + v_2\sigma_2 + v_3\sigma_3$.

Chapter 8

The Wave Equation

In this chapter we discuss the equation that governs light, sound, and most other waves in nature. The wave equation is a partial differential equation for a *wave function* $f(x,t)$, which is a function of position x and time t. The equation itself is

$$\frac{\partial^2 f(x,t)}{\partial x^2} - \frac{1}{v^2}\frac{\partial^2 f(x,t)}{\partial t^2} = 0, \tag{8.1}$$

where v is a constant that depends on our specific application. We shall soon see that v is the speed at which waves propagate. For sound waves in air, $f(x,t)$ is the density of matter at position x, at time t, and v is approximately 300 meters per second. For light waves, $f(x,t)$ is the strength of a component of the electric field and v is about 300,000,000 meters per second. Although light and sound are very different things, they are both described by equation (8.1).

Equation (8.1), taken literally, describes wave propagation in a 1-dimensional medium, such as a string or a wire. Of course, in physical space position is described by three variables, x, y, and z, not by one. Nonetheless, equation (8.1), with just one spatial variable, is relevant to wave phenomena in any dimension. When dealing with physical examples in $\mathbb{R}^3$, the variable x is understood to mean distance along the direction in which the wave is propagating.

A key feature of the wave equation (8.1) is that it is linear. If $f_1(x,t)$ and $f_2(x,t)$ are two solutions to (8.1), and c_1 and c_2 are scalars, then $c_1 f_1(x,t) + c_2 f_2(x,t)$ is also a solution. Indeed, one sometimes speaks of the wave *operator*

$$\frac{\partial^2}{\partial x^2} - \frac{1}{v^2}\frac{\partial^2}{\partial t^2}, \tag{8.2}$$

acting on an appropriate space of functions of x and t. The set of solutions to the equation (8.1) is precisely the kernel of the operator (8.2) and is a vector space in its own right. We may therefore apply linear algebra to the space of solutions to (8.1). One approach to solving (8.1) is to find a basis for the space of solutions, together with a method for decomposing an arbitrary solution in this basis. This will lead us, once again, to Fourier analysis.

Although x and t appear similarly in (8.1), their roles can be quite different. Typically, we think of $f(x,t)$ as a function of x that evolves in time, not the other way around. That function of x lives in some vector space, typically $L^2(\mathbb{R})$, or $L^2(U)$, where U is a subset of $\mathbb{R}$. In other words, x is restricted to lie in U, while t always runs from $-\infty$ to ∞. A typical problem is to be given initial conditions $f(x,0)$ and $\frac{\partial f}{\partial t}(x,0)$ (for all $x \in U$), and to have to compute $f(x,t)$ for all $x \in U$ and all $t \in \mathbb{R}$.

In previous chapters, whenever we spoke of a vector space of functions, such as L^2, we took our variable to be t. Here, however, t means time, and we are interested in functions of space (that evolve in time). Except where noted otherwise, L^2 and other function spaces will always refer to the variable x, not to t. Also, it is natural to refer to an element of such a function space as a function, rather than as a vector. (It really is both, of course.) If A is an operator on a space of functions, we call the eigenvectors of A *eigenfunctions*. Thus $\boldsymbol{\xi} = \sin(x)$, an element of $L^2[0,\pi]$, is an eigenfunction of $A = d^2/dx^2$ with eigenvalue -1.

8.1 Waves on the Line

In this section we consider solutions to (8.1) on the entire line. That is, we let $U = \mathbb{R}$. If $h_1(x)$ and $h_2(x)$ are twice-differentiable functions of x, then
$$f(x,t) = h_1(x - vt) + h_2(x + vt) \tag{8.3}$$
is a solution to (8.1). To see that, note, that by the chain rule,
$$\frac{\partial^2 f(x,t)}{\partial x^2} = h_1''(x - vt) + h_2''(x + vt), \tag{8.4}$$
while
$$\frac{\partial^2 f(x,t)}{\partial t^2} = v^2 h_1''(x - vt) + v^2 h_2''(x + vt). \tag{8.5}$$
By taking limits of solutions, we can drop the assumption that h_1 and h_2 are differentiable.

8.1. Waves on the Line

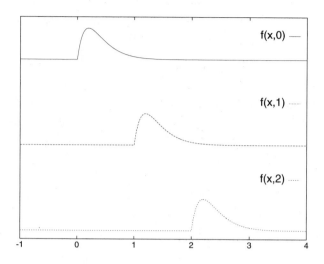

Figure 8.1: A forward traveling wave $f(x,t) = h_1(x - vt)$

The function $h_1(x-vt)$ is a wave propagating forward with speed v. See Figure 8.1. To see this, suppose that $h_1(x)$ is peaked near $x = a$. Then $h_1(x - vt)$ is peaked near $x - vt = a$, that is $x = a + vt$. Thus the crest of the wave (and, indeed, the whole wave form) is moving forward with speed v. Meanwhile, if $h_2(x)$ is peaked near $x = b$, then $h_2(x + vt)$ is peaked near $x + vt = b$, or $x = b - vt$, so the function $h_2(x + vt)$ is a wave propagating backward with speed v. Be careful with these signs, as it is easy to get them mixed up! *A function of $x - vt$ moves forward, while a function of $x + vt$ moves backward.*

A wave of the form $h_1(x - vt)$ or $h_2(x + vt)$ is called a *traveling wave*. It is very easy to observe, and to create, traveling waves in nature. You can create one by jerking the end of a piece of rope, by vibrating the end of a slinky, by dropping a stone in a pool of water, or just by speaking. When you speak, the wave form of your voice travels from you to your listeners. What is remarkable is that *every* solution to (8.1) is a sum of traveling waves.

Theorem 8.1 *Every solution to the wave equation (8.1) on $U = \mathbb{R}$ is of the form (8.3), namely the sum of a forward traveling wave and a backward traveling wave.*

Proof: Define the differential operators

$$D_\pm = \frac{\partial}{\partial x} \pm \frac{1}{v}\frac{\partial}{\partial t}. \tag{8.6}$$

The wave operator (8.2) is just $D_+D_- = D_-D_+$, and the wave equation (8.1) reduces to

$$D_+D_-f(x,t) = 0. \tag{8.7}$$

Since D_+ and D_- commute, they can be diagonalized simultaneously, and the eigenfunctions of $D_\pm$ are also the eigenfunctions of D_+D_-. By the results of Section 4.7, every solution $f(t)$ to (8.7) can be written as the sum of two terms, $f_+(x,t)$ and $f_-(x,t)$, with

$$D_+f_+(x,t) = D_-f_-(x,t) = 0. \tag{8.8}$$

Now define variables $x_\pm = x \pm vt$, so that $x = (x_+ + x_-)/2$ and $t = (x_+ - x_-)/2v$. Any function of x and t can be rewritten as a function of x_+ and x_-, and vice-versa. By the chain rule,

$$\frac{\partial}{\partial x_\pm} = \frac{\partial x}{\partial x_\pm}\frac{\partial}{\partial x} + \frac{\partial t}{\partial x_\pm}\frac{\partial}{\partial t} = \frac{1}{2}\frac{\partial}{\partial x} \pm \frac{1}{2v}\frac{\partial}{\partial t} = \frac{1}{2}D_\pm. \tag{8.9}$$

Since $D_+f_+ = 0$, f_+ is independent of x_+ and so can be written as a function of x_- alone, while f_- can be written as a function of x_+. That is, f_+ is a forward traveling wave and f_- is a backward traveling wave. ∎

Given this description, we can now solve the initial value problem on the real line. Suppose we are given the initial data

$$f(x,0) = f_0(x), \qquad \frac{\partial f}{\partial t}(x,0) = g_0(x), \tag{8.10}$$

where f_0 and g_0 are fixed functions of x. We need to compute $f(x,t)$ for all time from this initial data. Since the solution must take the form (8.3), this means finding the functions $h_1(x)$ and $h_2(x)$. Plugging (8.3) into (8.10) we get

$$h_1(x) + h_2(x) = f_0(x); \qquad -vh_1'(x) + vh_2'(x) = g_0(x). \tag{8.11}$$

Integrating the second equation of (8.11) gives

$$h_1(x) - h_2(x) = \frac{-1}{v}\int g_0(x)dx. \tag{8.12}$$

Combining this with the first equation of (8.11) gives

$$h_1(x) = \frac{vf_0(x) - \int g_0(x)dx}{2v}; \qquad h_2(x) = \frac{vf_0(x) + \int g_0(x)dx}{2v}. \tag{8.13}$$

Evaluating the integral $\int g_0(x)dx$ involves picking a constant of integration. Any value will do. Adding to this constant increases h_2 and decreases h_1 by the same amount, leaving the sum (8.3) and our actual solution $f(x,t)$ unchanged.

Notice that when $g_0 = 0$, then $h_1 = h_2$, and the wave propagates equally in both directions. This is what happens when you pluck a guitar string, pulling it out of shape and then letting go with zero velocity. Also notice that, when $f_0 = 0$, then $h_1 = -h_2$, and again the wave propagates with equal magnitude in both directions. This is what happens when you strike a note on a piano, imparting velocity (but not position) to the piano wire. To get waves to preferentially travel in one direction or the other, you must have both f_0 and g_0 nonzero.

Exercises

Consider the wave equation (8.1) with $v = 1$. For each of these initial conditions, compute $f(x,t)$ for all (x,t) and sketch the solutions at $t = 0$, $t = 1$, and $t = 5$.

1. $f(x,0) = \begin{cases} 1 - |x| & \text{if } |x| < 1 \\ 0; & \text{otherwise,} \end{cases}$ and $\dot{f}(x,0) = 0$.

This describes a string being plucked and then let go.

2. $f(x,0) = 0$ and $\dot{f}(x,0) = \begin{cases} 20 & \text{if } |x| < 0.1; \\ 0 & \text{otherwise.} \end{cases}$

This describes a string being hit by a narrow hammer, as on a piano.

3. $f(x,0) = \sin(x)$ and $\dot{f}(x,0) = 0$. Simplify your answer as much as possible.

4. $f(x,0) = \sin(x)$ and $\dot{f}(x,0) = \cos(x)$.

5. $f(x,0) = \exp(-x^2)$ and $\dot{f}(x,0) = 2x\exp(-x^2)$.

6. $f(x,0) = \exp(-x^2)$ and $\dot{f}(x,0) = -2x\exp(-x^2)$.

7. $f(x,0) = 3\exp(-x^2)$ and $\dot{f}(x,0) = 2x\exp(-x^2)$.

8. $f(x,0) = e^x$ and $\dot{f}(x,0) = 0$.

8.2 Waves on the Half Line; Dirichlet and Neumann Boundary Conditions

Waves bounce. Light bounces off shiny surfaces, sound bounces off walls, and water waves bounce off the ends of swimming pools. To understand the reflection of waves, we study the wave equation on a

230   Chapter 8. The Wave Equation

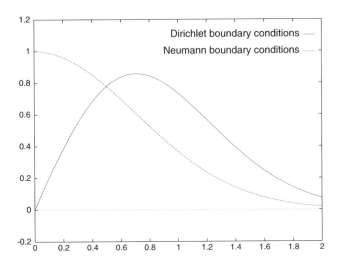

Figure 8.2: Dirichlet and Neumann boundary conditions

domain with a boundary, the half line $U = [0, \infty)$. As we shall see, the form of the reflection depends on what conditions we put at the boundary $x = 0$.

Boundary conditions

The most commonly studied boundary conditions, called *Dirichlet* boundary conditions, require that our function remain zero at the boundary:

$$f(0, t) = 0 \text{ for all time } t. \tag{8.14}$$

For waves on a string, this corresponds to the end of the string being tied down. See Figure 8.2.

Some problems involve a quantity that is free to vary at the boundary, like the height of the water at the edge of a swimming pool. For these problems, the relevant boundary condition, called *Neumann*, is that the normal derivative of the function be zero. That is,

$$\frac{\partial f}{\partial x}(0, t) = 0 \text{ for all time } t. \tag{8.15}$$

There are other, more complicated boundary conditions that occasionally come up, such as requiring that the derivative at the origin be a fixed multiple of the value. We will not consider these mixed boundary conditions.

8.2. Waves on the Half Line

The physical setup determines whether we apply the Dirichlet condition (8.14), the Neumann condition (8.15), or something more complicated. We never apply both (8.14) and (8.15) at the same time! The wave equation with both sets of boundary conditions would not have any nontrivial solutions at all.

The wave equation with Dirichlet conditions

We begin by solving the wave equation on the half line with Dirichlet boundary conditions. Let $h_1(x)$ and $h_2(x)$ be any two functions on $[0, \infty)$. Extend them to be functions on all of $\mathbb{R}$ by defining $h_i(x) = 0$ for $x < 0$. We then consider the function

$$f(x,t) = h_1(x-vt) + h_2(x+vt) - h_1(-x-vt) - h_2(-x+vt). \quad (8.16)$$

Notice that $f(x,t)$ is defined for all values of x, although only the region $x \geq 0$ is physically meaningful. It is not hard to see that (8.16) is a solution to the wave equation (8.1), since each of the four terms is a traveling wave. Furthermore,

$$f(0,t) = h_1(-vt) + h_2(vt) - h_1(-vt) - h_2(vt) = 0, \quad (8.17)$$

so it satisfies the Dirichlet condition (8.14).

The first two terms in (8.16) are easy to understand, as they are the same as in the whole line solution (8.3). The first term, $h_1(x-vt)$, is a forward traveling wave, moving away from the origin, while the second term, $h_2(x+vt)$, is a backward traveling wave, moving towards the origin.

The last two terms are past and future reflections of the first two, and are obtained by simply sending $x \to -x$ and flipping the overall sign. Since $h_2(x+vt)$ is an incoming wave, it will have a reflection in the future, namely $-h_2(-x+vt)$. (Notice that $h_2(-x+vt) = 0$ unless $t > x/v$, so this term is only relevant to the future, not to the present or the past.) Similarly, $h_1(x-vt)$ is an outgoing wave and must be the reflection of a past incoming wave; $-h_1(-x-vt)$ is that past incoming wave. Notice also that $-h_1(-x-vt)$ is actually a function of $x+vt$, and so is moving backward (inward), while $-h_2(-x+vt)$ is a function of $x-vt$, and so is moving outward (forward). Reflected waves emerge with the opposite sign of the original incoming waves. $-h_1(-x-vt)$ and $+h_2(x+vt)$ are incoming waves, while $+h_1(x-vt)$ and $-h_2(-x+vt)$ are their outgoing reflections.

Figure 8.3 demonstrates this in the case that $h_1 = 0$ and h_2 is a triangle wave. The dotted line represents the solution in the

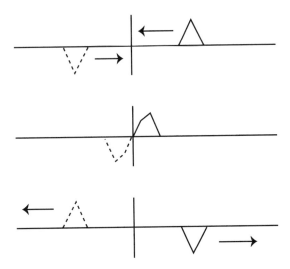

Figure 8.3: Reflection off a Dirichlet boundary

unphysical region $x < 0$. At $t = 0$, the term $-h_2(-x+vt)$ is nonzero only for x negative, and in the physical region $x \geq 0$ we merely have the incoming wave $h_2(x+vt)$. Some time later, the incoming wave $h_2(x+vt)$ and the reflected wave $-h_2(-x+vt)$ are interfering near $x = 0$. Later still, the incoming wave $h_2(x+vt)$ has moved to the unphysical $x < 0$ region, and in the physical $x \geq 0$ region we have only the reflection $-h_2(-x+vt)$.

The method we have used to solve the wave equation on the half line is called the *method of images*. To account for the effect of the boundary, we add to our solutions the effects of fictitious sources at $x < 0$. It should not be too surprising that this works. When you stand in a room with a mirrored wall, it looks as if there are two copies of everything, one in front of the mirror and one behind it. The light reflecting off the mirror might as well have come from honest sources behind it. There is no way (short of breaking the glass) to distinguish between the light patterns in a room with a mirrored wall and the light in two back-to-back rooms with identical objects facing each other. The only peculiarity of Dirichlet boundary conditions is that the mirror image sources appear with opposite signs.

The mirrored room is more than just an analogy. It is an actual example, as light waves are governed by the wave equation (8.1). A metal mirror reflects light by forcing the tangential electric field to be zero — that is, by imposing a Dirichlet boundary condition. The

8.2. Waves on the Half Line

light waves in such a room take the form (8.16), with the first two terms representing direct images of objects in the room, and the last two terms representing reflected images.

To get the functions h_1 and h_2 from initial data, we proceed as in Section 8.1, plugging the expression (8.16) into the initial conditions (8.10), and solving for h_1 and h_2. The analysis is identical to the whole line solution, except that there is now a natural constant of integration for $\int g_0$. The result is

$$h_1(x) = \frac{vf_0(x) - \int_0^x g_0(x')dx'}{2v}; \quad h_2(x) = \frac{vf_0(x) + \int_0^x g_0(x')dx'}{2v}, \quad (8.18)$$

for $x \geq 0$ and $h_1(x) = h_2(x) = 0$ for $x < 0$. The constant of integration is chosen so that $h_1(0) = h_2(0) = 0$. In general, the expression (8.16) will be continuous in x and t, as long as h_1 and h_2 are continuous on $[0, \infty)$, and $h_1(0) = -h_2(0)$.

Theorem 8.2 *Every solution to the wave equation (8.1) on the half line $U = [0, \infty)$ with Dirichlet boundary conditions is of the form (8.16).*

Proof: Given a solution $f(x,t)$ for $x \geq 0$ with Dirichlet boundary conditions, we can extend it to a solution on the whole line by defining

$$f(-x, t) = -f(x, t). \quad (8.19)$$

Since this is solution on the whole line (see Exercise 3), it must take the form $\tilde{h}_1(x - vt) + \tilde{h}_2(x + vt)$, by Theorem 8.1. We can add a constant to $\tilde{h}_1$ (and subtract it from $\tilde{h}_2$) to make $\tilde{h}_1(0) = 0$. Now, since $f(0, t) = 0$ for all time, we must have $\tilde{h}_2(0 + vt) = -\tilde{h}_1(0 - vt)$. In other words, $\tilde{h}_1$ and $\tilde{h}_2$ are not independent functions. Rather,

$$\tilde{h}_1(x) = -\tilde{h}_2(-x) \quad (8.20)$$

for all x. We define

$$h_1(x) = \begin{cases} \tilde{h}_1(x) & \text{if } x \geq 0; \\ 0 & \text{if } x \leq 0; \end{cases} = \begin{cases} -\tilde{h}_2(-x) & \text{if } x \geq 0; \\ 0 & \text{if } x \leq 0. \end{cases}$$

$$h_2(x) = \begin{cases} -\tilde{h}_1(-x) & \text{if } x \geq 0; \\ 0 & \text{if } x \leq 0; \end{cases} = \begin{cases} \tilde{h}_2(x) & \text{if } x \geq 0; \\ 0 & \text{if } x \leq 0, \end{cases} \quad (8.21)$$

so that, for all x,

$$\tilde{h}_1(x) = h_1(x) - h_2(-x) \quad \text{and} \quad \tilde{h}_2(x) = h_2(x) - h_1(-x). \quad (8.22)$$

Rewriting $\tilde{h}_1(x - vt) + \tilde{h}_2(x + vt)$ in terms of h_1 and h_2 then gives the form (8.16). ∎

Neumann boundary conditions

In the case of Neumann boundary conditions, the procedure and results are almost identical to the Dirichlet case, except for signs. We leave the derivation as a series of exercises. As before, let $h_1(x)$ and $h_2(x)$ be arbitrary functions that are zero for $x < 0$. The most general solution to the wave equation with Neumann boundary conditions at $x = 0$ is

$$f(x,t) = h_1(x-vt) + h_2(x+vt) + h_1(-x-vt) + h_2(-x+vt). \quad (8.23)$$

Notice the difference in sign between (8.23) and (8.16). When waves are reflected off a Neumann boundary, they do *not* flip sign.

The formula (8.18) for h_1 and h_2 in terms of initial data is still valid for $x > 0$. However, there may well be a discontinuity at $x = 0$, since $f_0(0)$ need not be zero. In such a case, we must define $h_1(0) = h_2(0) = f_0(0)/4$, rather than $f_0(0)/2$, in order to keep $f(x,t)$ continuous at $x = vt$ and $x = -vt$.

Exercises

1. On the half line $[0, \infty)$ with Dirichlet boundary conditions, imagine that at time $t = 0$ we have an incoming triangle wave and an outgoing square wave, both positive, as shown below. Describe and sketch what the wave function looked like in the distant past, and what it will look like in the distant future.

2. Derive the results (8.18) in the half line case.

3. Show that the function on the whole line defined by (8.19) is a solution to the wave equation on the whole line. (We already know it is a solution for $x > 0$. You must show that the wave equation is satisfied at $x < 0$ and at $x = 0$.)

4. Show that (8.23) is a solution of the wave equation on the half line with Neumann boundary conditions at $x = 0$.

5. Show that every solution to the wave equation on the half line with Neumann boundary conditions at $x = 0$ is of the form (8.23).

6. Repeat Exercise 1 for Neumann boundary conditions.

In Exercises 7–11, solve the wave equation on the half line with Dirichlet boundary conditions, $v = 1$, and the specified initial conditions. In each case sketch the solution at $t = 0$, $t = 1$, $t = 10$, and $t = -10$.

7. $f(x,0) = \begin{cases} 1 - |x - 10| & \text{if } |x - 10| < 1 \\ 0; & \text{otherwise,} \end{cases}$ and $\dot{f}(x,0) = 0$.

8. $f(x,0) = 0$ and $\dot{f}(x,0) = \begin{cases} 20 & \text{if } |x - 1| < 0.1; \\ 0 & \text{otherwise.} \end{cases}$

9. $f(x,0) = 0$ and $\dot{f}(x,0) = \begin{cases} x^2 & \text{if } 0 \leq x < 1; \\ 0 & \text{otherwise.} \end{cases}$

10. $f(x,0) = x\exp(-x^2)$ $\dot{f}(x,0) = 0$.

11. $f(x,0) = x\exp(-x^2)$ and $\dot{f}(x,0) = (2x^2 - 1)\exp(-x^2)$ (for $x \geq 0$). What is $f(3,5)$? What is $f(5,3)$?

12–14. Repeat Exercises 7–9 for Neumann boundary conditions.

8.3 The Vibrating String

We now turn to an example of great musical importance, the vibrating string. Imagine a piece of string of length L, stretched tight, and tacked down at both ends. The string can vibrate, and the vibrations are governed by the wave equation (8.1), but the ends of the string cannot move. Mathematically, we are solving the wave equation on the interval $U = [0, L]$ with Dirichlet boundary conditions at both ends. In other words,

$$f(0,t) = f(L,t) = 0 \quad \text{for all } t. \tag{8.24}$$

In this section we solve the equation using traveling waves and the method of images. In the next section we solve it using eigenvectors and eigenvalues, leading directly to Fourier series. We then see why the two approaches are equivalent.

If you stand in a room with a mirror on one wall, you will see one reflection of the room. In Section 8.2, with one boundary, we needed one reflection each of our original waves $h_1(x - vt)$ and $h_2(x + vt)$ to solve the wave equation on the half line. However, if you stand in a room with mirrors on opposite walls, you will see an infinite number of reflections. To apply the method of images in this section, we will need an infinite number of terms to account for the infinite number of times a traveling wave can bounce back and forth along the string.

Figure 8.4: An infinite sum of traveling waves

We begin by listing all the mirror images of the point $x = a$. Reflecting this point across $x = 0$ or $x = L$ gives $x = -a$ or $x = 2L - a$, respectively. Reflecting twice, once across each boundary, gives $x = a \pm 2L$. Reflecting three times (say, across $x = 0$, then across $x = L$, then across $x = 0$), gives $4L - a$ and $-2L - a$. In general, reflecting an odd number of times gives points of the form $2nL - a$, where n can be negative, zero, or positive, while reflecting an even number of times gives points of the form $2nL + a$.

This list of points suggests an extension of (8.16). Let h_1 and h_2 be arbitrary functions of x that vanish for $x < 0$ or $x > L$. To get $f(x,t)$ we will sum $h_1(x - vt) + h_2(x + vt)$ over all the mirror image points of x, flipping the sign at each reflection. That is, we count all the odd reflections negatively and all the even reflections positively. The resulting sum,

$$f(x,t) = \sum_{n=-\infty}^{\infty} h_1(x + 2nL - vt) + \sum_{n=-\infty}^{\infty} h_2(x + 2nL + vt)$$
$$- \sum_{n=-\infty}^{\infty} h_1(-x + 2nL - vt) - \sum_{n=-\infty}^{\infty} h_2(-x + 2nL + vt),$$
(8.25)

is our desired general solution. See Figure 8.4.

There is no difficulty with convergence since, for fixed x and t, at most one term in each sum is nonzero. Each term satisfies the wave equation (8.1), so the sum also satisfies (8.1). We then check the boundary conditions. At $x = 0$, each term in the first or second sum is exactly canceled by the corresponding term in the third or fourth sum. At $x = L$, the nth term in the first or second sum is canceled by the $n + 1$st term in the third or fourth sum. Thus (8.25) is a solution to the vibrating string problem.

To get h_1 and h_2 from initial data, we proceed exactly as in Sections 8.1 and 8.2, with essentially identical results. For $x < 0$ or $x > L$ we have $h_1(x) = h_2(x) = 0$, while for $0 \le x \le L$ we have

$$h_1(x) = \frac{vf_0(x) - \int_0^x g_0(x')dx'}{2v}; \qquad h_2(x) = \frac{vf_0(x) + \int_0^x g_0(x')dx'}{2v}.$$
(8.26)

8.3. The Vibrating String

Theorem 8.3 *Every solution to the wave equation on the interval $[0, L]$ with Dirichlet boundary conditions at $x = 0$ and at $x = L$ is of the form (8.25).*

Proof: Given a solution $f(x,t)$ for $0 \leq x \leq L$ with Dirichlet boundary conditions, we can extend it to a solution on the whole line by defining

$$f(x + 2nL, t) = f(x,t); \qquad f(2nL - x, t) = -f(x,t). \qquad (8.27)$$

Since this is a solution on the whole line (see Exercise 3), it must take the form $\tilde{h}_1(x - vt) + \tilde{h}_2(x + vt)$. Now, since $f(0,t) = 0$ for all time, we must have $\tilde{h}_2(vt) = -\tilde{h}_1(-vt)$, and so

$$\tilde{h}_2(x) = -\tilde{h}_1(-x). \qquad (8.28)$$

Also, since $f(L,t) = 0$, we must have $\tilde{h}_1(L - vt) = -\tilde{h}_2(L + vt)$, and so

$$\tilde{h}_2(L + x) = -\tilde{h}_1(L - x). \qquad (8.29)$$

Combining these last two results we have that

$$\tilde{h}_2(2L+x) = \tilde{h}_2(L+(L+x)) = -\tilde{h}_1(L-(L+x)) = -\tilde{h}_1(-x) = \tilde{h}_2(x). \qquad (8.30)$$

Thus $\tilde{h}_2$ is a periodic function with period $2L$. By (8.28), so is $\tilde{h}_1$. We then define, as in Section 8.2,

$$h_1(x) = \begin{cases} \tilde{h}_1(x) & \text{if } 0 \leq x \leq L; \\ 0 & \text{otherwise;} \end{cases} = \begin{cases} -\tilde{h}_2(-x) & \text{if } 0 \leq x \leq L; \\ 0 & \text{otherwise.} \end{cases}$$

$$h_2(x) = \begin{cases} -\tilde{h}_1(-x) & \text{if } 0 \leq x \leq L; \\ 0 & \text{otherwise;} \end{cases} = \begin{cases} \tilde{h}_2(x) & \text{if } 0 \leq x \leq L; \\ 0 & \text{otherwise.} \end{cases}$$

(8.31)

Rewriting $\tilde{h}_1(x - vt) + \tilde{h}_2(x + vt)$ in terms of h_1 and h_2 then gives (8.25). ∎

Exercises

1. Suppose that at time $t = 0$ we have a single triangle wave, centered at $x = L/2$, moving to the right (forward). Describe qualitatively what the solution will do at future times and what it did at past times. Sketch the wave function at various times to support your verbal description.

2. Show that the function defined by (8.27) is a solution to the wave equation on the whole line.

3. Show that the solution (8.25) is continuous if three conditions are met: 1) h_1 and h_2 are continous on $[0, L]$, 2) $h_1(0) = -h_2(0)$, and 3) $h_1(L) = -h_2(L)$. Show that, if f_0 and g_0 are continuous, that these conditions are indeed met by the functions h_1 and h_2 of (8.26).

4. How should formulas (8.25) and (8.26) be changed if we want our solutions to satisfy Neumann boundary conditions (instead of Dirichlet) at $x = 0$ and $x = L$?

5. On the interval $[0, 1]$, solve for $f(x, t)$, given $f_0(x) = \sin(\pi x)$, and $g_0(x) = 0$. Simplify as much as possible.

6. On the interval $[0, 1]$, solve for $f(x, t)$, given $f_0(x) = 0$, and $g_0(x) = \sin(\pi x)$. Simplify as much as possible.

8.4 Standing Waves and Fourier Series

In this section we solve the problem of a vibrating string by a completely different method. We use the fact that, for any fixed time t, our wave function $f(x, t)$ is an element of the vector space $L^2([0, L])$. When thinking of it as a vector, we write the wavefunction as **f**. From this point of view, the wave equation (8.1) is no longer a scalar-valued partial differential equation. Rather, it is a vector-valued ordinary differential equation

$$\frac{d^2 \mathbf{f}}{dt^2} = v^2 A \mathbf{f}, \tag{8.32}$$

where the operator

$$A = \frac{d^2}{dx^2}, \tag{8.33}$$

with Dirichlet boundary conditions, acts on $L^2([0, L])$.

We learned how to deal with second-order vector-valued ordinary differential equations in Section 5.3. First we must find the eigenvalues and eigenfunctions of A. To each eigenfunction $\boldsymbol{\xi}$ of A with

8.4. Standing Waves and Fourier Series

eigenvalue λ, there are two linearly independent solutions to (8.32), namely

$$\mathbf{f} = \boldsymbol{\xi} e^{\pm vt\sqrt{\lambda}}. \tag{8.34}$$

These can be recombined into hyperbolic sines and cosines of $vt\sqrt{\lambda}$ if $\lambda > 0$, or into ordinary sines and cosines of $vt\sqrt{-\lambda}$ if $\lambda < 0$.

Theorem 8.4 *The eigenfunctions of $A = d^2/dx^2$ on the interval $[0, L]$ with Dirichlet boundary conditions are the functions $\boldsymbol{\xi}_n(x) = \sin(n\pi x/L)$, with eigenvalues $\lambda_n = -n^2\pi^2/L^2$. These eigenfunctions form an orthogonal basis for $L^2[0, L]$.*

Proof: We saw in Section 6.9 that the functions $\boldsymbol{\xi}_n$ are eigenfunctions of A and are orthogonal in $L^2[0, L]$. What remains is to show that there are no other eigenfunctions, and the $\boldsymbol{\xi}_n$'s form a basis for $L^2[0, L]$.

First note that A is Hermitian:

$$\begin{aligned}\langle \mathbf{f} | A\mathbf{g} \rangle &= \int_0^L \overline{f(x)} g''(x) dx = \overline{f(x)} g'(x) \Big|_0^L - \int_0^L \overline{f'(x)} g'(x) dx \\ &= \left(\overline{f(x)} g'(x) - \overline{f'(x)} g(x) \right) \Big|_0^L + \int_0^L \overline{f''(x)} g(x) dx \\ &= 0 + \langle A\mathbf{f} | \mathbf{g} \rangle, \end{aligned} \tag{8.35}$$

since, with our boundary conditions, $\bar{f}(0) = g(0) = \bar{f}(L) = g(L) = 0$. Since Hermitian operators are diagonalizable, once we have found all our eigenfunctions we know we have a basis.[1]

The eigenvalue equation $A\boldsymbol{\xi} = \lambda \boldsymbol{\xi}$ is an ordinary differential equation

$$\frac{d^2\boldsymbol{\xi}(x)}{dx^2} = \lambda \boldsymbol{\xi}(x), \tag{8.36}$$

with boundary conditions

$$\boldsymbol{\xi}(0) = \boldsymbol{\xi}(L) = 0. \tag{8.37}$$

Since eigenvalues of Hermitian operators are real, we only need to consider (8.36) with λ positive, zero, or negative.

For $\lambda > 0$, the general solution to (8.36) is

$$\boldsymbol{\xi}(x) = c_1 \sinh(x\sqrt{\lambda}) + c_2 \cosh(x\sqrt{\lambda}). \tag{8.38}$$

---

[1] Theorem 7.6 only showed that Hermitian operators on finite dimensional vector spaces are diagonalizable. However, the spectral theorem (Theorem 9.5) extends this result to infinite dimensional spaces such as $L^2([0, L])$.

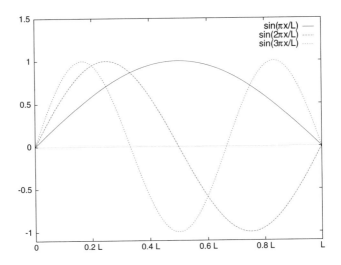

Figure 8.5: First 3 eigenfunctions of d^2/dt^2 on $[0, L]$

Since we need $\boldsymbol{\xi}(0) = 0$, we must have $c_2 = 0$. Since $\sinh(L\sqrt{\lambda}) \neq 0$, we must then also have $c_1 = 0$, so $\boldsymbol{\xi}(x) = 0$ for all x. However, eigenfunctions by definition must be nonzero, so $\boldsymbol{\xi}$ is not an eigenfunction and $\lambda > 0$ is not an eigenvalue. Similarly, 0 is not an eigenvalue, since the only solutions to $A\boldsymbol{\xi} = 0$ are $\boldsymbol{\xi}(x) = c_1 + c_2 x$, which cannot meet the boundary conditions unless $c_1 = c_2 = 0$.

Our eigenvalues must therefore all be negative. When $\lambda < 0$, the general solution to (8.36) is

$$\boldsymbol{\xi}(x) = c_1 \sin(x\sqrt{-\lambda}) + c_2 \cos(x\sqrt{-\lambda}). \tag{8.39}$$

As before, the condition $\boldsymbol{\xi}(0) = 0$ implies $c_2 = 0$. Let $k = \sqrt{-\lambda}$. $\boldsymbol{\xi}(L)$ is then equal to $c_1 \sin(kL)$, which must equal zero. If $\sin(kL) \neq 0$, then we must also have $c_1 = 0$ so we do not have an eigenfunction, but if $\sin(kL) = 0$, then $\sin(kx)$ *is* an eigenfunction. We therefore need $kL = n\pi$, so $k = n\pi/L$, and $\lambda = -n^2\pi^2/L^2$. ∎

As a result of this theorem, the most general solution to the wave equation (8.32, or equivalently 8.1) is

$$\begin{aligned} f(x,t) &= \sum_{n=1}^{\infty} \boldsymbol{\xi}_n(x)(a_n \cos(n\pi vt/L) + b_n \sin(n\pi vt/L)) \\ &= \sum_{n=1}^{\infty} \sin(n\pi x/L)(a_n \cos(n\pi vt/L) + b_n \sin(n\pi vt/L)). \end{aligned} \tag{8.40}$$

8.4. Standing Waves and Fourier Series

A single term in this sum is called a *standing wave* since, rather than moving back and forth, it maintains a fixed shape and merely varies in size. The first three standing waves are shown in Figure 8.5.

To compute the constants a_n and b_n we evaluate the function (8.40) and its time derivative at $t = 0$.

$$f(x,0) = f_0(x) = \sum_{n=1}^{\infty} a_n \boldsymbol{\xi}_n(x)$$
$$\frac{\partial f}{\partial t}(x,0) = g_0(x) = \frac{\pi v}{L} \sum_{n=1}^{\infty} n b_n \boldsymbol{\xi}_n(x). \quad (8.41)$$

The constants a_n and b_n are thus closely related to the n-th Fourier coefficients of the initial position $f_0(x)$ and velocity $g_0(x)$. Specifically,

$$a_n = \frac{\langle \boldsymbol{\xi}_n | \mathbf{f}_0 \rangle}{\langle \boldsymbol{\xi}_n | \boldsymbol{\xi}_n \rangle} = \frac{2}{L} \int_0^L f_0(x) \sin\left(\frac{n\pi x}{L}\right) dx,$$
$$b_n = \frac{L}{n\pi v} \frac{\langle \boldsymbol{\xi}_n | \mathbf{g}_0 \rangle}{\langle \boldsymbol{\xi}_n | \boldsymbol{\xi}_n \rangle} = \frac{2}{n\pi v} \int_0^L g_0(x) \sin\left(\frac{n\pi x}{L}\right) dx. \quad (8.42)$$

The remainder of this section is a discussion of the solution (8.40) and the coefficients (8.42).

It is convenient to define the quantities

$$k_n = n\pi/L; \quad \omega_n = v k_n = n\pi v/L. \quad (8.43)$$

These are called the *wave number* and *angular frequency* of the n-th mode $\boldsymbol{\xi}_n$, since $\boldsymbol{\xi}_n$ is proportional to $\sin(k_n x)$ and oscillates in time like the sine or cosine of $\omega_n t$. The *wavelength* of the n-th mode is the distance in space needed to enclose one complete cycle, namely $2\pi/k_n = 2L/n$. The *period* is the time needed to oscillate through one complete cycle, namely $2\pi/\omega_n = 2L/nv$. The *frequency* is the number of cycles per unit time. Since a cycle is 2π radians, the frequency (denoted ν_n) is

$$\nu_n = \omega_n/2\pi = 1/(\text{period}) = nv/2L. \quad (8.44)$$

The $n = 1$ mode is called the *fundamental* mode and the modes with $n > 1$ are called *overtones*.

When a string vibrates with frequency ν_n, it sets the air around it vibrating with the same frequency and the vibrational waves of the air propagate to your ear. Your ear detects the vibrations and registers the frequency as a musical tone. Higher frequencies are higher tones; lower frequencies are lower tones.

Hearing and music

Notice that all the frequencies of a string are proportional to v/L. To tune a string instrument, that is to get it to play different notes, you must either change v or change L. To change v, you adjust the tension on the string. Tighter strings have higher wave speeds, and therefore higher frequencies than looser strings, so string instruments typically come equipped with cranks at the end of the neck for adjusting the tension on each string. This is how musicians tune their instruments before playing.

It is impractical, however, to adjust the tension in the middle of a piece. Instead, string instrument players use their fingers to adjust the effective lengths of their strings. If you hold a string down at a point $a < L$, you have imposed a Dirichlet condition at $x = a$, and effectively are playing a string of length a. By moving their fingers, musicians can get the same string to have different natural frequencies at different times without having to change the tension.

A single mode, also called a pure tone, sounds like a whistle and is not very pleasant. Typically, pleasant sounds are a combination of pure tones, and it is the overtones that distinguish the sound of a guitar playing a note from the sound of a piano playing the same note. The fundamental modes are the same, and might have the same amplitudes a_1 and b_1, but one instrument might generate more second harmonic (bigger a_2 and b_2) than the other, while the other instrument might generate a bigger third or fourth harmonic. The human ear doesn't register these differences in the distribution of overtones as a difference in pitch, but rather as a difference in the quality of the sound.

Because the frequencies of a string (and other simple sound-making devices) come in the ratios $1 : 2 : 3 : 4 : \cdots$, these simple ratios appear constantly in nature, and our ears are trained from birth to appreciate them. The ratio 2:1 corresponds to notes an octave apart. The ratio 3:2 is a musical *fifth*, the ratio 4:3 is a *fourth*, 5:4 is a *major third*, and 6:5 is a *minor third*. All these ratios are pleasant, or *consonant*.

Ratios involving 7, however, turn out to be unpleasant, or *dissonant*. Musical instruments are designed to produce as little 7th harmonic as possible. The way you play an instrument also affects the mix of overtones. A string that is plucked near the end sounds different from the same string plucked in the middle. Exercise 1 is essentially to determine the right place to pluck a guitar.

8.4. Standing Waves and Fourier Series

Standing vs. traveling waves

In Section 8.3 we saw that the most general solution to the vibrating string was a sum of traveling waves. In this section we saw that the most general solution is a sum of standing waves. We can convert from one to the other. For example, the standing wave $2\sin(k_n x)\sin(\omega_n t)$ can be rewritten as a sum of traveling waves

$$\begin{aligned} 2\sin(k_n x)\sin(\omega_n t) &= \cos(k_n x - \omega_n t) - \cos(k_n x + \omega_n t) \\ &= \cos(k_n(x-vt)) - \cos(k_n(x+vt)), \quad (8.45) \end{aligned}$$

since $\omega_n = vk_n$. To write a sum of traveling waves $\tilde{h}_1(x-vt)+\tilde{h}_2(x+vt)$ as a sum of standing waves we must write $\tilde{h}_i(x)$ as a Fourier series. However, $\tilde{h}_1$ and $\tilde{h}_2$ do not individually satisfy the Dirichlet boundary conditions at 0 and L. Instead, they are periodic functions of x with period $2L$. To convert traveling waves into standing waves we must first learn how to Fourier analyze periodic functions. That is the content of the next section.

Exercises

1. Find a point a between 0 and L such that the triangle wave

$$f_0(x) = \begin{cases} x/a & \text{if } 0 \le x \le a; \\ (L-x)/(L-a) & \text{if } a \le x \le L \end{cases} \quad (8.46)$$

is orthogonal to ξ_7. How many such points are there? To find the points you must first use Fourier series to find a (transcendental) equation for a, and then use Newton's method (or technology) to find the solutions to this equation.

For each of the initial conditions of Exercises 2–4, find a solution to the wave equation (with $L = v = 1$) for all time, and sketch the solution at time $t = 0, \pi/4, \pi/2,$ and π. You may wish to use technology to assist with the sketching.

2. $f(x,0) = \begin{cases} x & \text{if } x \le 1/2 \\ 1-x & \text{if } x > 1/2, \end{cases}$ and $\dot{f}(x,0) = 0$.

3. $f(x,0) = 0$ and $\dot{f}(x,0) = \begin{cases} x & \text{if } x \le 1/2 \\ 1-x & \text{if } x > 1/2. \end{cases}$

4. $f(x,0) = x - x^2$ and $\dot{f}(x,0) = 0$.

In reality, a string is not a continuous medium, but is made up of a (very large but) finite number of atoms, each bound by chemical forces to its neighbors. This suggests modeling a vibrating string as a large number of blocks, coupled by springs, as in Section 5.3. The following problems show how to obtain the vibrating string from the coupled block problem by taking a limit as the number of blocks goes to infinity.

5. Consider an array of N coupled blocks, all of mass m, and all springs of constant k. Write down the coefficient matrix for this problem.

6. Show that, for any integer j, the vector $\mathbf{x} = (\sin(j\pi/(N+1)), \sin(2j\pi/(N+1)), \ldots, \sin(Nj\pi/(N+1)))^T$ is an eigenvector of the coefficient matrix A of Exercise 5. What is the corresponding eigenvalue? [This problem is essentially the same as Exercise 7 of Section 5.3.]

7. Let $\Delta x = L/N$, and imagine $N-1$ blocks on a string of length L, with equilibrium positions $x = \Delta x, 2\Delta x, \ldots, (N-1)\Delta x$. Suppose the masses are all a/N, and that the spring constants are all Nb, where a and b are constants, independent of N. Let $\boldsymbol{\xi}_n(x) = \sin(n\pi x/L)$. Let $x_j = \boldsymbol{\xi}_n(j\Delta x)$. Show that, for every value of N, $(x_1, x_2, \ldots, x_{N-1})^T$ is an eigenvector of the coefficient matrix A. What is the limit of the eigenvalue as $N \to \infty$? How does this limit depend on the constants a and b?

The upshot of these exercises is that the low-frequency modes of a vibrating string are essentially the same as those of a large number of coupled oscillators. The effect of having a finite number of oscillators is only to put an upper limit on the frequency (as well as to modify the frequencies of some modes that are close to the limit). Lattice models are often used in physics and engineering to model continuum behavior, especially on a computer.

8.5 Periodic Boundary Conditions

In this section we consider functions on a circle, periodic functions on the line, and functions on the interval $[0, L]$ with periodic boundary conditions, show how to decompose each type of function into an infinite sum of sines and cosines, and solve the wave equation in each setting. In fact, the three settings are all equivalent, and understanding one is tantamount to understanding all three. We first define our terms.

8.5. Periodic Boundary Conditions

Periodic functions

Definition *A function $f(x)$ on the real line is said to be* periodic with period L *if, for every x, $f(x+L) = f(x)$.*

Definition *A function $f(x)$ on the interval $[0, L]$ is said to satisfy* periodic boundary conditions *if $f(0) = f(L)$ and $f'(0) = f'(L)$.*

Consider a function on the unit circle. Such a function can be viewed as a function of the angle θ. However, since θ and $\theta + 2\pi$ represent the same point on the circle, we must have $f(\theta) = f(2\pi + \theta)$. In other words, a function on the unit circle can always be represented as a periodic function on the line with period 2π. We can also restrict our angles to always lie between 0 and 2π, inclusive. In that case, our function on the circle becomes a function on $[0, 2\pi]$ with periodic boundary conditions.

Similarly, other periodic functions can be associated with functions on the circle of circumference L. We associate the real number x with the point on the circle a distance x counterclockwise of a given reference point. Since the circle has circumference L, the coordinates x and $x + L$ refer to the same point, and a function on the circle is the same as a periodic function of x with period L. If we wish, we can restrict x to lie between 0 and L, in which case we have a function on $[0, L]$ with periodic boundary conditions.

To put it another way, we can obtain a circle by taking the real line and identifying any two points that differ by a multiple of L. This amounts to wrapping the line around and around the circle of circumference L and identifying points that land on the same spot. Alternatively, we can start with the interval $[0, L]$ and identify the endpoints to get a circle.

This shows how to go from functions on a circle to periodic functions on $\mathbb{R}$, or to functions on $[0, L]$ with periodic boundary conditions. To go from a periodic function to a function on $[0, L]$, just restrict the domain to $[0, L]$. Since the original function was periodic, the new function satisfies the boundary conditions. Finally, if f is a function on $[0, L]$ with periodic boundary conditions, we can extend it to a periodic function as follows. Every point $x \in \mathbb{R}$ can be written as $nL + a$ for a unique integer n and $a \in [0, L)$. Just define

$$f(x) = f(a). \tag{8.47}$$

Since $f(L) = f(0)$, the two definitions of $f(L)$ agree, and the new function is clearly periodic.

In short, we have the equivalence

$$\left\{\begin{array}{c}\text{Functions}\\ \text{on circle of}\\ \text{radius } L\end{array}\right\} = \left\{\begin{array}{c}\text{Periodic}\\ \text{functions on } \mathbb{R}\\ \text{with period } L\end{array}\right\} = \left\{\begin{array}{c}\text{Functions on } [0,L]\\ \text{with periodic}\\ \text{boundary conditions}\end{array}\right\}. \quad (8.48)$$

We give this space a name, $L^2(S_L^1)$, and an inner product

$$\langle f|g\rangle = \int_0^L \overline{f(x)}g(x)dx. \quad (8.49)$$

Strictly speaking, the space $L^2(S_L^1)$ is constructed in much the same way as $L^2(\mathbb{R})$. We begin with the space of smooth periodic functions, on which (8.49) is an inner product. We then complete the space, adding on limits of smooth functions in the norm defined by (8.49), and take the quotient of this space by the subspace of zero norm (i.e., the functions that are zero except on a set of measure zero). The result is essentially the space of square-integrable functions on our circle of circumference L, with the understanding that two functions are identified if they differ only on a set of measure zero.

Fourier series

The functions $\sin(2n\pi x/L)$ and $\cos(2n\pi x/L)$ are clearly periodic with period L. Therefore any linear combination of these functions is also periodic. Conversely, every periodic function is a (possibly infinite) linear combination of these sines and cosines.

Theorem 8.5 (Fourier series for periodic functions) *Any periodic square-integrable function $f(x)$ of period L can be represented as an infinite sum*

$$f(x) = \frac{a_0}{2} + \sum_{n=1}^{\infty} a_n \cos(2n\pi x/L) + b_n \sin(2n\pi x/L), \quad (8.50)$$

where

$$a_n = \frac{2}{L}\int_0^L f(x)\cos(2n\pi x/L)dx; \quad b_n = \frac{2}{L}\int_0^L f(x)\sin(2n\pi x/L)dx. \quad (8.51)$$

Alternatively, $f(x)$ can be represented as a sum of complex exponentials

$$f(x) = \sum_{n=-\infty}^{\infty} c_n \exp(2\pi i n x/L), \quad (8.52)$$

8.5. Periodic Boundary Conditions

where

$$c_n = \frac{1}{L}\int_0^L \exp(-2\pi i n x/L) f(x) dx. \tag{8.53}$$

Proof: The functions $\mathbf{b}_n(x) = \exp(2\pi i n x/L)$ are orthogonal:

$$\langle \mathbf{b}_n | \mathbf{b}_m \rangle = \int_0^L \exp(2\pi i (m-n)x/L) dx = \begin{cases} L & \text{if } n = m; \\ 0 & \text{otherwise.} \end{cases} \tag{8.54}$$

I claim that the $\mathbf{b}_n$'s are a basis for $L^2(S_L^1)$.

Given this claim (which will be proven last), (8.52) is just an expansion of $\mathbf{f}$ in the basis $\mathcal{B}$, and equation (8.53) is just the usual formula $c_n = \langle \mathbf{b}_n | \mathbf{f} \rangle / \langle \mathbf{b}_n | \mathbf{b}_n \rangle$ for the coefficients of an orthogonal expansion. Since $\mathbf{b}_n$ is the sum of a sine and a cosine:

$$\mathbf{b}_n = \cos(2\pi n x/L) + i\sin(2\pi n x/L), \tag{8.55}$$

we can rewrite (8.52) as an expansion in sines and cosines, obtaining (8.50) with

$$a_n = c_n + c_{-n}; \qquad b_n = -i(c_n - c_{-n}). \tag{8.56}$$

We have used the fact that $\cos(-\theta) = \cos(\theta)$ and that $\sin(-\theta) = -\sin(\theta)$ to write everything in terms of sines and cosines of non-negative multiples of $2\pi x/L$. Finally, plugging the formula (8.53) for c_n into (8.56) gives the formula (8.51) for a_n and b_n. Note that when $n > 0$, $\mathbf{b}_n$ and $\mathbf{b}_{-n}$ each contribute to the sum (8.52). However, $\mathbf{b}_0$ and $\mathbf{b}_{-0}$ are the same function, which only contributes once. Since $a_0 = c_0 + c_{-0} = 2c_0$, the leading term in (8.50) is $a_0/2$, rather than a_0.

To show that the $\mathbf{b}_n$'s form a basis, we find a Hermitian operator whose eigenfunctions are precisely the $\mathbf{b}_n$'s. Let $A = -i d/dx$. Then

$$\begin{aligned}\langle \mathbf{f} | A\mathbf{g} \rangle &= \int_0^L \overline{f(x)}(-ig'(x))dx = -i\overline{f(x)}g(x)\Big|_0^L + i\int_0^L \overline{f'(x)}g(x)dx \\ &= -i\overline{f(x)}g(x)\Big|_0^L + \int_0^L \overline{-if'(x)}g(x)dx \\ &= -i\overline{f(x)}g(x)|_0^L + \langle A\mathbf{f} | \mathbf{g} \rangle. \end{aligned} \tag{8.57}$$

But f and g are periodic, so the two terms in $-i\overline{f(x)}g(x)|_0^L$ cancel, and $\langle \mathbf{f} | A\mathbf{g} \rangle = \langle A\mathbf{f} | \mathbf{g} \rangle$. In other words, A is Hermitian.

We now look for the eigenvalues and eigenfunctions of A. The eigenvalue equation $A\boldsymbol{\xi} = \lambda \boldsymbol{\xi}$ becomes

$$-i\frac{d\boldsymbol{\xi}(x)}{dx} = \lambda \boldsymbol{\xi}(x) \tag{8.58}$$

248   Chapter 8. The Wave Equation

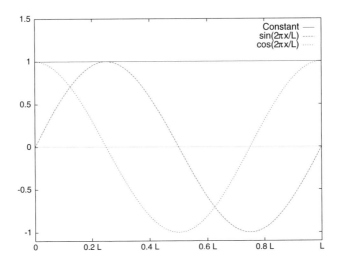

Figure 8.6: Three eigenfunctions of d^2/dx^2 on S_L^1

with periodic boundary conditions. The solutions to (8.58) are all multiples of the function

$$\xi(x) = \exp(i\lambda x). \tag{8.59}$$

The periodic boundary conditions demand $\xi(L) = \xi(0)$, so we must have $\exp(i\lambda L) = 1$. But then $\lambda L = 2n\pi$ for some integer n, so $\lambda = 2\pi n/L$, and $\xi = \mathbf{b}_n$. Since A is Hermitian, it is diagonalizable, so its eigenfunctions, namely the $\mathbf{b}_n$'s, form a basis for $L^2(S_L^1)$. ∎

It is also possible to obtain the expansion (8.50) in sines and cosines directly, without involving complex exponentials or the operator $-id/dx$. The functions 1, $\sin(2\pi nx/L)$, and $\cos(2\pi nx/L)$ are themselves eigenfunctions of the operator d^2/dx^2 on S_L^1, and the expansion (8.50) can be considered an eigenfunction expansion for d^2/dx^2. See Figure 8.6. This approach is explored further in Exercises 10–12.

Properties of Fourier coefficients

The coefficients c_n are called the *Fourier coefficients* of $\mathbf{f}$ and are denoted $\hat{f}_n$. In other words,

$$f(x) = \sum_n \hat{f}_n \mathbf{b}_n(x) = \sum_n \hat{f}_n \exp(2\pi inx/L). \tag{8.60}$$

Some of the many useful properties of the Fourier coefficients are summarized in

8.5. Periodic Boundary Conditions

Theorem 8.6 *Let $f(x)$ and $g(x)$ be periodic square-integrable functions with period L. Then*

$$\langle \mathbf{f} | \mathbf{g} \rangle = L \sum_{n=-\infty}^{\infty} \overline{\hat{f}_n} \hat{g}_n. \tag{8.61}$$

The Fourier coefficients of $\mathbf{f}$ are bounded by the L^2 norm of $\mathbf{f}$:

$$|\hat{f}_n| \leq \sqrt{\frac{1}{L} \int_0^L |f(x)|^2 dx}. \tag{8.62}$$

If $\mathbf{f}$ is differentiable, the Fourier coefficients of $\mathbf{f}'$ are

$$\hat{f'}_n = \frac{2\pi i n}{L} \hat{f}_n. \tag{8.63}$$

If $\mathbf{f}$ is k-times differentiable, then

$$|\hat{f}_n| \leq \frac{L^k}{(2\pi n)^k} \sqrt{\frac{1}{L} \int_0^L |f^{(k)}(x)|^2 dx}, \tag{8.64}$$

where $\mathbf{f}^{(k)}$ denotes the k-th derivative of $\mathbf{f}$. Conversely, rapid decay of Fourier coefficients implies smoothness: If the series $\sum_{n=-\infty}^{\infty} |\hat{f}_n|$ converges, then $\mathbf{f}$ is continuous, while if the series $\sum_{n=-\infty}^{\infty} |n^k \hat{f}_n|$ converges, then $\mathbf{f}$ is k-times differentiable.

Proof: Equations (8.61) and (8.63) are left as exercises. Taking (8.61) with $\mathbf{f} = \mathbf{g}$ we get

$$\frac{1}{L} \int_0^L |f(x)|^2 dx = \langle \mathbf{f} | \mathbf{f} \rangle / L = \sum_{i=-\infty}^{\infty} \hat{f}_i^2 \geq \hat{f}_n^2, \tag{8.65}$$

which is tantamount to (8.62). Applying (8.62) to $f^{(k)}$ and using (8.63) gives (8.64). What remains is to show that decay of Fourier coefficients implies smoothness.

Suppose that $\sum_{n=-\infty}^{\infty} |\hat{f}_n|$ converges. We construct a sequence of continuous functions $F_m(x)$ that converge to $f(x)$. Let

$$F_m(x) = \sum_{n=-m}^{m} \hat{f}_n \exp(2\pi i n x / L). \tag{8.66}$$

Being a finite sum of continuous functions, $\mathbf{F}_m$ is continuous. Also, for m large, $F_m(x)$ is close to $f(x)$ at every point, since

$$|f(x) - F_m(x)| = |(\sum_{n=-\infty}^{-m-1} + \sum_{n=m+1}^{\infty}) \hat{f}_n \exp(2\pi i n x / L)|$$

$$\leq \sum_{n=-\infty}^{-m-1} |\hat{f}_n| + \sum_{n=m+1}^{\infty} |\hat{f}_n|. \quad (8.67)$$

Since the sum $\sum_{n=-\infty}^{\infty} |\hat{f}_n|$ converges, the sum of the terms with $n > |m|$ can be made as small as we wish by picking m large enough. That is, given a number $\epsilon > 0$, we can find a number M such that, for all $m > M$ and all x, $|f(x) - F_m(x)| < \epsilon$. In other words, the sequence of functions F_m converges *uniformly* to f. Since the uniform limit of continuous functions is continuous, this implies that f is continuous.

Finally, if $\sum n^k |\hat{f}_n|$ converges, then $\sum |\hat{f}_n^{(k)}|$ converges, so $\mathbf{f}^{(k)}$ is continuous, and $\mathbf{f}$ is k-times differentiable. ∎

Equation (8.61) tells us that Fourier series makes $L^2(S_L^1)$ look like a space of infinite sequences, just like ℓ_2, except with the indices running from $-\infty$ to ∞ instead of from 1 to ∞. In this correspondence, the derivative operator is, up to an overall constant factor $2\pi i/L$, just multiplication by n. This tells us that the degree of differentiability of a function $\mathbf{f}$ is extremely closely related to the rate of decay of $\hat{f}_n$ with n. The remainder of Theorem 8.6 tells us how much decay is implied by a given degree of differentiability, and just how much decay is needed to imply that degree of differentiability.

We can now understand what sort of function has coefficients that go as n^{-k}. For such a function $\mathbf{f}$, the sum $\sum_n |n^{k-1} \hat{f}_n|$ fails to converge, but only barely, since $\sum_n |n^p||\hat{f}_n|$ converges for all real exponents $p < k - 1$. You might expect such functions to just barely fail to be $k - 1$ times differentiable. In fact, a function whose coefficients decay as n^{-k} is typically $k - 1$-times differentiable except at a finite number of points where the $k - 1$st derivative takes a jump. For example, a square wave, which is continuous everywhere but at a finite number of points, has coefficients that decay like n^{-1}. A triangle wave is differentiable except at the corners, where the first derivative takes a jump, and its Fourier coefficients go as n^{-2}.

Examples

With this in mind we revisit the examples of Section 6.9, working on the interval $[0, 1]$. In Section 6.9 we expanded these functions as a linear combination of $\sin(n\pi x)$. Here we expand them as linear combinations of $\exp(2\pi i n x)$.

8.5. Periodic Boundary Conditions

1. Suppose $f(x)$ is a square wave:
$$f(x) = \begin{cases} 1 & \text{if } x < 1/2; \\ -1 & \text{if } x \geq 1/2. \end{cases} \quad (8.68)$$
Then our coefficients are
$$\begin{aligned} \hat{f}_n &= \int_0^{1/2} \exp(-2\pi i n x) dx - \int_{1/2}^1 \exp(-2\pi i n x) dx \\ &= \begin{cases} 0 & \text{if } n \text{ is even}; \\ -2i/n\pi & \text{if } n \text{ is odd}. \end{cases} \end{aligned} \quad (8.69)$$
$f(x)$ is continuous except at two jumps ($x = 0$ and $x = 1/2$), the coefficients decay as n^{-1}, and $\sum |\hat{f}_n|$ barely fails to converge.

2. Next, suppose that $g(x)$ is a triangle wave:
$$g(x) = \begin{cases} x & \text{if } x < 1/2; \\ 1-x & \text{if } x \geq 1/2. \end{cases} \quad (8.70)$$
Notice that $f(x)$ of Example 1 is the derivative of $g(x)$, so, by (8.63), $\hat{f}_n = 2\pi i n \hat{g}_n$. This determines all values of $\hat{g}_n$ except $\hat{g}_0$, which equals $\int_0^1 g(x) dx = 1/4$. Thus
$$\hat{g}_n = \begin{cases} 0 & \text{if } n \text{ is even and nonzero}; \\ -1/n^2\pi^2 & \text{if } n \text{ is odd }; \\ 1/4 & \text{if } n = 0. \end{cases} \quad (8.71)$$
This can also be obtained by integrating $g(x)\exp(-2\pi i n x)$ directly. In this example $g(x)$ is continuous and is differentiable except at two points where g' takes a jump, and $\hat{g}_n$ decays as n^{-2}.

3. Suppose that $h(x) = x - x^2$. Our coefficients are then
$$\begin{aligned} \hat{h}_n &= \int_0^1 (x - x^2) \exp(2\pi i n x) dx \\ &= \begin{cases} 1/6 & \text{if } n = 0; \\ -1/2\pi^2 n^2 & \text{otherwise}. \end{cases} \end{aligned} \quad (8.72)$$
In this example the coefficients go as n^{-2}. This may be surprising, since $h(x)$ looks infinitely differentiable. However, when you extend $h(x)$ to a periodic function by declaring $h(x+n) = h(x)$, you develop a discontinuity in the first derivative at integer values of x. To put it another way, although $h(1) = h(0)$, $h'(1) = -1 \neq h'(0)$. This discontinuity in the first derivative naturally leads to $1/n^2$ behavior in $\hat{f}_n$.

Exercises

1. Derive the formula (8.61) for the inner product in terms of the Fourier coefficients.

2. Derive the formula (8.63) for the Fourier coefficients of the derivative of a function.

3. Show that, if $f(x)$ is real for every x, then $\hat{f}_n = \overline{\hat{f}_{-n}}$.

4. Show that, if $\mathbf{f}$ is an even function ($f(-x) = f(x)$), then $\hat{f}_{-n} = \hat{f}_n$, while if $\mathbf{f}$ is an odd function, then $\hat{f}_{-n} = -\hat{f}_n$.

5. Let $g(x) = f(-x)$. How are the Fourier coefficients $\hat{g}_n$ and $\hat{f}_n$ related?

6. In example 3, use the fact that $h(0) = \sum_{n=-\infty}^{\infty} \hat{h}_n$ to derive a formula for $\sum_{n=1}^{\infty} n^{-2}$.

7. Similarly, in example 2, plug $x = 0$ into the expansion for $g(x)$ to derive a formula for $\sum_{n=0}^{\infty} (2n+1)^{-2}$.

8. Find a function $f(x)$ whose Fourier coefficients, for $n \neq 0$, are proportional to $1/n^4$.

9. Use the results of Exercise 8 to evaluate the series $\sum_{n=1}^{\infty} n^{-4}$.

10. Show that the operator d^2/dx^2 on S_L^1 is Hermitian.

11. Find the eigenvalues of d^2/dx^2 on S_L^1.

12. For each nonzero eigenvalue λ of d^2/dx^2, find two different bases for E_λ, one involving real functions (sines and cosines), the other involving complex exponentials.

In Exercises 13–17, describe the rate of decay of the Fourier coefficients for each of these periodic functions defined on $[0, 1]$. You do not have to actually compute the coefficients.

13. $f(x) = x(x - 1/2)(x - 1)$.

14. $f(x) = x^2(x-1)^2$.

15. $f(x) = \sin(\pi x)$.

16. $f(x) = 3\sin(\pi x) - \sin(3\pi x)$.

17. $f(x) = 2x + \cos(\pi x)$.

For each of the periodic functions of Exercises 18–22, compute the Fourier coefficients.

18. $f(x) = \sin(\pi x)$.

19. $f(x) = \sin((2k+1)\pi x)$ for an arbitrary integer k.

20. $f(x) = 3\sin(\pi x) - \sin(3\pi x)$.

21. $f(x) = (1 + ce^{i\pi x})^{-1}$, where $|c| < 1$.

22. $f(x) = (1 + ce^{i\pi x})^{-1}$, where $|c| > 1$.

23. Compute the Fourier coefficients of $f(x) = \sum_{n=-\infty}^{\infty} e^{-(x-n)^2/2}$. You may use the fact that $\int_{-\infty}^{\infty} e^{-x^2/2} e^{ikx} dx = \sqrt{2\pi} e^{-k^2/2}$.

24. Using the results of Exercise 23, show that $\sum_{n=-\infty}^{\infty} e^{-n^2/2} = \sqrt{2\pi} \sum_{n=-\infty}^{\infty} e^{-2\pi^2 n^2}$. This is a special case of the *Poisson resummation formula*.

25. Prove the identity $\sum_{n=-\infty}^{\infty} e^{-n^2\pi/\sigma^2} = \sigma \sum_{n=-\infty}^{\infty} e^{-n^2\pi\sigma^2}$, where σ is an arbitrary positive constant.

8.6 Equivalence of Traveling Waves and Standing Waves

In this section we show that, for the vibrating string problem, the traveling wave solution (8.25) is equivalent to the standing wave solution (8.40). Indeed, we will show how to convert each one to the other form.

We start with the easier direction, converting standing waves into traveling waves. We extend the identity (8.45) to cover both sines and cosines of t, bearing in mind that $\omega_n = vk_n$:

$$\begin{aligned} 2\sin(k_n x)\sin(\omega_n t) &= \cos(k_n x - \omega_n t) - \cos(k_n x + \omega_n t) \\ &= \cos(k_n(x - vt)) - \cos(k_n(x + vt)) \\ 2\sin(k_n x)\cos(\omega_n t) &= \sin(k_n x - \omega_n t) + \sin(k_n x + \omega_n t) \\ &= \sin(k_n(x - vt)) + \sin(k_n(x + vt)). \end{aligned} \quad (8.73)$$

Substituting this into the standing wave solution (8.40) we have

$$f(x,t) = \frac{1}{2} \sum_{n=1}^{\infty} \Big[a_n \sin(k_n(x-vt)) + b_n \cos(k_n(x-vt)) \\ + a_n \sin(k_n(x+vt)) - b_n \cos(k_n(x+vt)) \Big]. \quad (8.74)$$

We define

$$\tilde{h}_1(x) = \frac{1}{2} \sum_{n=1}^{\infty} a_n \sin(k_n x) + b_n \cos(k_n x);$$

$$\tilde{h}_2(x) = \frac{1}{2} \sum_{n=1}^{\infty} a_n \sin(k_n x) - b_n \cos(k_n x) = -\tilde{h}_1(-x), \quad (8.75)$$

allowing us to rewrite

$$f(x,t) = \tilde{h}_1(x - vt) + \tilde{h}_2(x + vt). \tag{8.76}$$

Defining h_1 and h_2 as in (8.31) and substituting into (8.76), we then obtain the traveling wave solution (8.25).

Notice the similarity between (8.75) and the general periodic Fourier decomposition (8.50). The coefficients a_n and b_n of the standing wave solution (8.40) are essentially the coefficients of the Fourier decomposition of $\tilde{h}_1$. The only difference, aside from an overall factor of $1/2$, is that the roles of a_n and b_n have been reversed. a_n and b_n are coefficients of the sine and cosine of $k_n x$ in (8.75), in that order, while they multiply the cosine and sine of $\omega_n t$ in (8.40).

This then tells us how to go from traveling waves to standing waves. We must find an appropriate periodic function and decompose it into Fourier modes. The individual terms of the expansion will be individual standing waves. Given a traveling wave solution (8.25), we define the function

$$\tilde{h}(x) = \sum_{n=-\infty}^{\infty} h_1(x + 2nL) - h_2(-x + 2nL), \tag{8.77}$$

and the auxiliary functions

$$\tilde{h}_1(x) = \tilde{h}(x); \qquad \tilde{h}_2(x) = -\tilde{h}(-x), \tag{8.78}$$

in terms of which our solution is simply

$$f(x,t) = \tilde{h}_1(x - vt) + \tilde{h}_2(x + vt). \tag{8.79}$$

The function $\tilde{h}$ is manifestly periodic, with period $2L$, so it admits a decomposition

$$\tilde{h}(x) = \tilde{a}_0/2 + \sum_{n=1}^{\infty} \tilde{a}_n \cos(n\pi x/L) + \tilde{b}_n \sin(n\pi x/L), \tag{8.80}$$

with

$$\tilde{a}_n = \frac{1}{L} \int_0^{2L} \tilde{h}(x) \cos(n\pi x/L) dx; \quad \tilde{b}_n = \frac{1}{L} \int_0^{2L} \tilde{h}(x) \sin(n\pi x/L) dx. \tag{8.81}$$

Notice that (8.80) and (8.81) differ from (8.50) and (8.51) in that the period is $2L$, not L. The n-th wave number is therefore $k_n = n\pi/L$, as in (8.43).

8.6. Equivalence of Traveling Waves and Standing Waves

By the addition of angles formulas

$$\sin(k(x \pm vt)) = \sin(kx)\cos(kvt) \pm \cos(kx)\sin(kvt);$$
$$\cos(k(x \pm vt)) = \cos(kx)\cos(kvt) \mp \sin(kx)\sin(kvt), \quad (8.82)$$

we have $\tilde{h}_1(x - vt)$ and $\tilde{h}(x - vt)$ equaling

$$\frac{\tilde{a}_0}{2} + \sum_{n=1}^{\infty} \tilde{a}_n \cos(k_n x)\cos(\omega_n t) + \tilde{a}_n \sin(k_n x)\sin(k_n t)$$
$$+ \tilde{b}_n \sin(k_n x)\cos(\omega_n t) - \tilde{b}_n \cos(k_n x)\sin(k_n t); \quad (8.83)$$

while $\tilde{h}_2(x + vt)$ and $-\tilde{h}(-x - vt)$ equal

$$-\frac{\tilde{a}_0}{2} - \sum_{n=1}^{\infty} \tilde{a}_n \cos(k_n x)\cos(\omega_n t) - \tilde{a}_n \sin(k_n x)\sin(k_n t)$$
$$- \tilde{b}_n \sin(k_n x)\cos(\omega_n t) - \tilde{b}_n \cos(k_n x)\sin(k_n t), \quad (8.84)$$

where $\omega_n = vk_n = n\pi v/L$. Adding these together we get our standing wave solution

$$f(x,t) = 2\sum_{n=1}^{\infty} \tilde{a}_n \sin(k_n x)\sin(k_n t) + \tilde{b}_n \sin(k_n x)\cos(\omega_n t). \quad (8.85)$$

This is precisely of the form (8.40), with $a_n = 2\tilde{b}_n$ and $b_n = 2\tilde{a}_n$.

You may wonder why we developed both the theory of traveling waves and the theory of standing waves, given that they are equivalent. Wouldn't just one of them do? The answer is that for some applications one approach is much better than the other, while for other applications it is the other way around. See Figure 8.7.

When the wavefunction is spread all over the string and is fairly smooth, as in most musical applications, then the Fourier series (8.80) converges extremely rapidly. In such cases, it is best to think of our solution as a sum of standing waves.

There are other applications where the wavefunction is zero except in a small region, and is not particularly smooth. Imagine clapping your hands once in a canyon. It would take very many terms to adequately describe the resulting wave as a sum of standing waves, but it is easily understood as a traveling wave: there is a pulse from your hand clap, then an echo off the canyon wall, then a second echo, and so on.

In signal processing, working with traveling waves is also called working in the *time domain*, while decomposing into standing waves

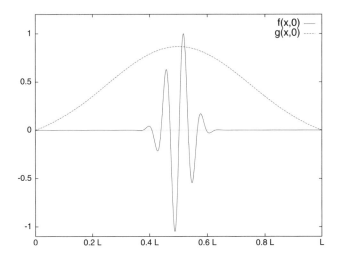

Figure 8.7: $f(x,t)$ is best resolved into traveling waves, $g(x,t)$ into standing waves

is called working in the *frequency domain*. To properly understand waves, you have to be able to work in both domains, and to go back and forth. As we have seen, Fourier series is the essential tool for this conversion.

Exercises

In Exercises 1–5, consider the wave equation on the interval $[0, 1]$ with velocity $v = 1$. For each of the following initial conditions, find the traveling wave solution and the standing wave solution, and show that they are equal.

1. $f_0(x) = \begin{cases} 0 & \text{if } x < 1/4; \\ x - 1/4 & \text{if } 1/4 \leq x < 1/2; \\ 3/4 - x & \text{if } 1/2 \leq x < 3/4; \\ 0 & \text{if } x \geq 3/4, \end{cases}$ $g_0(x) = 0.$

2. $f_0(x) = \begin{cases} 0 & \text{if } x < 1/4; \\ x - 1/4 & \text{if } 1/4 \leq x < 1/2; \\ 3/4 - x & \text{if } 1/2 \leq x < 3/4; \\ 0 & \text{if } x \geq 3/4, \end{cases}$

$g_0(x) = \begin{cases} 0 & \text{if } x < 1/4; \\ -1 & \text{if } 1/4 \leq x < 1/2; \\ 1 & \text{if } 1/2 \leq x < 3/4; \\ 0 & \text{if } x \geq 3/4. \end{cases}$

3. $f_0(x) = \sin(\pi x)$, $g_0(x) = \pi \cos(\pi x)$.

4. $f_0(x) = 0$, $g_0(x) = \sin(\pi x)$.

5. $f_0(x) = 0$, $g_0(x) = \begin{cases} x & \text{if } x < 1/2; \\ 1-x & \text{if } x \geq 1/2. \end{cases}$

8.7   The Different Types of Fourier Series

We have seen three types of Fourier series for functions on the interval $[0, L]$. The first, developed in Section 6.9, decomposes a function on the interval $[0, L]$, technically an element of $L^2[0, L]$, as a sum of sine waves with wave numbers $n\pi/L$. The second, developed in Section 8.5, decomposes a periodic function with period L (an element of $L^2(S_L^1)$) as a sum of sines and cosines with wave numbers $2\pi n/L$. In this second form of Fourier series, the wave numbers are twice as large, but there are two modes per wave number instead of one. Finally, there is the expansion of periodic functions (elements of $L^2(S_L^1)$), as a sum of complex exponentials $\exp(\pm 2\pi i n/L)$.

The two spaces $L^2[0, L]$ and $L^2(S_L^1)$ are actually the same, with the same inner product. It is clear that every element of $L^2(S_L^1)$, that is every periodic square-integrable function, is a square-integrable function and so an element of $L^2([0, L])$. Conversely, any function on $[0, L]$ can be modified at a single point, changing $f(L)$ to make $f(L) = f(0)$, and then extended into a periodic function by the rule $f(x + nL) = f(x)$. Changing the function at one point does not change the element of $L^2[0, 1]$, since we identify functions that differ on a set of measure zero (and a single point is certainly of measure zero). This shows that $L^2[0, L]$ and $L^2(S_L^1)$ are subsets of each other, and are therefore the same.

What we have done is exhibit three orthogonal bases for the same space. One orthogonal basis is the set of functions $\sin(n\pi x/L)$. A second orthogonal basis is the set of functions $\sin(2n\pi x/L)$ together with the functions $\cos(2n\pi x/L)$. Closely related to the second basis is the third basis $\{\exp(2\pi i n x/L)\}$, with n now ranging from $-\infty$ to ∞. These bases are in turn obtained as the eigenfunctions of three different operators on $L^2[0, L]$.

The first operator is d^2/dx^2 with Dirichlet boundary conditions, whose eigenvalues are $-n^2\pi^2/L^2$, and whose eigenfunctions are the functions $\sin(n\pi x/L)$. This basis is naturally useful for problems involving second derivatives with Dirichlet boundary conditions, such as the vibrating string. See Figure 8.8.

Chapter 8. The Wave Equation

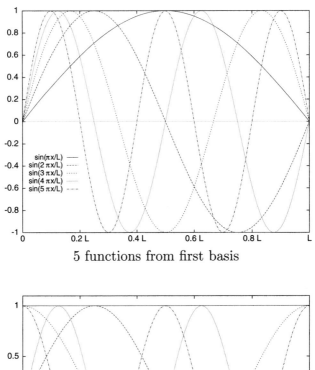

5 functions from first basis

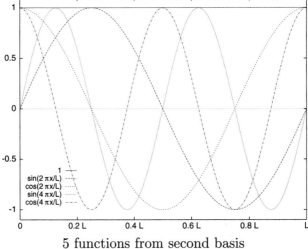

5 functions from second basis

Figure 8.8: Alternate bases for $L^2[0, L]$

8.7. The Different Types of Fourier Series

The second operator is d^2/dx^2 with periodic boundary conditions. The eigenvalues are $\lambda_n = -4n^2\pi^2/L^2$, and each eigenspace E_{λ_n} is 2-dimensional, with basis $\{\sin(2n\pi x/L), \cos(2n\pi x/L)\}$. Naturally, this basis is particularly useful for problems involving second derivatives and periodic functions.

The third operator is $-id/dx$ with periodic boundary conditions. The eigenvalues are $2n\pi/L$, and the eigenfunctions are $\exp(2\pi inx/L)$, with n ranging from $-\infty$ to ∞. This basis is particularly useful for problems involving first derivatives, as well as higher derivatives. Indeed, we saw that the Fourier coefficients of $\mathbf{f}'$ are closely related to those of $\mathbf{f}$, and this allowed us to directly relate the smoothness of $\mathbf{f}$ to the speed at which the Fourier series converges. Note that the second operator is minus the square of the third, so the eigenfunctions of the third operator are also eigenfunctions of the second.

One can easily come up with additional Hermitian operators (such as the operator d^2/dx^2 with Neumann boundary conditions, whose eigenfunctions are $\cos(2n\pi x/L)$). For any such operator, the eigenfunctions will form an orthogonal basis for L^2, and we can decompose an arbitrary function in the basis, using the inner product to find the coefficients. Although such "generalized Fourier series" expansions can sometimes be useful, we will not pursue them here.

Decay rates of Fourier coefficients

In some applications, such as the vibrating string, the conditions of the problem tell you whether to apply periodic, Dirichlet, or other boundary conditions, which in turn tells you which type of Fourier series to use. But what if you are given a function and just want to find a convenient series representation, preferably one whose coefficients decay rapidly. Which type of Fourier series should you use?

If the function is discontinuous in the middle, or generally very rough, then it doesn't matter very much which series you use. In such cases, neither series will converge very quickly. The behavior in the middle is the key, and the boundary conditions are not very important.

If the function is smooth in the middle, however, then it can matter quite a bit which series you use. The following rules explain how quickly each type of series will converge. Use whichever type of series gives faster convergence. If the rates of convergence are the same, then take your pick.

- For the first kind of Fourier series (in terms of $\sin(n\pi x/L)$), you want your function to be zero at the endpoints. Ideally, the even-order derivatives of your function should also be zero at the endpoints. (The odd-order derivatives do not have to be zero, as long as they exist.) If the function, the 2nd derivative, the 4th derivative, and so on through the $2k$-th derivative are all zero at both $x = 0$ and $x = L$, but the $2k + 2$nd derivative is nonzero at an endpoint, then the Fourier coefficents go as $n^{-(2k+3)}$.

- For the second or third kind of Fourier series (either with complex exponentials or sines and cosines of $2n\pi x/L$), you do not need values and derivatives to be zero at the endpoints. However, you do need values and derivatives, both odd-order and even-order, to agree at the two endpoints. If $f(0) = f(L)$, $f'(0) = f'(L)$, and so on through the k-th derivative, but the $k+1$st derivatives at the two endpoints do not agree, then the Fourier coefficients go as $n^{-(k+2)}$.

Example 1: We have already expanded the function $f(x) = x - x^2$ on $[0, 1]$ in the first and third kind of Fourier series (Example 3 of Sections 6.9 and 8.5, respectively). Since the function is zero at the endpoints but its second derivative is not, the first kind of Fourier series has coefficients that go as n^{-3}. As for the second or third kind of Fourier series, we have $f(0) = f(1) = 0$, but $1 = f'(0) \neq f'(1) = -1$. Therefore the second kind of Fourier series has coefficients that go as n^{-2}. In this example, the first kind of Fourier series is clearly better.

Example 2: The function $f(x) = 2x^3 - 3x^2 + x$ on the interval $[0, 1]$ is smooth, has $f(0) = f(1) = 0$, has $f'(0) = f'(1) = 1$, has $f''(0) = -6$ and $f''(1) = 6$. By the first rule, since $f''(0) \neq 0$ and $f''(1) \neq 0$, the first kind of Fourier series has coefficients that go as n^{-3}. By the second rule, since $f''(0) \neq f''(1)$, the second (or third) kind of Fourier series also has coefficients that go a n^{-3}. Neither choice is particularly better than the other.

Example 3: Consider the function $f(x) = 2x^3 - 3x^2 + x + 1$. Since $f(0) \neq 0$, the first kind of Fourier series has coefficients that go as n^{-1}. However, since $f(0) = f(1)$ and $f'(0) = f'(1)$, the second kind of Fourier series has coefficients that go as n^{-3}.

To see where these rules come from, take a smooth function on $[0, L]$ and extend it to a periodic function on $\mathbb{R}$. This function will

8.7. The Different Types of Fourier Series

be smooth except at one or two points per period, and the rate of convergence of the Fourier series depends on the behavior at those one or two points.

For the first kind of Fourier series, we construct the periodic function

$$\tilde{f}(x) = \sum_{n=-\infty}^{\infty} f(x+2nL) - f(-x+2nL). \tag{8.86}$$

Note that, for $0 < x < L$, $f(x) = \tilde{f}(x)$. The function $\tilde{f}$ is odd, $\tilde{f}(-x) = -\tilde{f}(x)$, so when we decompose it into a Fourier series (of the second type, with periodic boundary conditions of period $2L$) we only get sine terms; the integral that computes a_n is identically zero. As a result we can write, for $0 < x < L$,

$$f(x) = \tilde{f}(x) = \sum_n b_n \sin(n\pi x/L). \tag{8.87}$$

In other words, the Fourier coefficients of **f** using the first kind of Fourier series are the coefficients of $\tilde{\mathbf{f}}$ using the second kind. But we already know how the Fourier coefficients of a periodic function depend on the smoothness of that function (Theorem 8.6). If the function is smooth except at a finite number of points and is r-times differentiable at those points but not $r+1$ times differentiable, then the coefficients go as $n^{-(r+2)}$.

All that remains, then, is to figure out how many times $\tilde{f}$ can be differentiated. $\tilde{f}$ is smooth except possibly at $x = nL$, so we examine the derivatives at $x = 0$ and $x = L$. Since $\tilde{f}(-x) = -\tilde{f}(x)$, the rth derivatives satisfy $\frac{d^r\tilde{f}}{dx^r}(-x) = (-1)^{r+1}\frac{d^r\tilde{f}}{dx^r}(x)$. Taking the limit as $x \to 0$, we find that

$$\lim_{x \to 0+} \frac{d^r \tilde{f}(x)}{dx^r} = (-1)^{r+1} \lim_{x \to 0-} \frac{d^r \tilde{f}(x)}{dx^r}. \tag{8.88}$$

Thus, for r even, the r-th derivative of $\tilde{\mathbf{f}}$ is continuous at zero, only if the r-th derivative of f at zero vanishes. On the other hand, for r odd, the limits as $x \to 0^{\pm}$ of $d^r\tilde{f}(x)/dx^r$ always agree. A similar analysis applies at $x = L$. Therefore, if the function **f**, the 2nd derivative, the 4th derivative, and so on through the $2k$-th derivative are all zero at both $x = 0$ and $x = L$, but the $2k + 2$nd derivative is nonzero at an endpoint, then the first $2k + 1$ derivatives of $\tilde{\mathbf{f}}$ exist at $x = 0$ and $x = L$, but the $2k + 2$nd derivative does not, so the Fourier coefficients go as $n^{-(2k+3)}$.

To prove the second rule, we simply extend f to a periodic function $\tilde{f}$ by the rule $\tilde{f}(x+nL) = f(x)$. The function $\tilde{f}$ is smooth except at $x = nL$. If $f(0) = f(L)$, then $\tilde{f}$ is continuous, if $f'(0) = f'(L)$ then $\tilde{f}$ is once differentiable, and in general if the first k derivatives of $\mathbf{f}$ at 0 and L agree, then $\tilde{f}$ is k times differentiable. If the $k+1$st derivatives do not agree, then the Fourier coefficients must go as $n^{-(k+2)}$.

Exercises

For each of the following functions on the interval $[0,1]$, give the qualitative behavior of the coefficients of the different types of Fourier series.

1. $f(x) = x(x - 1/2)(x - 1)$.
2. $f(x) = x^2(x - 1)^2$.
3. $f(x) = \sin(\pi x)$.
4. $f(x) = 3\sin(\pi x) - \sin(3\pi x)$.
5. $f(x) = 2x + \cos(\pi x)$.
6. $f(x) = (1 + ce^{i\pi x})^{-1}$, where $|c| < 1$.
7. $f(x) = (1 + ce^{i\pi x})^{-1}$, where $|c| > 1$.

Chapter 9

Continuous Spectra and the Dirac Delta Function

Up until now, we have been decomposing vectors in bases of eigenvectors of linear operators. If $\mathbf{v}$ is a vector and A is an operator, we write

$$\mathbf{v} = \sum a_n \mathbf{b}_n, \qquad (9.1)$$

where

$$A\mathbf{b}_n = \lambda_n \mathbf{b}_n. \qquad (9.2)$$

If our vector space has an inner product, and if A is Hermitian (or unitary), then the eigenvectors can be chosen orthogonal, in which case the coefficients a_n can be computed by

$$a_n = \frac{\langle \mathbf{b}_n | \mathbf{v} \rangle}{\langle \mathbf{b}_n | \mathbf{b}_n \rangle}. \qquad (9.3)$$

More formally, we can write the identity operator, or A itself, as a sum of products of bras and kets:

$$I = \sum_n \frac{|\mathbf{b}_n\rangle\langle\mathbf{b}_n|}{\langle \mathbf{b}_n | \mathbf{b}_n \rangle}; \qquad A = \sum_n \lambda_n \frac{|\mathbf{b}_n\rangle\langle\mathbf{b}_n|}{\langle \mathbf{b}_n | \mathbf{b}_n \rangle}. \qquad (9.4)$$

If, furthermore, the eigenvectors are chosen orthonormal, then the denominators in (9.4) all equal one, and we have the simpler expansions

$$I = \sum_n |\mathbf{b}_n\rangle\langle\mathbf{b}_n|; \qquad A = \sum_n \lambda_n |\mathbf{b}_n\rangle\langle\mathbf{b}_n|. \qquad (9.5)$$

Most of the book, up to this point, has been an exploration of the power of these expansions, both for diagonalizable operators on finite dimensional spaces (where n ranges from 1 to the dimension of the space), and for many operators on infinite dimensional function spaces (where n ranges from 1 to ∞, or from $-\infty$ to ∞).

I must now confess that there are important examples of Hermitian and unitary operators for which the expansions (9.1–9.5) simply do not work as written. In these examples one must modify the expansions, replacing sums with integrals or with generalized integrals. Such generalized expansions can be extremely useful, and Chapters 10 and 11 are devoted to these applications.

In this chapter we learn how to handle such expansions. In Section 9.1 we define the spectrum of a linear operator, see that the spectrum can sometimes be continuous, and learn about generalized eigenfunctions. Normalizing these generalized eigenfunctions involves the Dirac delta function, which we explore in Section 9.2. These first two sections are heuristic, with little attempt at mathematical rigor.

Generalized eigenfunctions are typically not elements of the original vector space of functions, but live in a larger space, a space of *distributions*. The delta function itself is an example of a distribution. The theory of distributions, briefly presented in Section 9.3, provides the rigorous justification for the formal manipulations we do with the generalized eigenfunctions and the Dirac delta function. In Section 9.4 we consider generalized eigenfunction expansions for Hermitian operators and precisely state the spectral theorem, which guarantees their existence.

9.1 The Spectrum of a Linear Operator

Before beginning our analysis, we consider two important operators. The first is the "position" operator X on $L^2(\mathbb{R})$, which multiplies any function of x by x. That is,

$$(Xf)(x) = xf(x). \tag{9.6}$$

It is easy to see that X is Hermitian (see Example 2 of Section 7.1 and Example 3 of Section 7.2), but X does not have any eigenfunctions in $L^2(\mathbb{R})$. If $X\mathbf{f} = \lambda \mathbf{f}$, then $(x - \lambda)f(x)$ is identically zero, so $f(x) = 0$ everywhere but at $x = \lambda$. A function that is nonzero only at a single point is equivalent (in $L^2(\mathbb{R})$) to the zero function, so $\mathbf{f}$ cannot be a true eigenfunction. Thus, if we are to generalize the notion of eigenfunctions for operators such as X, we must allow "eigenfunctions" that are not elements of $L^2(\mathbb{R})$. We will soon see

9.1. The Spectrum of a Linear Operator

that X does indeed have a generalized eigenfunction with eigenvalue λ, namely the delta function $\delta(x - \lambda)$.

Our second operator is the "momentum" operator $K = -id/dx$, either on $L^2(S_L^1)$, or on $L^2(\mathbb{R})$. We have already seen that K is Hermitian on either space, and that an eigenfunction of eigenvalue k must be a multiple of $\exp(ikx)$. On $L^2(S_L^1)$, $\exp(ikx)$ is indeed an eigenfunction, as long as k is a multiple of $2\pi/L$. As we have seen, Fourier series (for periodic functions) is precisely the expansion (9.1) of an arbitrary element of $L^2(S_L^1)$ in eigenfunctions of K. On $L^2(\mathbb{R})$, however, K has no true eigenfunctions, since $\boldsymbol{\xi} = \exp(i\lambda x)$ is not square integrable,

$$\langle \boldsymbol{\xi} | \boldsymbol{\xi} \rangle = \int_{-\infty}^{\infty} |e^{ikx}|^2 dx = \int_{-\infty}^{\infty} 1 dx = \infty, \tag{9.7}$$

and so is not an element of $L^2(\mathbb{R})$. An arbitrary element of $L^2(\mathbb{R})$ cannot be written as a sum of (true) eigenfunctions of K, since K does not have any (true) eigenfunctions. $\exp(ikx)$ is a *generalized* eigenfunction, and the Fourier transform (Chapter 10) is the expansion of an arbitrary element of $L^2(\mathbb{R})$ as an *integral* (not a sum) of generalized eigenfunctions of K.

We have been speaking informally of generalized eigenvalues and generalized eigenfunctions. It is time to make those notions more precise.

Definition *The* spectrum *of an operator A, denoted $\sigma(A)$, is the set of all numbers λ for which $A - \lambda I$ is not continuously invertible. An element of $\sigma(A)$ is called a* generalized eigenvalue *of A. A solution $\boldsymbol{\xi}$ to the eigenvalue equation $A\boldsymbol{\xi} = \lambda\boldsymbol{\xi}$, where λ is a generalized eigenvalue of A, is called a* generalized eigenfunction *of A.*

If A is an $n \times n$ matrix, or more generally an operator on a finite dimensional vector space, then $\sigma(A)$ is precisely the set of eigenvalues of A. λ being in $\sigma(A)$ is equivalent to $A - \lambda I$ not being invertible, which is equivalent to $\det(A - \lambda I) = 0$, which is equivalent to λ being an eigenvalue. However, for operators on infinite dimensional spaces, it is possible to be in the spectrum without being a (true) eigenvalue.

Example 1: On $L^2(\mathbb{R})$, 0 is in the spectrum of X, although 0 is not an eigenvalue. To see that 0 is in the spectrum, let $f(x) = \exp(-x^2)$. **f** is in $L^2(\mathbb{R})$, but if $X^{-1}\mathbf{f}$ existed, it would have to be the function $\exp(-x^2)/x$. However, $\exp(-x^2)/x$ is not square integrable due to

266 Chapter 9. Continuous Spectra and the Dirac Delta Function

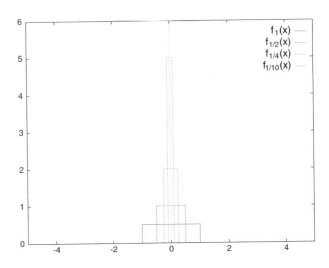

Figure 9.1: The functions $\mathbf{f}_{1/n}$

the singularity at $x = 0$. Since $X^{-1}\mathbf{f}$ is not in $L^2(\mathbb{R})$, X^{-1} is not a well-defined operator from $L^2(\mathbb{R})$ to itself, so $X = X - 0I$ is not invertible and $0 \in \sigma(X)$.

Example 2: On $L^2(\mathbb{R})$, 0 is in the spectrum of K. To see this, suppose K^{-1} exists, and take $f(x) = \exp(-x^2)$, as before. If $\mathbf{g} = K^{-1}\mathbf{f}$ exists, then

$$\lim_{a \to \infty}(g(a) - g(-a)) = \int_{-\infty}^{\infty} g'(x)dx = i\int_{-\infty}^{\infty} f(x)dx = i\sqrt{\pi}. \quad (9.8)$$

Since $g(x)$ cannot approach zero at both ∞ and $-\infty$, $\mathbf{g}$ cannot be square integrable, so K^{-1} is not a well-defined operator on all of $L^2(\mathbb{R})$.

The following is a useful criterion for determining when a number is in the spectrum of an operator:

Theorem 9.1 *If there exists a sequence of nonzero vectors $\mathbf{f}_n$, such that $\lim_{n \to \infty} \langle (A - \lambda I)\mathbf{f}_n | (A - \lambda I)\mathbf{f}_n \rangle / \langle \mathbf{f}_n | \mathbf{f}_n \rangle = 0$, then λ is in the spectrum of the operator A.*

Proof: If the sequence exists and $(A - \lambda I)^{-1}$ exists and is continuous, then there exists a sequence of vectors $\mathbf{g}_n = (A - \lambda I)\mathbf{f}_n / \sqrt{\langle \mathbf{f}_n | \mathbf{f}_n \rangle}$, converging to zero, such that $(A - \lambda I)^{-1}\mathbf{g}_n$ is a unit vector for every n. Taking the limit as $n \to \infty$ we get that $0 = (A - \lambda I)^{-1}0$ is a unit vector, which is a contradiction. Thus the existence of the sequence implies that λ is in the spectrum. ∎

9.1. The Spectrum of a Linear Operator

The converse requires an additional assumption, that A be Hermitian.

Theorem 9.2 *If A is a Hermitian operator on an inner product space V, and if $\lambda \in \sigma(A)$, then there exists a sequence of vectors $\mathbf{f}_n$, such that the limit $\lim_{n\to\infty} \langle (A-\lambda I)\mathbf{f}_n | (A-\lambda I)\mathbf{f}_n \rangle / \langle \mathbf{f}_n | \mathbf{f}_n \rangle$ equals zero.*

Proof: Exercises 6-10. ∎

We revisit Example 1 using this criterion. Let

$$\delta_\epsilon(x) = \begin{cases} \frac{1}{2\epsilon} & \text{if } -\epsilon \leq x \leq \epsilon; \\ 0 & \text{otherwise,} \end{cases} \quad (9.9)$$

and let $\mathbf{f}_n = \delta_{1/n}$. The functions $\mathbf{f}_n$ are tall and skinny, with total integral equal to one. See Figure 9.1. An easy calculation shows that

$$\frac{\langle X\mathbf{f}_n | X\mathbf{f}_n \rangle}{\langle \mathbf{f}_n | \mathbf{f}_n \rangle} = \frac{1}{12n^2}, \quad (9.10)$$

which goes to zero as $n \to \infty$. Thus 0 is in the spectrum of X.

Exercises

1. Let A be an operator on a (possibly infinite dimensional) inner product space V. Show that every eigenvalue of A lies in $\sigma(A)$. [In other words, the set of eigenvalues of A, sometimes called the *point spectrum*, is a subset of $\sigma(A)$.]

2. Find a sequence $\mathbf{f}_n$ of functions for which $\langle K\mathbf{f}_n | K\mathbf{f}_n \rangle / \langle \mathbf{f}_n | \mathbf{f}_n \rangle$ approaches zero as $n \to \infty$.

3. Let λ be a complex number that is not real. Without using Theorem 9.2, show that $X - \lambda I$ is invertible.

4. Show that $\sigma(X) = \mathbb{R}$. You may use Theorem 9.2.

5. Show that $\sigma(K) = \mathbb{R}$. You may use Theorem 9.2.

For Exercises 6–9, assume A is a continuous operator on an inner product space V. These problems explore Theorem 9.2, and show that the nonexistence of the sequence $\mathbf{f}_n$ implies the existence of a left-inverse for $(A - \lambda I)$. For Hermitian operators, this then implies the existence of a right-inverse; hence a 2-sided inverse, implying that λ is not in the spectrum. For non-Hermitian operators this is not the case, and Exercises 11–13 give a counterexample.

6. Show that there exists a real constant M such that, for all $\mathbf{f} \in V$, $\langle A\mathbf{f} | A\mathbf{f} \rangle \leq M^2 \langle \mathbf{f} | \mathbf{f} \rangle$.

7. Suppose that there is a continuous operator B such that AB is the identity. Using the result of Exercise 6, show that there exists a positive constant m such that, for all $\mathbf{f} \in V$, $\langle A\mathbf{f}|A\mathbf{f}\rangle \geq m^2\langle\mathbf{f}|\mathbf{f}\rangle$.

8. Suppose that there exists a positive constant m such that, for all $\mathbf{f} \in V$, $\langle A\mathbf{f}|A\mathbf{f}\rangle \geq m^2\langle\mathbf{f}|\mathbf{f}\rangle$. Show that there exists a continuous operator B such that BA is the identity.

9. Suppose A is Hermitian and there exists an operator B such that BA is the identity. Show that AB is also equal to the identity.

10. Prove Theorem 9.2.

11. Let $\{\mathbf{b}_1, \mathbf{b}_2, \ldots\}$ be an orthonormal basis for an inner product space V, and suppose that $A\mathbf{b}_n = \mathbf{b}_{n+1}$ for all n. Compute $A^\dagger \mathbf{b}_n$.

12. With A as in Exercise 9, show that $A^\dagger A$ equals the identity, but that $AA^\dagger$ does not.

13. With A as in Exercise 9, show that $0 \in \sigma(A)$ even though $\langle A\mathbf{f}|A\mathbf{f}\rangle = \langle\mathbf{f}|\mathbf{f}\rangle$ for all $\mathbf{f} \in V$.

9.2 The Dirac δ Function

The sequence of functions $\mathbf{f}_n = \delta_{1/n}$ of Section 9.1 does not actually converge to an element of $L^2(\mathbb{R})$. If it did converge, then that limiting function would be a true eigenvector of the X operator. But we already know that there are no true eigenvectors, so convergence in $L^2(\mathbb{R})$ is out of the question.

In this section, we take the limit anyway, and call it the *Dirac delta function*, denoted $\delta(x)$. In other words,

$$\delta(x) = \lim_{\epsilon \to 0} \delta_\epsilon(x). \tag{9.11}$$

Although the delta function is not an element of $L^2(\mathbb{R})$, it *is* an element of a well-defined space, the space of *distributions*, to be discussed in Section 9.3. While it is not a true eigenfunction of X, it is a generalized eigenfunction.

In what sense can we take the limit (9.11)? It does not make sense pointwise; the limit is zero everywhere but at $x = 0$, and does not exist at $x = 0$. We cannot take the limit in $L^2(\mathbb{R})$, since $\langle \delta_\epsilon | \delta_\epsilon \rangle = \int \delta_\epsilon^2(x)dx = 1/(2\epsilon)$, which blows up. However, we *can* take the limit under an integral sign.

9.2. The Dirac δ Function

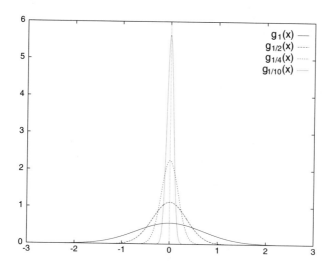

Figure 9.2: Smooth approximations to $\delta(x)$

Let $f(x)$ be any smooth function, and consider the integral

$$\int_{-\infty}^{\infty} dx\, \delta_\epsilon(x) f(x) = \frac{1}{2\epsilon} \int_{-\epsilon}^{\epsilon} f(x) dx$$
$$= \frac{1}{2\epsilon} \int_{-\epsilon}^{\epsilon} f(0) + f'(0)x + f''(0)x^2/2 + \cdots dx$$
$$= f(0) + f''(0)\epsilon^2/4 + f''''(0)\epsilon^4/48 + \cdots \quad (9.12)$$

Taking the limit as $\epsilon \to 0$ we get

$$\langle \delta | f \rangle = \int \delta(x) f(x) dx = f(0). \quad (9.13)$$

In the theory of distributions, (9.13) is the defining property of the δ function.

Although the functions δ_ϵ are convenient for some purposes, it is sometimes useful to think of $\delta(x)$ as a limit of smooth functions. Let $g(x)$ be any smooth function that decays sufficiently rapidly at $\pm\infty$, such that $\int_{-\infty}^{\infty} g(x) dx = 1$. For example, one might take $g(x) = (1/\sqrt{\pi}) \exp(-x^2)$. Let $g_\epsilon(x) = (1/\epsilon) g(x/\epsilon)$. See Figure 9.2. g_ϵ is a sharply peaked function, with width of order ϵ, height of order $1/\epsilon$, and total integral 1. Then, for any smooth function **f**, I claim that

$$\lim_{\epsilon \to 0} \int g_\epsilon(x) f(x) dx = f(0). \quad (9.14)$$

(The proof of (9.14) is left to the exercises.) Thus the δ function may be viewed as the limit $\lim_{\epsilon \to 0} g_\epsilon$, not just as the limit of δ_ϵ.

270   Chapter 9. Continuous Spectra and the Dirac Delta Function

The "function" $\delta(x)$ is peaked at the origin. To get a function that is peaked at $x = a$, we simply take $\delta(x - a)$. To see that this works, we let $y = x - a$ and compute

$$\int_{-\infty}^{\infty} \delta(x - a) f(x) dx = \int_{-\infty}^{\infty} \delta(y) f(y + a) dy = f(a). \tag{9.15}$$

In terms of bras and kets, we often denote the function $\delta(x - a)$ by $|X, a\rangle$ since, as we shall see, $|X, a\rangle$ is a generalized eigenfunction of X with generalized eigenvalue a. Thus (9.15) may be rewritten in bracket notation as

$$\langle X, a | f \rangle = f(a). \tag{9.16}$$

Theorem 9.3 *The Dirac δ function has the following properties:*

- $\delta(-x) = \delta(x)$.

- *If c is a nonzero constant, then $\delta(cx) = \delta(x)/|c|$.*

- *If $\rho(x)$ is an increasing or decreasing function of x such that $\rho(a) = 0$, then*

$$\delta(\rho(x)) = \delta(x - a)/|\rho'(a)|. \tag{9.17}$$

- *If $\rho(x)$ is a function of x with zeroes at $a_1, \ldots, a_n$, then*

$$\delta(\rho(x)) = \sum_{i=1}^{n} \frac{\delta(x - a_i)}{|\rho'(a_i)|}. \tag{9.18}$$

- *It is permissible to integrate by parts:*

$$\int \delta'(x) f(x) dx = -\int \delta(x) f'(x) dx = -f'(0). \tag{9.19}$$

Proof: Since the δ function is defined as a limit of the g_ϵ functions, or of the δ_ϵ functions, inside an integral, we must prove each result inside an integral. To see that the δ function is even, we let $y = -x$, and compute

$$\begin{aligned}\int_{-\infty}^{\infty} \delta(-x) f(x) dx &= \int_{\infty}^{-\infty} \delta(y) f(-y)(-dy) = \int_{-\infty}^{\infty} \delta(y) f(-y)(dy) \\ &= f(-0) = f(0). \end{aligned} \tag{9.20}$$

9.2. The Dirac δ Function

Since $\delta(-x)$, multiplied by any function $f(x)$ and integrated gives $f(0)$, $\delta(-x)$ must be the same as $\delta(x)$. Similarly, if $c > 0$, we let $y = cx$ and compute

$$\int_{-\infty}^{\infty} \delta(cx)f(x)dx = \int_{-\infty}^{\infty} \delta(y)f(\frac{y}{c})\frac{dy}{c} = \frac{f(0)}{c}, \quad (9.21)$$

so $\delta(cx) = \delta(x)/c$. If $c < 0$, then $\delta(cx) = \delta(-cx) = \delta(x)/(-c)$. In either case, $\delta(cx) = \delta(x)/|c|$. For the third item, suppose ρ is an increasing function. Just set $y = \rho(x)$ and compute

$$\int \delta(\rho(x))f(x)dx = \int \delta(y)f(\rho^{-1}(y))\frac{dy}{\rho'(x)} = \frac{f(\rho^{-1}(0))}{\rho'(\rho^{-1}(0))}$$
$$= \frac{f(a)}{\rho'(a)} = \int f(x)\frac{\delta(x-a)}{\rho'(a)}dx. \quad (9.22)$$

If ρ is decreasing, then $-\rho$ is increasing, so $\delta(\rho(x)) = \delta(-\rho(x)) = \delta(x-a)/(-\rho'(a)) = \delta/|\rho'(a)|$. Now suppose ρ is a general function, with several roots. The product $f(x)\delta(\rho(x))$ is zero away from the roots of ρ, so we can restrict the integral $\int f(x)\delta(\rho(x))dx$ to a small interval around each root. In each such integral, ρ is either increasing or decreasing, and the contribution from that interval is $f(a_i)/|\rho'(a_i)|$. Adding the contributions gives (9.18). Finally, if $\delta(x)$ is the limit of functions $g_\epsilon(x)$, then $\delta'(x)$ is the limit of $g'_\epsilon(x)$. But $g'_\epsilon(x)$ is a smooth function, so integration by parts is permissible, so

$$\int \delta'(x)f(x)dx = \lim_{\epsilon \to 0} \int g'_\epsilon(x)f(x)dx = -\lim_{\epsilon \to 0} \int g_\epsilon(x)f'(x)dx$$
$$= -\int \delta(x)f'(x)dx = -f'(0). \blacksquare \quad (9.23)$$

For example, $\int \delta(x^2 - 4)f(x)dx = (f(2) + f(-2))/4$, since the roots of $\rho(x) = x^2 - 4$ are ± 2 with $\rho'(2) = 4$ and $\rho'(-2) = -4$.

We are now ready to do a generalized eigenfunction expansion for the operator X. Compare the following to the discrete expansions (9.1–9.5).

Theorem 9.4 *Any function $f(x)$ can be written as linear combinations of delta functions*

$$|\mathbf{f}\rangle = \int_{-\infty}^{\infty} da\, f(a)|X,a\rangle, \quad (9.24)$$

where the function $|X,a\rangle = \delta(x-a)$ is a generalized eigenfunction of the X operator

$$X|X,a\rangle = a|X,a\rangle, \quad (9.25)$$

and the coefficients are
$$f(a) = \langle X, a | \mathbf{f} \rangle. \tag{9.26}$$

The identity operator and the operator X can be written as

$$I = \int_{-\infty}^{\infty} da \, |X,a\rangle\langle X,a|; \quad X = \int_{-\infty}^{\infty} da \, a|X,a\rangle\langle X,a|. \tag{9.27}$$

Proof: Reversing the roles of x and a in (9.15) we have

$$f(x) = \int_{-\infty}^{\infty} da \, f(a)\delta(a-x) = \int_{-\infty}^{\infty} da \, f(a)\delta(x-a), \tag{9.28}$$

which is equivalent to (9.24). The coefficient (9.26) was already derived as (9.16). Combining these we have

$$\begin{aligned} I|\mathbf{f}\rangle = |\mathbf{f}\rangle &= \int_{-\infty}^{\infty} da \, |X,a\rangle f(a) = \int_{-\infty}^{\infty} da \, |X,a\rangle\langle X,a|\mathbf{f}\rangle \\ &= \left(\int_{-\infty}^{\infty} da \, |X,a\rangle\langle X,a| \right) |\mathbf{f}\rangle. \end{aligned} \tag{9.29}$$

Since this applies to all functions $\mathbf{f}$, the identity operator takes the form given in (9.27). Since $X = XI$, the form (9.27) for X follows by applying (9.25) to the form (9.27) for I. All that remains is to establish (9.25).

We will show that $(X - a)|X,a\rangle = 0$. We do this by showing that $\langle \mathbf{f}|X - a|X,a\rangle = 0$ for every function $\mathbf{f}$. This is equivalent to $\langle X,a|X - a|\mathbf{f}\rangle = 0$, since X (and therefore $X - a$) is Hermitian. But

$$\langle X, a|(X-a)\mathbf{f}\rangle = \int_{-\infty}^{\infty} dx \, \delta(x-a)(x-a)f(x) = (a-a)f(a) = 0. \blacksquare \tag{9.30}$$

The parallel between (9.1–9.4) and (9.24–9.27) is direct. Where before we had a sum over true eigenvectors, we now have an integral over generalized eigenvectors. Our discrete condition of orthonormality had been

$$\langle \mathbf{b}_i | \mathbf{b}_j \rangle = \delta_{ij}, \tag{9.31}$$

where

$$\delta_{ij} = \begin{cases} 1 & \text{if } i = j; \\ 0 & \text{otherwise} \end{cases} \tag{9.32}$$

is called the *Kronecker δ function*. It is replaced by the normalization

$$\langle X, a | X, b \rangle = \delta(a-b), \tag{9.33}$$

9.2. The Dirac δ Function

involving the Dirac δ function. The Kronecker and Dirac δ functions play analogous roles for sums and integrals, in that

$$\sum_j \delta_{ij} v_j = v_i; \qquad \int db\, \delta(a-b) f(b) = f(a). \qquad (9.34)$$

This is why both functions are traditionally denoted by the same letter δ.

To see that the normalization (9.33) holds for the generalized eigenfunctions $|X, a\rangle$, consider the identity

$$\int db\, \delta(a-b) f(b) = f(a) = \langle X, a|f\rangle = \int_{-\infty}^{\infty} db\, \langle X, a|X, b\rangle \langle X, b|f\rangle$$

$$= \int db\, \langle X, a|X, b\rangle f(b). \qquad (9.35)$$

Since this holds for every possible f, we must have $\langle X, a|X, b\rangle = \delta(a-b)$. Henceforth we will use the word "orthonormal" to describe a continuum of functions satisfying the condition (9.33), as well as for a discrete set of functions satisfying the condition (9.31).

Exercises

1. Compute $\int_{-\infty}^{\infty} x^2 \delta(3x - 6) dx$.

2. Compute $\int_{-\infty}^{\infty} (x^2 + 1)\delta(x^3 - x) dx$.

3. Compute $\int_0^{\infty} 2x^5 \delta(3x^2 - 6) dx$ in two ways, first by using (9.18), and then by doing a change of variables to convert the problem into Exercise 1.

4. Prove the identity (9.14).

5. Show that the indefinite integral of the delta function $\delta(x)$ is the Heaviside function $f(x) = \begin{cases} 0 & \text{if } x < 0; \\ 1 & \text{if } x \geq 0. \end{cases}$ This suggests that the derivative of $f(x)$ is $\delta(x)$. This last statement will be given a precise meaning in Section 9.3.

6. Let $0 < a < 1$ and let $f(x) = \sum_{m=-\infty}^{\infty} \delta(x + m - a)$. This is a periodic function with period one. Compute its Fourier coefficients $\hat{f}_n$.

7. Let $0 < a < b < 1$ and consider the periodic function, with period 1, whose restriction to $[0, 1]$ is $\begin{cases} 0 & \text{if } x < a; \\ 1 & \text{if } a \leq x < b; \\ 0 & \text{if } x \geq b. \end{cases}$ Compute

274 Chapter 9. Continuous Spectra and the Dirac Delta Function

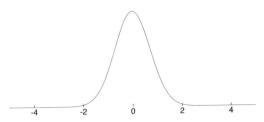

Figure 9.3: A test function is smooth, with compact support

the Fourier coefficients $\hat{f}_n$ and compare to the results of Exercise 6. (Note that the derivative of f is a difference of δ functions.)

8. Consider the periodic function $f(x) = x - [x]$, where $[x]$ is the greatest integer less than or equal to x. Compute the Fourier coefficients of f and relate to the results of Exercises 6 and 7.

9. Consider the periodic function **f** such that $f(x) = x - x^2$ for $0 \leq x \leq 1$ and $f(x + n) = f(x)$ for every integer n. Evaluate the first and second derivatives of **f**. Use this to compare the Fourier coefficients of $\hat{f}_n$ to the results of the Exercises 6–8.

9.3 Distributions

In this section we define what distributions are and examine their properties. To understand what can or cannot be done with δ functions you need to understand what δ functions really are, and that means understanding distributions. We begin with test functions.

Definition *A test function is an infinitely differentiable function on $\mathbb{R}$ that equals zero outside a bounded set. In other words, if f is a test function, then there exists a constant M such that $f(x) = 0$ for all $|x| > M$. The space of all test functions is denoted $C_0^\infty(\mathbb{R})$.*

It is not hard to see that the sum of two test functions is a test function, and that a scalar multiple of a test function is a test function. Thus $C_0^\infty(\mathbb{R})$ is a vector space, a subspace of $L^2(\mathbb{R})$. It is a little difficult to explicitly write down a formula for an element of $C_0^\infty(\mathbb{R})$, as these functions cannot be represented as infinite Taylor series. (If you do a Taylor series around $x = -M - 1$, you get the zero function, which is not correct.) However, the space of test functions is actually quite large, and we construct some examples in the exercises.

9.3. Distributions

Definition A distribution *is a linear map from* $C_0^\infty(\mathbb{R})$ *to the scalars. In other words, the space of distributions, denoted* $\mathcal{D}(\mathbb{R})$, *is the dual space to* $C_0^\infty(\mathbb{R})$.

Here are some examples of distributions:

1. The map that sends the function $\mathbf{f} \in C_0^\infty(\mathbb{R})$ to the number $\int_{-\infty}^\infty f(x)dx$ is linear and well-defined. In other words, integration is a distribution.

2. If $\mathbf{g}$ is a locally integrable function, or a square-integrable function, then the map $T_\mathbf{g}(\mathbf{f}) = \int_{-\infty}^\infty \overline{g(x)} f(x) dx$ is a distribution. We write this formally as $T_\mathbf{g}(\mathbf{f}) = \langle \mathbf{g} | \mathbf{f} \rangle$, even though $\mathbf{g}$ may not be an element of $L^2(\mathbb{R})$. Example 1 was a special case of this, with $g(x) = 1$. In this way, the space of locally integrable functions and the space $L^2(\mathbb{R})$ each correspond to subspaces of $\mathcal{D}(\mathbb{R})$.

3. The linear map $T_\delta(f) = f(0)$ is called the δ distribution. It is not as smooth as the previous examples, in that it only depends on the value of f at a single point, rather than a weighted average.

4. The linear map $T_{\delta'}(f) = -f'(0)$ is a distribution that is even more singular than the δ distribution.

The δ distribution is a perfectly well-defined distribution. It is not of the form of Example 2, in that it does not correspond to a true function, but it is the limit of distributions of the form T_g. Indeed we showed that $\lim_{\epsilon \to 0} T_{\delta_\epsilon}(\mathbf{f}) = T_\delta(\mathbf{f})$ for every $\mathbf{f}$, which means that $\lim_{\epsilon \to 0} T_{\delta_\epsilon} = T_\delta$. In other words, while the *functions* δ_ϵ don't have a well-defined limit as $\epsilon \to 0$, the *distributions* T_{δ_ϵ} do have a limit — the distribution T_δ.

Example 2 gives an extremely large class of distributions. In fact, it can be shown that every distribution is either in this class, or is a limit of distributions of this class. We adopt a notation that pretends that all distributions are of this form. If the distribution $T_\mu = \lim T_{g_i}$ for a sequence of functions g_i, then we write

$$T_\mu(f) = \int \overline{\mu(x)} f(x) dx, \qquad (9.36)$$

where the "function" $\mu(x)$ means the limit of the functions $g_i(x)$. In other words, the right-hand side of (9.36) is to be understood as shorthand for the left-hand side, not as a Riemann integral of an actual function. If the limit of the g_i's happens to be an actual function, then the right-hand side is equal to an ordinary Riemann integral, and we can manipulate $\mu(x)$ as an ordinary function. If the g_i's do not have a limit as functions, then $\mu(x)$ is not a true function, but the expression $\mu(x)$ still makes sense underneath an integral sign, where it really means "apply the distribution T_μ".

The δ "function" is therefore just a notational technique for writing the distribution T_δ as an integral. Since the δ distribution is truly the limit of the distributions T_{δ_ϵ}, or of the distributions T_{g_ϵ}, we speak of the δ function as being the limit of the functions δ_ϵ or g_ϵ. The true test of these limits, and of all our other manipulations involving the δ function, is whether they give the right answer when multiplied by an arbitrary test function and integrated (hence the name "test function").

From now on, we identify functions with the corresponding distributions. The function **g** and the distribution associated with **g** will both be denoted **g**, and we use the cumbersome notation $T_\mathbf{g}$ only for emphasis. With this identification, the space of functions *is* a subspace of the space of distributions, and the space of distributions is an extension of the space of functions. For this reason, distributions are sometimes called "generalized functions".

We next consider how to differentiate distributions. If $\mu = \lim g_i$, where the g_i's are differentiable functions, then we define the distribution μ' to be the limit of the g_i's. That is, for any test function **f**,

$$\begin{aligned}\langle \mu'|\mathbf{f}\rangle &= \int \overline{\mu'(x)} f(x)dx = \lim \int \overline{g_i'(x)} f(x)dx \\ &= \lim \int -\overline{g_i(x)} f'(x)dx = -\int \overline{\mu(x)} f'(x)dx \\ &= -\langle \mu|f'\rangle. \end{aligned} \qquad (9.37)$$

Using the integration-by-parts formula (9.37), it makes sense to take derivatives of distributions, whether the underlying functions are differentiable or not.

For example, consider the Heaviside function

$$h(x) = \begin{cases} 1 & \text{if } x \geq 0; \\ 0 & \text{otherwise.} \end{cases} \qquad (9.38)$$

This is not a differentiable function, but we may take its derivative as a distribution. If **f** is a test function, then

$$\langle H'|\mathbf{f}\rangle = -\langle H|\mathbf{f}'\rangle = -\int_0^\infty f'(x)dx = f(0) = \langle \delta|\mathbf{f}\rangle. \quad (9.39)$$

In other words, the derivative of the Heaviside function is the delta function, as previously worked out formally in Exercise 5 of Section 9.2. This can also be seen by approximating $H(x)$ by a smooth function such as $e^{x/\epsilon}/(1 + e^{x/\epsilon})$. The derivative of this smooth function is very small away from $x = 0$ and has total integral one, and so is an approximate δ function.

Exercises

1. Consider the function

$$\alpha(x) = \begin{cases} \exp(-1/x^2) & \text{if } x > 0; \\ 0 & \text{otherwise.} \end{cases} \quad (9.40)$$

Show that $\alpha(x)$ is infinitely differentiable. [This is trivial away from $x = 0$. At $x = 0$ you must apply the definition of the derivative and a clever use of L'Hôpital's rule.]

2. Let $\rho(x) = \alpha(x)\alpha(1-x)dx'$. Show that ρ is a smooth function that is zero for $x \leq 0$ or $x \geq 1$, but is positive for $0 < x < 1$. Functions like ρ are sometimes called *bump functions*.

3. Let $\beta(x) = \frac{1}{N}\int_0^x \rho(x')dx'$, where $N = \int_0^1 \rho(x')dx'$. Show that β is a smooth function with $\beta(x) = 0$ for $x \leq 0$, and $\beta(x) = 1$ for $x \geq 1$. Functions like β are called *cutoff functions*.

4. Let f be any smooth function, and let $M > 0$. Consider the function $g(x) = f(x)\beta(M-x)\beta(M+x)$. Show that $g(x)$ is a test function.

5. Compute the derivative of the approximate δ function δ_ϵ. Apply this to a test function **f**, and then take the limit as $\epsilon \to 0$. Is the result the same as δ' applied to **f**?

9.4 Generalized Eigenfunction Expansions; The Spectral Theorem

Given a Hermitian operator A, whose spectrum ranges over a continuum, we would like to construct a decomposition similar to the decomposition (9.24–9.27) for the X operator. For each generalized eigenvalue λ, we need to find a corresponding generalized eigenfunction $|A, \lambda\rangle$ such that

$$I = \int d\lambda \, |A, \lambda\rangle\langle A, \lambda|. \tag{9.41}$$

From this the other expansions follow immediately:

$$A = AI = \int d\lambda \, A|A, \lambda\rangle\langle A, \lambda| = \int d\lambda \, \lambda |A, \lambda\rangle\langle A, \lambda|, \tag{9.42}$$

$$|\mathbf{f}\rangle = I|\mathbf{f}\rangle = \int d\lambda \, |A, \lambda\rangle\langle A, \lambda|\mathbf{f}\rangle = \int d\lambda \, g(\lambda)|A, \lambda\rangle, \tag{9.43}$$

where
$$g(\lambda) = \langle A, \lambda|\mathbf{f}\rangle. \tag{9.44}$$

Since (generalized) eigenfunctions with different eigenvalues are orthogonal, $\langle \boldsymbol{\xi}_\lambda | \boldsymbol{\xi}_{\lambda'} \rangle$ must equal $c(\lambda)\delta(\lambda - \lambda')$ for some function $c(\lambda)$. However, the decomposition (9.41) implies that

$$|A, \lambda\rangle = I|A, \lambda\rangle = \int d\lambda' \, |A, \lambda'\rangle\langle A, \lambda'|A, \lambda\rangle = c(\lambda)|A, \lambda\rangle, \tag{9.45}$$

so $c(\lambda) = 1$ for all λ and we have the normalization

$$\langle A, \lambda|A, \lambda'\rangle = \delta(\lambda - \lambda'), \tag{9.46}$$

which is a generalization of (9.33).

The decomposition (9.41) also implies the existence of a family of projection operators

$$P_a = \int_{-\infty}^{a} d\lambda \, |A, \lambda\rangle\langle A, \lambda|. \tag{9.47}$$

The operator P_a projects onto the span of the eigenfunctions with eigenvalue $\lambda \leq a$. If U is a (measurable) subset of $\mathbb{R}$, we can also define the projection

$$P_U = \int_U d\lambda \, |A, \lambda\rangle\langle A, \lambda|. \tag{9.48}$$

P_U projects onto the span of the eigenfunctions with $\lambda \in U$.

For example, suppose that $A = X$, the position operator. Then

$$(P_a f)(x) = \begin{cases} f(x) & \text{if } x \leq a; \\ 0 & \text{if } x > a, \end{cases} \tag{9.49}$$

9.4. Generalized Eigenfunction Expansions

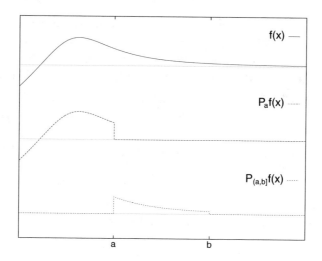

Figure 9.4: Projections for the position operator

and in general

$$(P_U f)(x) = \begin{cases} f(x) & \text{if } x \in U; \\ 0 & \text{otherwise.} \end{cases} \qquad (9.50)$$

See Figure 9.4.

Working with projection operators, rather than the eigenfunctions themselves, has two great advantages. The first is that the projections are well-defined operators on $L^2(\mathbb{R})$, while the generalized eigenfunctions are not elements of $L^2(\mathbb{R})$. This makes it possible to state and prove precise theorems. The second advantage is that projections allow for a unified approach that handles both discrete and continuous spectra. This is important, because some operators have a spectrum that consists both of discrete points (true eigenvalues) and a continuous range (generalized eigenvalues). For example, in quantum mechanics the operator that describes the combined energy of an electron and a proton has both continuous and discrete spectrum. The discrete eigenvalues correspond to possible energies of the electron and proton bound together as a hydrogen atom, while the continuous part of the spectrum describes the possible energies of the electron and proton wandering free of one another.

For an operator with discrete spectrum (i.e., true eigenvalues and eigenvectors), we can still define P_a to be the projection onto the span of the eigenfunctions with eigenvalue $\lambda \leq a$. That is,

$$P_a = \sum_{\lambda_i \leq a} P_{\mathbf{b}_i} = \sum_{\lambda_i \leq a} |\mathbf{b}_i\rangle\langle\mathbf{b}_i|. \qquad (9.51)$$

280 Chapter 9. Continuous Spectra and the Dirac Delta Function

If an operator has both discrete and continuous spectrum, then the projection P_a is again well-defined, and is the sum of two terms. The first term is the sum (9.51) over the true eigenvalues, and the second term is the integral (9.47) over the continuous part of the spectrum. Regardless of the nature of the spectrum, the projections P_a always exist. Similarly, one can always define the projection P_U for any measurable subset U of the real line.

In terms of projections, the decomposition (9.41) of the identity simply becomes

$$I = \lim_{a \to \infty} P_a, \qquad (9.52)$$

while the decomposition (9.42) of A becomes

$$A = \int_{-\infty}^{\infty} \lambda \, dP_\lambda. \qquad (9.53)$$

The integral (9.53) is not a Riemann integral. Rather, it is an operator-valued Riemann-Stieltjes integral, and that warrants some explanation.

Riemann-Stieltjes integrals

An ordinary integral, also called a Riemann integral, is the limit of a sum. The expression

$$\int_a^b f(x)\,dx \qquad (9.54)$$

means the following: Pick a large number N, set $\Delta x = (b-a)/N$, and pick $N+1$ numbers $x_0 = a$, $x_1 = a + \Delta x$, $x_2 = a + 2\Delta x, \ldots, x_N = b$. Pick a point z_i in each interval $(x_{i-a}, x_i]$, and look at the sum

$$\sum_{i=1}^N f(z_i)(x_i - x_{i-1}). \qquad (9.55)$$

Take the limit of the sum (9.55) as $N \to \infty$. If the limit exists and is independent of the choice of points z_i, then the Riemann integral (9.54) is defined to be that limit. Finally, a Riemann integral from $-\infty$ to ∞ is defined to be the limit as $a \to -\infty$ and $b \to \infty$ of (9.54).

Now, suppose we have a second function α. The Riemann-Stieltjes integral

$$\int_a^b f(x)\,d\alpha(x) \qquad (9.56)$$

9.4. Generalized Eigenfunction Expansions

is constructed as follows. Pick N, the points x_i and the points z_i as before, and consider the sum

$$\sum_{i=1}^{N} f(z_i)(\alpha(x_i) - \alpha(x_{i-1})). \tag{9.57}$$

If the limit of (9.57) as $N \to \infty$ exists and is independent of the choices made, the Riemann-Stieltjes integral (9.56) is defined to be that limit.

In the Riemann sum (9.55), each interval had weight equal to its length, while in the Riemann-Stieltjes sum (9.57), each interval has weight equal to the change in α. If α is a smooth function, then the change in α is approximately $\alpha'(x)$ times the change in x, and the Riemann-Stieltjes integral (9.56) reduces to a Riemann integral:

$$\int_a^b f(x)d\alpha(x) = \int_a^b f(x)\frac{d\alpha(x)}{dx}dx. \tag{9.58}$$

However, the Riemann-Stieltjes integral makes sense even if α is not differentiable. It can even make sense if α is discontinuous.

Suppose, for example, that $\alpha(x)$ is a step function, equalling zero for $x < 0$, and one for $x \geq 0$. Then $\alpha(x_i) - \alpha(x_{i-1})$ is either equal to zero or one. It is equal to one if $0 \in (x_{i-1}, x_i]$ and zero otherwise. Thus, if $a < 0 < b$, the sum (9.57) reduces to a single term, and equals $f(z_i)$, where z_i is the point chosen from the interval containing zero. As $N \to \infty$, the intervals get narrower and z_i has to approach zero, so the integral (9.56) equals $f(0)$.

Similarly, if α is a sum of step functions, with jumps of size w_i at the points y_i, then the Riemann-Sieltjes integral (9.56) equals $\sum_i w_i f(y_i)$. If α is differentiable everywhere except at some jump discontinuities, then the Riemann-Sieltjes integral (9.56) is a weighted sum of $f(x)$ at the jump points, plus an ordinary Riemann integral of $f(x)\frac{d\alpha(x)}{dx}dx$ over the remainder.

In the sum (9.57), and hence in the integral (9.56), $f(x)$ and $\alpha(x)$ do not have to be scalars. $f(x)$ and α might be vectors, or operators, as long as the product $f(x)(\alpha(x_i) - \alpha(x_{i-1}))$ makes sense. In the decomposition (9.53) of an operator A, the variable is called λ, not x, $f(\lambda) = \lambda$ is a scalar, and $\alpha(\lambda) = P_\lambda$ is an operator. The mysterious dP_λ term is

$$dP_\lambda = P_{\lambda+d\lambda} - P_\lambda = P_{(\lambda, \lambda+d\lambda]}, \tag{9.59}$$

282 Chapter 9. Continuous Spectra and the Dirac Delta Function

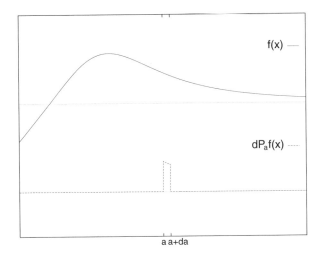

Figure 9.5: The infinitesimal projection dP_a

the projection onto the infinitesimal interval $(\lambda, \lambda + d\lambda]$. See Figure 9.5. For continuous spectrum, P_λ is a continuous (operator-valued) function of λ, and dP_λ is the same as $|A, \lambda\rangle\langle A, \lambda| d\lambda$, while for discrete spectrum, P_λ is a discontinuous function of λ, and dP_λ equals $|\mathbf{b}_i\rangle\langle\mathbf{b}_i|$ if the i-th eigenvalue lies in the interval $(\lambda, \lambda + d\lambda]$, and is zero if no eigenvalues lie in the interval.

To summarize, the integral (9.53) is obtained by the following limiting procedure. Suppose first that we are integrating from $\lambda = a$ to $\lambda = b$. Pick $N+1$ equally spaced points $\lambda_0 = a, \lambda_1, \ldots, \lambda_N = b$, and consider the sum

$$\sum_{i=1}^{N} \lambda_i P_{(\lambda_{i-1}, \lambda_i]} = \sum_{i=1}^{N} \lambda_i (P_{\lambda_i} - P_{\lambda_{i-1}}), \qquad (9.60)$$

just like (9.57) only with operators instead of scalars, and with the particular choice $z_i = \lambda_i$. Taking the limit as $N \to \infty$ gives the integral $\int_a^b \lambda \, dP_\lambda$. The integral (9.53) is the limit as $b \to \infty$ and $a \to -\infty$ of $\int_a^b \lambda \, dP_\lambda$. The result is the (operator-valued Riemann) integral of $\lambda |A, \lambda\rangle\langle A, \lambda| d\lambda$ over the continuous spectrum of A plus the sum of $\lambda_i |\mathbf{b}_i\rangle\langle\mathbf{b}_i|$ over the discrete spectrum.

So far in this section we have assumed that generalized eigenfunctions exist and have the desired properties. From these we have constructed projections and derived the "spectral resolution" (9.53). The spectral theorem says that such a resolution exists for every Hermitian operator on every appropriate inner product space.

9.4. Generalized Eigenfunction Expansions

Theorem 9.5 (the Spectral Theorem) *Let A be a Hermitian operator on a complete inner product space V. Then the spectrum of A is a subset of the real line. There exists a family of operators P_a with the following properties:*

- *P_a is an orthogonal projection. (This is equivalent to the equations $P_a^\dagger = P_a^2 = P_a$.)*

- *If $a < b$, $P_a P_b = P_b P_a = P_a$.*

- *For every $\mathbf{v} \in V$, $\lim_{a \to \infty}(P_a \mathbf{v} - \mathbf{v}) = 0$ and $\lim_{a \to -\infty}(P_a \mathbf{v}) = 0$. (This is a precise way of saying that $\lim_{a \to \infty} P_a = I$, and that $\lim_{a \to -\infty} P_a = 0$.)*

- *$A = \int_{-\infty}^{\infty} \lambda \, dP_\lambda$.*

When V is a finite dimensional space, the spectral theorem reduces to Theorems 7.4 and 7.6. The reality of the eigenvalues follows from $\sigma(A)$ being a subset of the real line (as opposed to the complex plane). The orthogonality of the eigenvectors follows from the first two properties of P_a, together with Exercise 5, below. The existence of the spectral resolution (9.53) implies that A is diagonalizable.

When V is an infinite dimensional space and the spectrum of A is discrete, the spectral theorem again shows that the eigenvalues are real, the eigenvectors are orthogonal, and that there exists a basis of eigenvectors. In particular, this justifies Fourier series.

When V is infinite dimensional and the spectrum of A is continuous, the spectral theorem implies that some sort of generalized eigenfunction expansion is possible. In the next chapter, we will apply this to the operator K to obtain Fourier integrals.

Exercises

1. Suppose $a > b$, and let $U = (a, b]$. Using (9.46), show that

$$P_a^2 = P_a^\dagger = P_a; \qquad P_a P_b = P_b P_a = P_b; \qquad P_U^\dagger = P_U^2 = P_U.$$

2. Let $V = \mathbb{R}^3$, and let $A = \begin{pmatrix} 0 & 0 & 0 \\ 0 & 1 & 0 \\ 0 & 0 & -1 \end{pmatrix}$. For every $a \in \mathbb{R}$, find the projection operator P_a.

3. For the example of Exercise 2, construct the integral (9.53) as a limit of sums (9.60) and show that the integral equals A.

284   Chapter 9.   Continuous Spectra and the Dirac Delta Function

4. If A is a Hermitian operator with discrete spectrum, show that the integral (9.53) means the same thing as the sum $\sum_n \lambda_n |\xi_n\rangle\langle\xi_n|$.

5. Assume a family of operators P_a satisfies items 1 and 2 of the spectral theorem. If $a < b$, show that $P_a(P_b - P_a) = (P_b - P_a)P_b = 0$. If, for every a, $P_a = \sum_{\lambda_i \leq a} |\mathbf{b}_i\rangle\langle\mathbf{b}_i|$, show that this implies that $\langle \mathbf{b}_1 | \mathbf{b}_2 \rangle = 0$.

6. Let $\alpha(x) = \begin{cases} 0 & \text{if } x < 0; \\ x^2 & \text{if } 0 \leq x < 1; \\ 1 & \text{if } 1 \leq x < 2; \\ 2 & \text{if } x \geq 2. \end{cases}$ Compute $\int_{x=-\infty}^{\infty} e^x d\alpha$.

7. Consider the matrix-valued function

$$P(x) = \begin{cases} \begin{pmatrix} 0 & 0 & 0 \\ 0 & 0 & 0 \\ 0 & 0 & 0 \end{pmatrix} & \text{if } x < 0; \\ \begin{pmatrix} 1 & 0 & 0 \\ 0 & 0 & 0 \\ 0 & 0 & 0 \end{pmatrix} & \text{if } 0 \leq x < 2; \\ \begin{pmatrix} 1 & 0 & 0 \\ 0 & 0 & 0 \\ 0 & 0 & 1 \end{pmatrix} & \text{if } 2 \leq x < 0; \\ \begin{pmatrix} 1 & 0 & 0 \\ 0 & 1 & 0 \\ 0 & 0 & 1 \end{pmatrix} & \text{if } x \geq 5. \end{cases}$$

Compute $\int_{x=-\infty}^{\infty} x\, dP$.

8. Consider the matrix-valued function

$$P(x) = \begin{cases} \begin{pmatrix} 0 & 0 \\ 0 & 0 \end{pmatrix} & \text{if } x < -1; \\ \begin{pmatrix} 1/2 & 1/2 \\ 1/2 & 1/2 \end{pmatrix} & \text{if } -1 \leq x < 2; \\ \begin{pmatrix} 1 & 0 \\ 0 & 1 \end{pmatrix} & \text{if } x \geq 2. \end{cases}$$

Compute $A = \int_{x=-\infty}^{\infty} x\, dP$. Compute $\int_{x=-\infty}^{\infty} x^2 dP$, and $\int_{x=-\infty}^{\infty} x^2 dP$, and compare to A^2 and A^3, respectively.

9. Let P_a be given by (9.49). Show that $X = \int_{a=-\infty}^{\infty} a\, dP_a$ and that $X^2 = \int_{a=-\infty}^{\infty} a^2 dP_a$.

10. Let $\alpha(x)$ be an increasing function (not necessarily continuous).

9.4. Generalized Eigenfunction Expansions

Show that the map $\mathbf{f} \to \int f(x)d\alpha$ is a distribution. What distribution corresponds to the step function $\alpha(x) = \begin{cases} 0 & \text{if } x < 0; \\ 1 & \text{if } x \geq 0 \end{cases}$?

Chapter 10

Fourier Transforms

In Chapters 6–8 we developed the theory of Fourier series. Given a function $f(x)$ on the interval $[0, L]$, we learned how to write $f(x)$ as a linear combination of sines, sines and cosines, or complex exponentials. These were eigenfunction expansions for operators on $L^2[0, L]$. Depending on the type of Fourier series, the operator was either d^2/dx^2 with Dirichlet or periodic boundary conditions, or $-id/dx$ with periodic boundary conditions. In each case, the spectrum of the operator was discrete, the eigenfunctions were the sines and cosines (or complex exponentials), and the operator being diagonalizable meant that the eigenfunctions formed a basis. Since the operator was Hermitian, the eigenfunctions were orthogonal, and we had simple formulas for computing the coefficients.

In this chapter we learn how to decompose functions on the entire real line $\mathbb{R}$, or on the half line $[0, \infty)$, into complex exponentials (or sines and cosines). This is a *generalized* eigenfunction expansion, since the relevant operators $(-id/dx$ or $-d^2/dx^2)$ have continuous spectra. As a result, we have to integrate over the eigenfunctions rather than sum. Otherwise, the formulas are almost identical. This decomposition is called a *Fourier integral*, and the set of coefficients is called the *Fourier transform* of the original function.

10.1 Existence of Fourier Transforms

The basic formulas of Fourier analysis are given in the following theorem.

Theorem 10.1 *Let* **f** *be a square-integrable function on* $\mathbb{R}$. *Then*

$$f(x) = \frac{1}{\sqrt{2\pi}} \int_{-\infty}^{\infty} \hat{f}(k) e^{ikx} dk, \qquad (10.1)$$

where

$$\hat{f}(k) = \frac{1}{\sqrt{2\pi}} \int_{-\infty}^{\infty} f(x) e^{-ikx} dx. \qquad (10.2)$$

The decomposition (10.1) is called a *Fourier integral*. Notice that the formula (10.2) for finding the coefficients $\hat{f}(k)$ looks almost identical to the Fourier integral (10.1) itself. This suggests treating the functions $f(x)$ and $\hat{f}(k)$ on an equal footing. The equations (10.1) and (10.2) transform $f(x)$ into $\hat{f}(k)$ and vice-versa, and the function $\hat{f}(k)$ is called the *Fourier transform* of the function $f(x)$. The number k is called the *wave number* of the plane wave e^{ikx}. When **f** is a function of time (t) rather than position (x), then the second variable is denoted ω rather than k, and is called the *angular frequency*.

Before proving the theorem, we restate it in terms of vectors and work a few examples. Consider the operator $K = -i d/dx$, and let $|K, k\rangle$ be the function $\exp(ikx)/\sqrt{2\pi}$. The vector $|K, k\rangle$ is a generalized eigenfunction of the operator K with eigenvalue k. Equation (10.1) says that we can decompose an arbitrary vector $|\mathbf{f}\rangle$ in $L^2(\mathbb{R})$ in the eigenstates of K:

$$|\mathbf{f}\rangle = \int_{-\infty}^{\infty} dk \, \hat{f}(k) |K, k\rangle, \qquad (10.3)$$

while (10.2) says that the coefficients are given by

$$\hat{f}(k) = \langle K, k|\mathbf{f}\rangle. \qquad (10.4)$$

Taken together, the theorem is equivalent to the statement that

$$I = \int_{-\infty}^{\infty} dk |K, k\rangle\langle K, k|. \qquad (10.5)$$

This is entirely analogous to Theorem 9.4, the decomposition (9.24) of a function into eigenfunctions of the X operator. When working with the X operator, we use the decomposition (9.24), and our coefficients are $f(a)$. When working with the K operator, it is more convenient to use the decomposition (10.3), with coefficients $\hat{f}(k)$. The Fourier integral (10.1), and its inverse (10.2), allow us to go back and forth between these two representations of the vector $|\mathbf{f}\rangle$.

10.1. Existence of Fourier Transforms

Example 1: Let $f(x) = \exp(-x^2/2)$. Then

$$\begin{aligned}
\hat{f}(k) &= \frac{1}{\sqrt{2\pi}} \int_{-\infty}^{\infty} e^{-ikx} e^{-x^2/2} dx = \frac{1}{\sqrt{2\pi}} \int_{-\infty}^{\infty} e^{-(x^2+2ikx)/2} dx \\
&= \frac{1}{\sqrt{2\pi}} \int_{-\infty}^{\infty} e^{-(x+ik)^2/2} e^{-k^2/2} dx \\
&= \frac{e^{-k^2/2}}{\sqrt{2\pi}} \int_{-\infty}^{\infty} e^{-y^2/2} dy,
\end{aligned} \qquad (10.6)$$

where $y = x + ik$. The value of this last integral is independent of k, and equals $\sqrt{2\pi}$. Thus

$$\hat{f}(k) = e^{-k^2/2}. \qquad (10.7)$$

In other words, the Fourier transform of a standard normal distribution is a standard normal distribution. An almost identical calculation shows that $\frac{1}{\sqrt{2\pi}} \int dk \hat{f}(k) e^{ikx}$ is equal to $e^{-x^2/2}$. That is, the theorem works when applied to the function $f(x) = e^{-x^2/2}$.

Example 2: Let $f(x)$ be the step function

$$f(x) = \begin{cases} 1 & \text{if } -1 < x < 1; \\ 0 & \text{otherwise.} \end{cases} \qquad (10.8)$$

Then

$$\hat{f}(k) = \frac{1}{\sqrt{2\pi}} \int_{-1}^{1} \exp(-ikx) dx = \frac{i}{k\sqrt{2\pi}} (e^{-ik} - e^{ikx}) = \frac{2\sin(k)}{k\sqrt{2\pi}}. \qquad (10.9)$$

Compare this with Example 1 of Section 6.9. When computing the Fourier *series* of a step function, the coefficients decay as n^{-1}. When doing Fourier *integrals*, the coefficients $\hat{f}(k)$ decay as k^{-1}. In general, the Fourier transform of a function with jump discontinuities will always decay as k^{-1} for large k.

The existence of Fourier transforms is so important that we give two proofs. The first proof obtains Fourier integrals as limits of Fourier series. In the second proof, we do an explicit calculation to show that equations (10.1) and (10.2) work when **f** is an approximate δ function. We then use the inner product to extend the result to all functions.

First proof of Theorem 10.1: We first prove the theorem for functions that vanish outside a bounded region, and then extend to general functions. Let $f(x)$ be a square-integrable function that vanishes for $|x| > M$. For each number $L > M$, we construct a Fourier series that equals $f(x)$ for $-L < x < L$. By taking the limit as $L \to \infty$, we get an expression that equals $f(x)$ everywhere. This expression will be our Fourier integral (10.1).

For each $L > M$, construct a periodic function $\tilde{f}$ with period $2L$ through the definitions

$$\begin{aligned}\tilde{f}(x) &= f(x) \text{ for } -L < x \le L, \\ \tilde{f}(x+2nL) &= f(x) \text{ for all } x.\end{aligned} \tag{10.10}$$

In other words, $\tilde{\mathbf{f}}$ is obtained by taking the graph of $\mathbf{f}$ between $x = -L$ and $x = L$ and repeating it. For x between $-L$ and L, this new function equals $f(x)$. For $|x| > L$ it typically does not.

Since $\tilde{f}(x)$ is a periodic function, it can be expanded in a Fourier series

$$\tilde{f}(x) = \sum_n c_n \exp(\pi i n x/L), \tag{10.11}$$

where

$$\begin{aligned}c_n &= \frac{1}{2L}\int_0^{2L} \tilde{f}(x)e^{-i\pi n x/L}dx = \frac{1}{2L}\int_{-L}^{L}\tilde{f}(x)e^{-i\pi n x/L}dx \\ &= \frac{1}{2L}\int_{-L}^{L} f(x)e^{-i\pi n x/L}dx = \frac{1}{2L}\int_{-\infty}^{\infty} f(x)e^{-i\pi n x/L}dx \\ &= \frac{\sqrt{2\pi}}{2L}\hat{f}\left(\frac{n\pi}{L}\right).\end{aligned} \tag{10.12}$$

Now let $k_n = \pi n/L$, and let $\Delta k = k_{n+1} - k_n = \pi/L$. Plugging the formula (10.12) for c_n into (10.11) we have

$$f(x) = \tilde{f}(x) = \frac{1}{\sqrt{2\pi}}\sum_n \hat{f}(k_n)e^{in\pi x/L}\Delta k \quad \text{for } -L < x < L. \tag{10.13}$$

For any fixed $L > M$, the relation (10.13) gives us our function $f(x)$ over a large interval $[-L, L]$. However, for $|x| > L$ the right-hand side equals $\tilde{f}(x)$, not $f(x)$. In order to get an expression for $f(x)$ everywhere we take the limit as $L \to \infty$. As $L \to \infty$, $\Delta k \to 0$, and the Riemann sum (10.13) approaches the integral (10.1). This proves the theorem for functions that vanish outside a bounded set.

Now suppose that f is a general square-integrable function. Pick a large number M, and define the function

$$f_M(x) = \begin{cases} f(x) & \text{if } |x| < M; \\ 0 & \text{if } |x| \ge M. \end{cases} \tag{10.14}$$

10.1. Existence of Fourier Transforms

This function is zero outside a bounded set, so, by our previous arguments, it can be represented as a Fourier integral. For $|x| < M$ we then have

$$f(x) = f_M(x) = \frac{1}{\sqrt{2\pi}} \int_{-\infty}^{\infty} dk\, \hat{f}_M(k) e^{ikx}. \tag{10.15}$$

To get an expression that applies to all x we take the limit as $M \to \infty$. Since

$$\begin{aligned}\lim_{M \to \infty} \hat{f}_M(k) &= \lim_{M \to \infty} \frac{1}{\sqrt{2\pi}} \int_{-M}^{M} f(x) e^{-ikx} dx \\ &= \frac{1}{\sqrt{2\pi}} \int_{-\infty}^{\infty} f(x) e^{-ikx} dx = \hat{f}(k),\end{aligned} \tag{10.16}$$

the limit of the integral (10.15) as $M \to \infty$ is the integral (10.1). ∎

Second proof of Theorem 10.1: We need to establish the decomposition (10.5) of the identity. We therefore define the operator

$$P_\infty = \int_{-\infty}^{\infty} dk\, |K,k\rangle\langle K,k|, \tag{10.17}$$

and show that $P_\infty = I$. Notice that P_∞ is Hermitian. We will show that $P_\infty|\mathbf{f}\rangle = |\mathbf{f}\rangle$ for all $\mathbf{f}$ by working from a simple example to the general case. Let $g(x) = e^{-x^2/2}/\sqrt{2\pi}$, and let $g_\epsilon(x) = g(\frac{x}{\epsilon})/\epsilon$. There are four cases.

1. $f(x) = g(x)$. This is Example 1, where we showed that (10.1) and (10.2) did indeed apply to the standard normal distribution, and hence that $P_\infty \mathbf{f} = \mathbf{f}$.

2. $f(x) = g_\epsilon(x) = g(\frac{x}{\epsilon})/\epsilon$. We compute, using the substitutions $y = \frac{x}{\epsilon}$ and $k' = k\epsilon$,

$$\begin{aligned}\hat{f}(k) &= \frac{1}{\sqrt{2\pi}} \int_{-\infty}^{\infty} \frac{e^{-ikx} g(\frac{x}{\epsilon}) dx}{\epsilon} = \frac{1}{\sqrt{2\pi}} \int_{-\infty}^{\infty} e^{-ik\epsilon y} g(y) dy \\ &= \hat{g}(k\epsilon) = \hat{g}(k').\end{aligned} \tag{10.18}$$

$$\begin{aligned}P_\infty f(x) &= \frac{1}{\sqrt{2\pi}} \int_{-\infty}^{\infty} dk\, \hat{f}(k) e^{ikx} = \frac{1}{\sqrt{2\pi}} \int_{-\infty}^{\infty} \frac{dk'}{\epsilon} \hat{g}(k') e^{ik'x/\epsilon} \\ &= \frac{1}{\epsilon} g(\frac{x}{\epsilon}) = f(x).\end{aligned} \tag{10.19}$$

3. $f(x) = g_\epsilon(x-a)$. We compute, using $y = k-a$,

$$\begin{aligned}
\hat{f}(k) &= \frac{1}{\sqrt{2\pi}} \int_{-\infty}^{\infty} e^{-ikx} g_\epsilon(x-a) dx \\
&= \frac{1}{\sqrt{2\pi}} \int_{-\infty}^{\infty} e^{-ika} e^{-iky} g_\epsilon(y) dy \\
&= e^{-ika} \hat{g}_\epsilon(k).
\end{aligned} \qquad (10.20)$$

$$\begin{aligned}
P_\infty f(x) &= \frac{1}{\sqrt{2\pi}} \int_{-\infty}^{\infty} e^{ikx} e^{-ika} \hat{g}_\epsilon(k) dk \\
&= \frac{1}{\sqrt{2\pi}} \int_{-\infty}^{\infty} e^{iky} \hat{g}_\epsilon(k) dk \\
&= g_\epsilon(y) = f(x).
\end{aligned} \qquad (10.21)$$

4. $f(x)$ is arbitrary. Since $\int_{-\infty}^{\infty} g(x) dx = 1$, we have $\delta(x-a) = \lim_{\epsilon \to 0} g_\epsilon(x-a)$. Thus for every number a,

$$\begin{aligned}
(P_\infty \mathbf{f})(a) &= \langle X, a | P_\infty | \mathbf{f} \rangle \\
&= \lim_{\epsilon \to 0} \langle g_\epsilon(x-a) | P_\infty | \mathbf{f} \rangle \\
&= \lim_{\epsilon \to 0} \langle P_\infty g_\epsilon(x-a) | \mathbf{f} \rangle \quad \text{since } P_\infty \text{ is Hermitian} \\
&= \lim_{\epsilon \to 0} \langle g_\epsilon(x-a) | \mathbf{f} \rangle \quad \text{by case 3} \\
&= \langle X, a | \mathbf{f} \rangle = f(a).
\end{aligned} \qquad (10.22)$$

Since $P_\infty f(a) = f(a)$ for every $\mathbf{f}$ and every a, we must have $P_\infty = I$. ∎

Exercises

Find the Fourier transforms of the following functions:

1. $f(x) = \begin{cases} 1 & \text{if } -1 < x < 0; \\ -1 & \text{if } 0 < x < 1; \\ 0 & \text{otherwise.} \end{cases}$

2. $f(x) = \begin{cases} x+1 & \text{if } -1 < x < 0; \\ 1-x & \text{if } 0 \leq x < 1; \\ 0 & \text{otherwise.} \end{cases}$ Compare to your answer to Exercise 1.

3. $f(x) = \begin{cases} x & \text{if } 0 < x < 1; \\ 0 & \text{otherwise.} \end{cases}$

4. $f(x) = \delta(x)$.
5. $f(x) = e^{-|x|}$.
6. $f(x) = xe^{-x^2/2}$. [Hint: let $y = x + ik$.]
7. $f(x) = x^2 e^{-x^2/2}$.
8. $f(x) = (x^2 + 1)^{-1}$.

10.2 Basic Properties of Fourier Transforms

In Section 10.1 we demonstrated the existence of Fourier transforms. In this section we explore their properties, and in Sections 10.3 and 10.4 we show how to use them to solve ordinary and partial differential equations.

Theorem 10.2 (Plancherel's Theorem) *The Fourier transform is an isometry from $L^2(\mathbb{R})$ to $L^2(\mathbb{R})$. That is, for all $\mathbf{f}, \mathbf{g} \in L^2(\mathbb{R})$,*

$$\int_{-\infty}^{\infty} dx\, \overline{f(x)} g(x) = \int_{-\infty}^{\infty} dk\, \overline{\hat{f}(k)} \hat{g}(k). \tag{10.23}$$

and

$$\int_{-\infty}^{\infty} dx\, |f(x)|^2 = \int_{-\infty}^{\infty} dk\, |\hat{f}(k)|^2. \tag{10.24}$$

Proof: This is a direct consequence of the decomposition (10.5) of the identity.

$$\int_{-\infty}^{\infty} dx\, \overline{f(x)} g(x) = \langle \mathbf{f} | \mathbf{g} \rangle = \langle \mathbf{f} | I | \mathbf{g} \rangle = \int_{-\infty}^{\infty} dk\, \langle \mathbf{f} | P, k \rangle \langle P, k | \mathbf{g} \rangle$$

$$= \int_{-\infty}^{\infty} dk\, \overline{\hat{f}(k)} \hat{g}(k). \tag{10.25}$$

The special case of $\mathbf{f} = \mathbf{g}$ is (10.24). ∎

Plancherel's theorem says that you can treat a function $\mathbf{f}$ equally well by considering $\hat{f}(k)$ as by considering $f(x)$. Calculations involving $\hat{f}(k)$ are sometimes called working *in momentum space*, while calculations involving $f(x)$ are called working *in position space*. The following theorem allows us to go back and forth by showing how applying simple operations to $f(x)$ affects $\hat{f}(k)$ and vice-versa. The results are essentially symmetric. If doing an operation to $f(x)$ causes a result to $\hat{f}(k)$, then, up to a sign, doing the operation to $\hat{f}(k)$ will cause the same result to $f(x)$.

Theorem 10.3 *Let* **f** *be a square-integrable function, let* $f_a(x) = f(x-a)$, *let* $g_b(x) = \exp(ibx)f(x)$, *let* $h_c(x) = f(cx)$, *let* $m(x) = f'(x)$, *and let* $n(x) = xf(x)$. *Then*

$$\hat{f}_a(k) = \exp(-ika)\hat{f}(k), \tag{10.26}$$

$$\hat{g}_b(k) = \hat{f}(k-b), \tag{10.27}$$

$$\hat{h}_c(k) = \hat{f}(k/c)/c. \tag{10.28}$$

If $m(x)$ *is square-integrable then*

$$\hat{m}(k) = ik\hat{f}(k), \tag{10.29}$$

and if $n(x)$ *is square-integrable then*

$$\hat{n}(k) = i\frac{d\hat{f}(k)}{dk}. \tag{10.30}$$

In other words, translating a function by a is the same as multiplying its Fourier transform by e^{-ika}, while translating the transform by b is the same as multiplying the original function by e^{ibx}. Rescaling a function horizontally by c^{-1} is equivalent to rescaling its Fourier transform horizontally by c, and dividing by c (and vice-versa). Taking the derivative of a function is the same as multiplying the Fourier transform by ik, while taking the derivative of the Fourier transform is the same as multiplying the function by $-ix$.

Proof: The identity (10.26) is a direct calculation using (10.2) and the substitution $y = x - a$:

$$\begin{aligned}\hat{f}_a(k) &= \frac{1}{\sqrt{2\pi}}\int_{-\infty}^{\infty} dx\, e^{-ikx} f(x-a) \\ &= \frac{1}{\sqrt{2\pi}}\int_{-\infty}^{\infty} dy\, e^{-ik(y+a)} f(y) = \exp(-ika)\hat{f}(k). \end{aligned} \tag{10.31}$$

The identities (10.27) and (10.28) are similar substitutions, and are left to the exercises. The relations (10.29) and (10.30) follow from the fact that taking a derivative of $\exp(ikx)$ with respect to x is the same as multiplying by ik, while taking the derivative of $\exp(-ikx)$ with respect to k is the same as multiplying by $-ix$.

$$m(x) = \frac{df(x)}{dx} = \frac{1}{\sqrt{2\pi}}\int_{-\infty}^{\infty} dk\, \hat{f}(k)\frac{de^{ikx}}{dx} = \frac{1}{\sqrt{2\pi}}\int_{-\infty}^{\infty} dk\, ik\hat{f}(k)e^{ikx}, \tag{10.32}$$

10.2. Basic Properties of Fourier Transforms

$$i\frac{d\hat{f}(k)}{dk} = \frac{id}{dk}\frac{1}{\sqrt{2\pi}}\int_{-\infty}^{\infty}f(x)e^{-ikx}dx = \frac{1}{\sqrt{2\pi}}\int_{-\infty}^{\infty}xf(x)e^{-ikx}dx$$
$$= \hat{n}(k). \quad \blacksquare \qquad (10.33)$$

Example 1: In the proof of Theorem 10.1, we had to compute the Fourier transforms of $g(x) = \exp(-x^2/2)/\sqrt{2\pi}$, of $g_\epsilon(x) = g(x/\epsilon)/\epsilon$, and of $g_\epsilon(x - a)$. Theorem 10.3 allows us to get by with a single calculation. Once we know that $\hat{g}(k) = \exp(-k^2/2)/\sqrt{2\pi}$, it follows from (10.28) that $\hat{g}_\epsilon(k) = \exp(-\epsilon^2 k^2/2)/\sqrt{2\pi}$. By (10.26) we then know that $\exp(-ika)\exp(-\epsilon^2 k^2/2)/\sqrt{2\pi}$ is the Fourier transform of $g_\epsilon(x - a)$.

Example 2: The function $g(x) = x\exp(-x^2/2)$ is minus the derivative of $f(x) = \exp(-x^2/2)$. By (10.29), $\hat{g}(k)$ equals $-ik\hat{f}(k)$, which equals $-ik\exp(-k^2/2)$. This could also have been computed from (10.30).

Example 3: Although we have only proven the existence of the Fourier transform for square-integrable functions, it is actually defined for many functions and generalized functions that are not square-integrable. For example, the Fourier transform of the Dirac δ function is

$$\hat{\delta}(k) = \frac{1}{\sqrt{2\pi}}\int_{-\infty}^{\infty}dx\,\exp(-ikx)\delta(x) = \frac{1}{\sqrt{2\pi}}. \qquad (10.34)$$

Similarly, the Fourier transform of the constant function $f(x) = 1$ is $\hat{f}(k) = \delta(k)\sqrt{2\pi}$.

Example 4: Consider the step function $f(x)$ defined in (10.8). The derivative of the step function (viewed as a distribution) is $\delta(x+1) - \delta(x-1)$. By (10.26) and example 3, the Fourier transform of f' is $(\exp(ik) - \exp(-ik))/\sqrt{2\pi} = 2i\sin(k)/\sqrt{2\pi}$. By (10.28), $\hat{f}(k) = 2\sin(k)/(k\sqrt{2\pi})$, as we previously found. Similarly, the derivative of the step function

$$\xi_{[a,b]} = \begin{cases} 1 & \text{if } a \leq x \leq b; \\ 0 & \text{otherwise} \end{cases} \qquad (10.35)$$

is $\delta(x - a) - \delta(x - b)$, whose Fourier transform is $(\exp(-ika) - \exp(-ikb))/\sqrt{2\pi}$. Applying (10.26), we then find that

$$\hat{\xi}_{[a,b]}(k) = \frac{\exp(-ika) - \exp(-ikb)}{ik\sqrt{2\pi}}. \qquad (10.36)$$

Example 5: Consider the triangle wave

$$t(x) = \begin{cases} 0 & \text{if } x < -1; \\ x+1 & \text{if } -1 \leq x < 0; \\ 1-x & \text{if } 0 \leq x < 1; \\ 0 & \text{if } x \geq 1. \end{cases} \qquad (10.37)$$

Since the derivative of the triangle wave is $\xi_{[-1,0]} - \xi_{[0,1]}$, we have

$$ik\hat{t}(k) = \hat{\xi}_{[-1,0]}(k) - \hat{\xi}_{[0,1]}(k) = \frac{e^{ik} + e^{-ik} - 2}{ik\sqrt{2\pi}} = 2\frac{\cos(k) - 1}{ik\sqrt{2\pi}}, \qquad (10.38)$$

and so

$$\hat{t}(k) = \sqrt{\frac{2}{\pi}} \frac{1 - \cos(k)}{k^2}. \qquad (10.39)$$

These examples indicate a general trend that we have already seen for Fourier series. The smoother a function is, the faster its Fourier transform decays with k. The triangle wave (10.37) is continuous but not differentiable, and its Fourier transform goes as k^{-2}. The step function (10.35) is well-defined but not continuous, and its Fourier transform goes as k^{-1}. The δ function is not even a function in the usual sense, and its transform goes as k^0. In general, if a function is r-times differentiable everywhere, with a finite number of discontinuities in the $r+1$st derivative, then its Fourier transform will typically go as $k^{-(r+2)}$. If a function is infinitely differentiable, then its Fourier transform will decay faster than any power of k (e.g., like $e^{-|k|}$ or e^{-k^2}).

There is a similar relation between the decay of $f(x)$ for large x and the smoothness of $\hat{f}(k)$. The relation (10.30) shows that multiplying $f(x)$ by x is equivalent to taking the derivative of $\hat{f}(k)$. If $f(x)$ decays rapidly, we can multiply by x many times, and so can take the derivative of $\hat{f}(k)$ many times. The general rule is exactly the reverse of the relation between the smoothness of f and the decay of $\hat{f}$. If $f(x)$ goes as $|x|^{-(r+2)}$ for large $|x|$, then $\hat{f}(k)$ will be r-times differentiable everywhere, and the $r+1$st derivative will typically jump at a finite number of points.

Example 6: The function $f(x) = e^{-|x|}$ has a Fourier transform

$$\begin{aligned}\hat{f}(k) &= \frac{1}{\sqrt{2\pi}} \int_{-\infty}^{\infty} dx\, e^{-ikx} e^{-|x|} \\ &= \frac{1}{\sqrt{2\pi}} \left(\int_{-\infty}^{0} dx\, e^{(1-ik)x} + \int_{0}^{\infty} dx\, e^{-(1+ik)x} \right)\end{aligned}$$

10.2. Basic Properties of Fourier Transforms

$$= \frac{1}{\sqrt{2\pi}} \left(\frac{1}{1-ik} + \frac{1}{1+ik} \right) = \frac{\sqrt{2/\pi}}{(1+k^2)}. \quad (10.40)$$

The function $f(x)$ is continuous and piecewise differentiable, with $f'(x)$ having a jump discontinuity at $x = 0$, so $\hat{f}(k)$ decays as k^{-2}. Since $f(x)$ decays faster than any power of x, $\hat{f}(k)$ is infinitely differentiable. The function $g(x) = 1/(x^2+1)$ has the opposite properties. It is infinitely differentiable, so its Fourier transform $\hat{g}(k)$ should decay faster than any power of k. However, $g(x)$ decays only as x^{-2} for large x, so $\hat{g}(k)$ should be continuous, but only piecewise differentiable, with jumps in its first derivative. In fact $\hat{g}(k) = \sqrt{\pi/2} e^{-|k|}$.

Example 7: The function $f(x) = e^{-x^2/2}$ is infinitely differentiable and decays faster than any power of x. Therefore its Fourier transform must decay faster than any power of k and be infinitely differentiable. Indeed, $\hat{f}(k) = e^{-k^2/2}$ does have those properties. Functions that are infinitely differentiable and decay faster than any power of x are called *smooth functions of rapid decrease* or *Schwarz functions*. The test functions of Section 9.3 are all Schwarz functions.

Table 10.2 gives a list of some common functions and their Fourier transforms, together with some of the rules of Theorems 10.3 and 10.4.

Exercises

1. Derive the identity (10.27).
2. Derive the identity (10.28).
3. The Fourier transform of a function **f** is $\hat{f}(k) = (1+|k|)^{-5}$. How many times differentiable is **f**? How quickly does $f(x)$ decay as $x \to \pm\infty$?
4. Compute the Fourier transform of $f(x) = x^5 e^{-x^2/2}$.
5. Find the Fourier transform of $f(x) = \frac{e^{2ix}}{(x-3)^2+1}$.
6. Let $f(x) = \begin{cases} e^x & \text{if } x \geq 0; \\ -e^{-x} & \text{if } x < 0. \end{cases}$ Find $\hat{f}(k)$.
7. In the previous problem, what is $f'(x)$? How does $ikf(k)$ compare to the Fourier transforms of $e^{-|x|}$ and of $\delta(x)$?

Name	Function	Fourier transform
Standard Gaussian	$e^{-x^2/2}$	$e^{-k^2/2}$
General Gaussian	$e^{-(x-\mu)^2/2\sigma^2}$	$\sigma e^{ik\mu} e^{-k^2\sigma^2/2}$
Standard Lorentzian	$(x^2+1)^{-1}$	$\sqrt{\pi/2}\, e^{-|k|}$
Exponential decay	$e^{-|x|}$	$\sqrt{2/\pi}/(k^2+1)$
Delta function	$\delta(x-a)$	$e^{ika}/\sqrt{2\pi}$
Complex exponential	e^{iax}	$\sqrt{2\pi}\,\delta(k-a)$
Step function	$\begin{cases} 1 & \text{if } a < x < b; \\ 0 & \text{otherwise.} \end{cases}$	$\dfrac{i(e^{ikb}-e^{ika})}{k\sqrt{2\pi}}$
Derivative	$f'(x)$	$ik\hat{f}(k)$
x times function	$xf(x)$	$i\dfrac{d\hat{f}(k)}{dk}$
Translate	$f(x-a)$	$e^{ika}\hat{f}(k)$
Boost	$e^{iax}f(x)$	$\hat{f}(k-a)$
Product	$f(x)g(x)$	$\hat{f}*\hat{g}(k)/\sqrt{2\pi}$
Convolution	$f*g(x)$	$\sqrt{2\pi}\,\hat{f}(k)\hat{g}(k)$

Table 10.1: Common Fourier transforms

10.3 Convolutions and Differential Equations

One of the primary uses of Fourier transforms is to solve differential equations. For example, suppose you are given a function $g(x)$ and asked to solve the differential equation

$$f''(x) - f(x) = g(x) \tag{10.41}$$

for $f(x)$. One approach is to take the Fourier transform of both sides. By (10.28), applied twice, the Fourier transform of $f''(x)$ is just $-k^2 \hat{f}(k)$, so we have

$$-(k^2 + 1)\hat{f}(k) = \hat{g}(k), \tag{10.42}$$

or equivalently

$$\hat{f}(k) = -\hat{g}(k)/(k^2 + 1). \tag{10.43}$$

To find $f(x)$ we take the Fourier tranform of $g(x)$, divide by (k^2+1), multiply by -1, and take the inverse Fourier transform of the result.

The tricky step is the last one. We know a function ($g(x)$ itself) whose Fourier transform is $\hat{g}(k)$, and we know a function ($e^{-|x|}$) whose Fourier transform is a constant times $1/(1+k^2)$, but how are we to find a function whose Fourier transform is the product of $\hat{g}(k)$ and $(1+k^2)^{-1}$? The keys are the following definition and theorem.

Definition *The convolution of two functions* **f** *and* **g**, *denoted* **f**∗**g**, *is the function*

$$f * g(x) = \int_{-\infty}^{\infty} dy \, f(x-y) g(y). \tag{10.44}$$

Similarly, the convolution of two functions $\hat{f}$ *and* $\hat{g}$ *is the function*

$$\hat{f} * \hat{g}(k) = \int_{-\infty}^{\infty} dk' \, \hat{f}(k-k') \hat{g}(k'). \tag{10.45}$$

Theorem 10.4 *Let* **f** *and* **g** *be functions such that* **f**, **g**, *the product* **fg**, *and the convolution* **f**∗**g** *are all square integrable. Then*

$$\widehat{f * g}(k) = \sqrt{2\pi} \hat{f}(k) \hat{g}(k); \qquad \widehat{fg}(k) = \frac{1}{\sqrt{2\pi}} \hat{f} * \hat{g}(k). \tag{10.46}$$

In other words, up to factors of $\sqrt{2\pi}$, the Fourier transform of a convolution is the product of the Fourier transforms, while the Fourier transform of a product is the convolution of the Fourier transforms.

Proof: We use the substitution $z = x - y$ to compute

$$\begin{aligned}
\widehat{f*g}(k) &= \frac{1}{\sqrt{2\pi}} \int_{-\infty}^{\infty} dx \, f*g(x) e^{-ikx} \\
&= \frac{1}{\sqrt{2\pi}} \int_{-\infty}^{\infty} dx \int_{-\infty}^{\infty} dy \, e^{-ikx} f(x-y) g(y) \\
&= \frac{1}{\sqrt{2\pi}} \int_{-\infty}^{\infty} dz \int_{-\infty}^{\infty} dy \, e^{-ik(y+z)} f(z) g(y) \\
&= \sqrt{2\pi} \left(\frac{1}{\sqrt{2\pi}} \int_{-\infty}^{\infty} dz \, f(z) e^{-ikz} \right) \left(\frac{1}{\sqrt{2\pi}} \int_{-\infty}^{\infty} dy \, g(y) e^{-iky} \right) \\
&= \sqrt{2\pi} \hat{f}(k) \hat{g}(k). \quad (10.47)
\end{aligned}$$

Similarly, we can use the Fourier integral (10.1) and the substitution $k'' = k - k'$ to show that the function whose Fourier transform is $\hat{f} * \hat{g}/\sqrt{2\pi}$ is **fg**. ∎

Theorem 10.4 gives us our solution to the differential equation (10.41). Let $h(x) = -e^{-|x|}/2$. Since we have already computed $\hat{h}(x) = -1/(\sqrt{2\pi}(1+k^2))$, the Fourier transform of $h*g$ is $-\hat{g}(k)/(1+k^2)$. Our solution is then

$$f(x) = h*g(x) = -\int_{-\infty}^{\infty} dy \, e^{-|x-y|} g(y)/2. \quad (10.48)$$

This formula will be revisited in Chapter 11.

Exercises

1. Show that $f*g(x) = g*f(x)$ for arbitrary functions **f** and **g**.

2. Let $f(x) = \exp(-x^2)$. Suppose g is a continuous function in $L^2(\mathbb{R})$. Show that **f∗g** is infinitely differentiable. (Operators such as **f∗**, that send continuous functions to infinitely differentiable functions, are known as *smoothing operators*.)

In Exercises 3–5, solve (10.41) for **f**, with **g** as follows:

3. $g(x) = e^{-|x|}$.

4. $g(x) = \sin(x)$.

5. $g(x) = \begin{cases} 1 & \text{if } x \geq 0; \\ 0 & \text{if } x < 0. \end{cases}$

6. Use Fourier transforms to convert the socond-order ordinary differential equation (ODE) $f''(x) + xf(x) = g(x)$ into a first-order ODE for $\hat{f}$. Use this setup to solve $f''(x) + xf(x) = 1$.

10.4 Partial Differential Equations

In this section we use Fourier transforms to study several important partial differential equations (PDEs) that arise in physics, namely the heat, Schrödinger, wave, and Klein-Gordon equations. In each case the Fourier transform converts the PDE into a family of ordinary differential equations that we already know how to solve.

The heat equation

Imagine that heat is distributed along a line. Let $f(x,t)$ be the temperature at point x, at time t. The equation that governs the distribution of heat is

$$\frac{\partial f(x,t)}{\partial t} = D\frac{\partial^2 f(x,t)}{\partial x^2}, \tag{10.49}$$

where D is a constant. (The letter D stands for "diffusion", and is related to the rate at which heat diffuses out from one area to another.) Equation (10.49) is called the *heat equation*, and applies to many physical systems (diffusion and Brownian motion, to name two) besides heat. A physical derivation of (10.49) is included in the exercises.

Equation (10.49) is of a form we already understand, $d\mathbf{f}/dt = A\mathbf{f}$, where in this case

$$A = D\frac{d^2}{dx^2}. \tag{10.50}$$

As in Chapter 5, we must diagonalize A, expand $\mathbf{f}$ in the eigenfunctions of A, multiply the coefficient of eigenvalue λ by $\exp(\lambda t)$, and then recombine to get $\mathbf{f}$ at time t. In this case, the diagonalization, expansion, and recombination are all taken care of by Fourier analysis.

Notice that the k-th Fourier mode, $|K,k\rangle = \exp(ikx)$, is an eigenfunction of A:

$$A|K,k\rangle = -Dk^2|K,k\rangle. \tag{10.51}$$

Expanding the two sides of (10.49) in eigenfunctions of A is the same as taking the Fourier transform of each side. The result is

$$\frac{\partial \hat{f}(k,t)}{\partial t} = -Dk^2 \hat{f}(k,t), \tag{10.52}$$

whose solution is simply

$$\hat{f}(k,t) = \exp(-Dk^2 t)\hat{f}(k,0). \tag{10.53}$$

Taking the inverse Fourier transform (10.1) of each side and applying Theorem 10.4 we get

$$f(x,t) = G_t * f_0(x), \qquad (10.54)$$

where $f_0(x) = f(x,0)$ and

$$G_t(x) = \frac{\exp(-x^2/4Dt)}{2\sqrt{\pi Dt}} \qquad (10.55)$$

is the inverse Fourier transform of $\exp(-Dk^2t)/\sqrt{2\pi}$. Notice that $G_t(x-y)$ is positive for all values of x and y, but is very small if $|x-y|$ is significantly bigger than $2\sqrt{Dt}$. In other words, heat diffuses out from the original distribution f_0, and while *some* heat gets arbitrarily far, arbitrarily soon, the bulk of the heat takes a time of order L^2/D to diffuse a distance L.

For example, suppose that $D = 1$, and that

$$f_0(x) = \begin{cases} 1 & \text{if } -1 < x < 1; \\ 0 & \text{otherwise.} \end{cases} \qquad (10.56)$$

Then

$$f(x,t) = \int_{-\infty}^{\infty} G_t(x-y)f_0(y)dy = \frac{1}{2\sqrt{\pi t}} \int_{-1}^{1} \exp(-(x-y)^2/4t)dy. \qquad (10.57)$$

If x is substantially bigger than $1 + 2\sqrt{t}$, or substantially smaller than $-1 - 2\sqrt{t}$, then this integral is extremely small. Figure 10.1 shows $f(x,t)$ for several values of t.

The Schrödinger equation

We next consider the Schrödinger equation, which governs nonrelativistic quantum mechanics. In quantum mechanics, the probability of finding a particle at point x, at time t is proportional to the absolute value squared of a "wave function" $\psi(x,t)$, that evolves according to the Schrödinger equation

$$i\hbar \frac{\partial \psi(x,t)}{\partial t} = \frac{-\hbar^2}{2m} \frac{\partial^2 \psi(x,t)}{\partial x^2} + V(x)\psi(x,t), \qquad (10.58)$$

where $V(x)$ is the potential energy, m is the mass of the particle, and $\hbar$ is Planck's constant divided by 2π. For a free particle, $V(x) = 0$ and the Schrödinger equation reduces to

$$\frac{\partial \psi(x,t)}{\partial t} = \frac{i\hbar}{2m} \frac{\partial^2 \psi(x,t)}{\partial x^2}. \qquad (10.59)$$

10.4. Partial Differential Equations

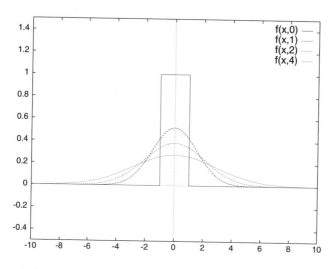

Figure 10.1: Diffusion from a step function

This is just like the heat equation (10.49), only with a pure imaginary diffusion constant $D = i\hbar/2m$, and we solve it in the same way. Taking the Fourier transform of both sides gives the differential equation

$$\frac{\partial \hat{\psi}(k,t)}{\partial t} = \frac{-i\hbar k^2}{2m}\hat{\psi}(k,t), \tag{10.60}$$

with solution

$$\hat{\psi}(k,t) = \exp(-ik^2 t/2m)\hat{\psi}(k,0), \tag{10.61}$$

hence

$$\psi(x,t) = \sqrt{\frac{m}{2\pi i\hbar t}} \int_{-\infty}^{\infty} \exp(im(x-y)^2/2\hbar t)\psi(y,0)dy. \tag{10.62}$$

This is identical in form to (10.54), only with $D = i\hbar/2m$. However, there is one key difference. While the heat kernel $G_t(x-y)$ is small in absolute value for $|x-y|$ large, the absolute value of $\exp(im(x-y)^2/2\hbar t)$ is equal to one for all values of x and y. As a result, a solution $\psi(x,t)$ to the Schrödinger equation may decay, for large x, far more slowly than a solution to the heat equation with the same initial conditions.

This is shown in Figure 10.2, which shows the solution to the Schrödinger equation (10.58), with $\hbar = 2m = 1$, and with initial condition

$$\psi(x,0) = \begin{cases} 1 & \text{if } -1 < x < 1; \\ 0 & \text{otherwise.} \end{cases} \tag{10.63}$$

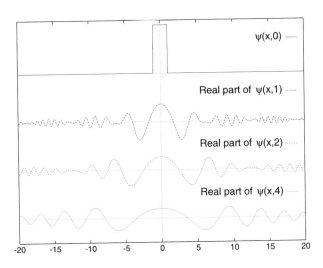

Figure 10.2: Solutions to the Schrödinger equation

The wave equation

We now return to the wave equation (8.1) on the whole line. The equation and its Fourier transform are

$$\frac{\partial^2 f(x,t)}{\partial t^2} = v^2 \frac{\partial^2 f(x,t)}{\partial x^2},$$
$$\frac{\partial^2 \hat{f}(k,t)}{\partial t^2} = -v^2 k^2 \hat{f}(k,t). \quad (10.64)$$

From this we see that $\hat{f}(k,t)$ must take the form $\hat{h}_1(k)\exp(-i\omega_k t) + \hat{h}_2(k)\exp(i\omega_k t)$, where $\omega_k = vk$, and where $\hat{h}_1(k)$ and $\hat{h}_2(k)$ are coefficients that are yet to be determined. Comparing to the initial conditions

$$f_0(x) = f(x,0) = \frac{1}{\sqrt{2\pi}} \int_{-\infty}^{\infty} \hat{f}(k,0)\exp(ikx)dk,$$
$$g_0(x) = \frac{\partial f}{\partial t}(x,0) = \frac{1}{\sqrt{2\pi}} \int_{-\infty}^{\infty} \frac{\partial \hat{f}}{\partial t}(k,0)\exp(ikx)dk, \quad (10.65)$$

we see that

$$\hat{f}_0(k) = \hat{h}_1(k) + \hat{h}_2(k),$$
$$\hat{g}_0(k) = -i\omega_k(\hat{h}_1(k) - \hat{h}_2(k)), \quad (10.66)$$

and hence that

$$\hat{h}_1(k) = \frac{\omega_k \hat{f}_0(k) + i\hat{g}_0(k)}{2\omega_k},$$

10.4. Partial Differential Equations

$$\hat{h}_2(k) = \frac{\omega_k \hat{f}_0(k) - i\hat{g}_0(k)}{2\omega_k}. \tag{10.67}$$

The functions $\mathbf{h_1}$ and $\mathbf{h_2}$ are the same as those of Section 8.1, that is, the forward and backward traveling waves. To see this, note that, since $\omega_k = vk$, we have

$$\begin{aligned}
f(x,t) &= \frac{1}{\sqrt{2\pi}} \int_{-\infty}^{\infty} (\hat{h}_1(k)\exp(-ikvt) + \hat{h}_2(k)\exp(ikvt))\exp(ikx)dk \\
&= \frac{1}{\sqrt{2\pi}} \int_{-\infty}^{\infty} (\hat{h}_1(k)\exp(ik(x-vt)) + \hat{h}_2(k)\exp(ik(x+vt)))dk \\
&= h_1(x-vt) + h_2(x+vt). \tag{10.68}
\end{aligned}$$

The Klein-Gordon equation

While the wave equation governs light, the Klein-Gordon equation

$$\hbar^2 \frac{\partial^2 \phi(x,t)}{\partial t^2} = \hbar^2 c^2 \frac{\partial^2 \phi(x,t)}{\partial x^2} - m^2 c^4 \phi(x,t), \tag{10.69}$$

governs massive noninteracting relativistic particles in quantum mechanics (the "free field"). Here, $\phi(x,t)$ is the wavefunction at position x and time t, m is the mass of the particle, and c is the speed of light. To simplify things, we choose units in which $m = \hbar = c = 1$, which reduces the Klein-Gordon equation to

$$\frac{\partial^2 \phi(x,t)}{\partial t^2} = \frac{\partial^2 \phi(x,t)}{\partial x^2} - \phi(x,t). \tag{10.70}$$

In vector form, this is $d^2\phi/dt^2 = A\phi$, where $A = d^2/dx^2 - I$. As with the previous equations, $|K, k\rangle$ is a generalized eigenfunction of A, this time with eigenvalue $-(k^2+1)$. Expanding ϕ in eigenfunctions of A (i.e., taking the Fourier transform of (10.70)), we obtain

$$\frac{\partial^2 \hat{\phi}(k,t)}{\partial t^2} = -(k^2+1)\hat{\phi}(k,t), \tag{10.71}$$

so $\hat{\phi}(k,t)$ can once again be written as

$$\hat{\phi}(k,t) = \tilde{h}_1(k)\exp(-i\omega_k t) + \tilde{h}_2(k)\exp(i\omega_k t), \tag{10.72}$$

only this time with

$$\omega_k = \sqrt{k^2+1}. \tag{10.73}$$

As with the wave equation, the initial field, and time derivative of the field are given by (10.65), and (10.66) gives $\hat{h}_i(k)$ in terms of $\hat{f}_0$ and $\hat{g}_0$.

The biggest difference between the wave and Klein-Gordon equations is that, in the Klein-Gordon equation the terms involving $\hat{h}_1(k)$ (resp $\hat{h}_2(k)$) are not forward (backward) traveling waves. We still have

$$\phi(x,t) = \frac{1}{\sqrt{2\pi}} \int_{-\infty}^{\infty} (\hat{h}_1(k)\exp(i(kx-\omega_k t)) + \hat{h}_2(k)\exp(i(kx+\omega_k t))) dk, \tag{10.74}$$

but, with ω_k no longer proportional to k, $kx \mp \omega_k t$ is no longer a function only of $x \mp vt$. For the wave equation, Fourier analysis was convenient, but a traveling wave solution could be constructed from initial data without it. For the Klein-Gordon equation, Fourier analysis is essential.

Exercises

1. Derive the heat equation as follows. Suppose we have a wire of constant thickness and constant heat capacity, so that the heat energy per unit length is the heat capacity c, times the temperature T. Suppose further that heat flows from hot to cold regions at a rate proportional to the difference in temperature. In other words, the rate at which heat is flowing past a point is $-\sigma \partial T/\partial x$, where σ is the (constant) heat conductivity. Compute the rate at which heat is flowing into the interval $[a, a+\Delta a]$. From this, derive a formula for the evolution of the average temperature on that interval. Finally, take the limit as $\Delta a \to 0$ to obtain the heat equation. How is the diffusion constant D, related to the heat capacity c, and the conductivity σ?

2. Solve the heat equation with $D=1$ and initial data $f(x,0) = e^{-x^2}$.

3. Solve the Schrödinger equation with $\hbar = 2m = 1$ and initial data $\psi(x,0) = e^{-x^2}$. Compare your answer to that of Exercise 2. Which solution spreads out faster in the long run? Which spreads faster initially?

4. Solve the Schrödinger equation with $\hbar = 2m = 1$ and initial data $\psi(x,0) = e^{iax-x^2}$, where a is an arbitrary constant. Where is the solution peaked at time t? How does the rate of spreading depend on a?

10.5. Bandwidth and Heisenberg's Uncertainty Principle

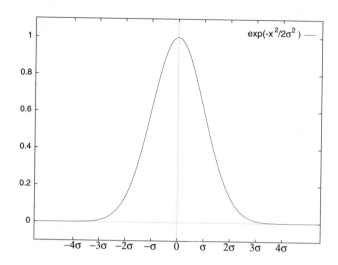

Figure 10.3: The width of a Gaussian is of order σ

5. Show that the formula (10.67) for $\hat{h}_1$ and $\hat{h}_2$ is equivalent to the formula (8.13) for $\mathbf{h}_1$ and $\mathbf{h}_2$.

6. Solve the wave and Klein-Gordon equations with initial data $f_0(x) = \begin{cases} 1 & \text{if } -1 < x < 1 \\ 0 & \text{otherwise} \end{cases}$ and $g_0(x) = 0$. Plot the results at times $t = 1$ and $t = 4$. You may need to do the integral (10.74) numerically.

10.5 Bandwidth and Heisenberg's Uncertainty Principle

Consider the function

$$f(x) = \exp(-x^2/2\sigma^2). \tag{10.75}$$

The graph of **f** is a bell curve, also known as a Gaussian, centered at $x = 0$, with width of order σ. I say "of order" σ, since we have not precisely defined what the width of a curve is. In this case, a slight majority of the area under the curve is between $x = -\sigma$ and $x = \sigma$, while an overwhelming majority (about 95%) is between $x = -2\sigma$ and $x = 2\sigma$. By any reasonable yardstick, the width of the curve is of order σ. See Figure 10.3.

The Fourier transform of **f** is

$$\hat{f}(k) = \sigma \exp(-k^2 \sigma^2/2). \tag{10.76}$$

308   Chapter 10. Fourier Transforms

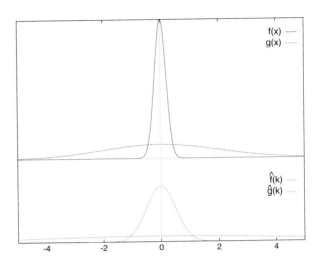

Figure 10.4: Narrower functions have wider Fourier transforms

This is also a bell curve, of width of order $1/\sigma$. If σ is small, then the graph of $\mathbf{f}$ is narrow, but the graph of $\hat{\mathbf{f}}$ is wide. If σ is large, then the graph of $\hat{\mathbf{f}}$ is narrow, but the graph of $\mathbf{f}$ is wide. See Figure 10.4. There is no way to get both graphs to be narrow. This is an example of

Heisenberg's Uncertainty Principle: *The width of the graph of a function, times the width of the graph of its Fourier transform, is at least of order 1.*

Without a definition of the width of a curve, we cannot make the uncertainty principle more precise. We will make such a definition and prove a precise restatement of the principle later in this section. However, those definitions are not suited for all applications, and the imprecise principle stated above is often much more useful than the precise theorem proven below. So, before seeking precision, we examine some of the qualitative implications of the uncertainty principle.

Suppose you want to send a message over a telephone line. Telephone lines are designed to deliver sound to human ears, most of which can only hear sounds of up to about 10KHz (10,000 cycles per second). Traditional analog telephone lines (called POTS, for "plain old telephone system") greatly attenuate signals with a frequency of over 4KHz, thereby filtering away annoying high-pitched background noise without harming voice transmission. However, this filtering severely limits the ability of POTS lines to carry data. A

10.5. Bandwidth and Heisenberg's Uncertainty Principle

signal that emerges from a POTS line will have a Fourier transform, whose width is at most about 8KHz, or about 50,000 radians/second. By the uncertainty principle, this transmitted signal will have a duration of at least 1/50,000 of a second. Two signals spaced less than 1/50,000 of a second apart will overlap and be essentially impossible to disentangle. Therefore, you cannot send or receive more than about 50,000 bits of information per second over a POTS line, no matter how good your equipment is. Inexpensive modems, using POTS lines, pretty much reached this theoretical limit in 1998.[1]

The key issue is *bandwidth*. Signals carried on frequencies between ω_{min} and ω_{max} trap the signal's Fourier transform within a band of width $\Delta\omega \sim \omega_{max} - \omega_{min}$. By the uncertainty principle, each bit of signal information occupies a time interval Δt, of length at least $1/\Delta\omega$, making it impossible to transmit information at a rate much faster than $\Delta\omega$.

Rapid transmission of data requires a large range of frequencies. Consider a fiber optic cable that carries visible light. Since visible light ranges in frequency from under 4×10^{14}Hz for red light to over 7×10^{14}Hz for blue light, the fiber optic bandwidth is over 3×10^{14}Hz, almost 100 billion times wider than a POTS line, or enough bandwidth to allow over a quadrillion bits of information per second. In principle, one optical fiber could carry far more information than all of the world's POTS lines put together!

The precise form of the uncertainty principle

To properly state the uncertainty principle, we must first define the width of a function precisely.

Definition *Let M be a Hermitian operator on an inner product space, and let $\mathbf{f}$ be a vector in that space. The* expectation *of M in the state $\mathbf{f}$ is $\langle \mathbf{f}|X|\mathbf{f}\rangle/\langle \mathbf{f}|\mathbf{f}\rangle$. The* variance *of M (in the state $\mathbf{f}$) is the expectation of $(M-\mu)^2$, where μ is itself the expectation of M. The* standard deviation *of M is the square root of the variance.*

When the inner product space is $L^2(\mathbb{R})$, the expectation, variance, and standard deviation of the position operator X are denoted

---

[1] Asymmetrical digital subscriber line (ADSL) technology promises much faster transmission using standard copper telephone wires. However, "copper wire" is not the same thing as POTS. ADSL uses very high frequencies, well outside the 0-4KHz POTS band, and requires lines to be cleared of filters and other impediments to high-frequency transmission.

μ_x, V_x, and Δx, respectively. Similarly, the expectation, variance, and standard deviation of the momentum operator K are denoted μ_k, V_k, and Δk.

For example, suppose that

$$f(x) = \begin{cases} 1 & \text{if } 0 < x < 1; \\ 0 & \text{otherwise.} \end{cases} \qquad (10.77)$$

Then we compute

$$\langle \mathbf{f}|\mathbf{f}\rangle = \int_{-\infty}^{\infty} |f(x)|^2 dx = \int_0^1 1 dx = 1$$

$$\langle \mathbf{f}|X|\mathbf{f}\rangle = \int_{-\infty}^{\infty} x|f(x)|^2 dx = \int_0^1 x dx = 1/2, \qquad (10.78)$$

so $\mu_x = 1/2$. The variance of the position operator is then

$$V_x = \frac{\int_{-\infty}^{\infty} \left(x - \frac{1}{2}\right)^2 |f(x)|^2 dx}{\int_{-\infty}^{\infty} |f(x)|^2 dx} = \int_0^1 \left(x - \frac{1}{2}\right)^2 dx = 1/12 \qquad (10.79)$$

and the standard deviation is

$$\Delta x = \sqrt{V_x} = \frac{1}{2\sqrt{3}}. \qquad (10.80)$$

In general, the expectation of the position is a weighted average of x, where the weight is proportional to $|f(x)|^2$. As such, it gives a measure of the middle of the wave. The variance is the average of $(x - \mu_x)^2$, and as such measures the average squared deviation from the middle. It is therefore an indicator of squared width. The standard deviation is then an indicator of width. Similarly, the expectation, variance, and standard deviation of the momentum operator give the middle, squared width, and width of the function $\hat{f}(k)$:

$$\mu_k = \frac{\int_{-\infty}^{\infty} k|\hat{f}(k)|^2 dk}{\int_{-\infty}^{\infty} |\hat{f}(k)|^2 dk} = \frac{\langle \mathbf{f}|K|\mathbf{f}\rangle}{\langle \mathbf{f}|\mathbf{f}\rangle},$$

$$V_x = \frac{\int_{-\infty}^{\infty} (k - \mu_k)^2 |\hat{f}(k)|^2 dk}{\int_{-\infty}^{\infty} |\hat{f}(k)|^2 dk} = \frac{\langle \mathbf{f}|(K - \mu_k)^2|\mathbf{f}\rangle}{\langle \mathbf{f}|\mathbf{f}\rangle}$$

$$\Delta k = \sqrt{V_k}. \qquad (10.81)$$

With these definitions we can now state the precise uncertainty principle:

10.5. Bandwidth and Heisenberg's Uncertainty Principle

Theorem 10.5 (the Uncertainty Principle) *Let* $\mathbf{f}$ *be any element of* $L^2(\mathbb{R})$. *Then the widths* Δx *of* $f(x)$ *and* Δk *of* $\hat{f}(k)$ *satisfy* $(\Delta x)(\Delta k) \geq 1/2$.

Proof: We prove the theorem in the special case that $\mu_x = \mu_k = 0$. The extension to the general case is left to the exercises. Rescale $\mathbf{f}$, so that $\langle \mathbf{f}|\mathbf{f}\rangle = 1$. Since $\mu_x = \mu_k = 0$, we then have $(\Delta x)^2 = \langle \mathbf{f}|X^2|\mathbf{f}\rangle$, and $(\Delta k)^2 = \langle \mathbf{f}|K^2|\mathbf{f}\rangle$. For every real number t we compute

$$\begin{aligned}
0 &\leq \langle (X+itK)\mathbf{f}|(X+itK)\mathbf{f}\rangle \\
&= \langle \mathbf{f}|(X-itK)(X+itK)|\mathbf{f}\rangle \qquad \text{since } X \text{ and } K \text{ are Hermitian} \\
&= \langle \mathbf{f}|X^2 + t^2K^2 + it(XK - KX)|\mathbf{f}\rangle \\
&= \langle \mathbf{f}|X^2 + t^2K^2 - t|\mathbf{f}\rangle \qquad \text{by Exercise 1} \\
&= (\Delta x)^2 + t^2(\Delta k)^2 - t. \qquad (10.82)
\end{aligned}$$

Since this is true for all t, it is in particular true for $t = 1/(2(\Delta k)^2)$, so

$$\begin{aligned}
0 &\leq (\Delta x)^2 + \frac{1}{4(\Delta k)^2} - \frac{1}{2(\Delta k)^2} \\
&= (\Delta x)^2 - \frac{1}{4(\Delta k)^2}. \qquad (10.83)
\end{aligned}$$

Multiplying by $(\Delta k)^2$, adding $1/4$ to each side, and taking a square root then gives $1/2 \leq (\Delta x)(\Delta k)$. ∎

AM, FM, and PM

Suppose you want to send a low-frequency signal, such as a human voice or a piece of music, over a medium that only transmits high frequencies (such as the atmosphere carrying radio waves). You need a way to encode the low-frequency data into a high-frequency wave. There are three standard technologies for doing this, called *amplitude modulation*, or AM, *frequency modulation*, or FM, and *phase modulation*, or PM.

The simplest method is amplitude modulation. Take the signal $f(t)$ and multiply it by a fixed carrier wave $\exp(ibt)$. The result can be viewed as a wave with fixed frequency b, but with varying amplitude $f(t)$. However, the Fourier transform of $h(t) = f(t)\exp(ibt)$ is not localized exactly at frequency b. By Theorem 10.3, the Fourier transform of $\mathbf{h}$ is just the Fourier transform of $\mathbf{f}$, shifted over by the carrier frequency b. Thus, if the Fourier transform of **f** lies between frequencies $-\Delta\omega$ and $+\Delta\omega$, then the Fourier transform of **h** lies between $b-\Delta\omega$ and $b+\Delta\omega$. For radio transmissions, as for telephones, the important sounds have frequencies less than 5KHz, so an AM radio broadcast takes up a band of frequencies approximately of size 10KHz. In fact, AM radio stations are assigned bands of width exactly 10KHz, centered at multiples of 10KHz. To minimize the possibility of interference, radio stations in the same city are usually not assigned adjacent bands.

The next simplest method is phase modulation. Here the broadcast wave is given constant strength, but varying phase. Specifically, the signal $f(t)$ is coded as $\exp(i[bt + f(t)])$. Phase modulation never caught on for commercial radio broadcasts, but is used for other forms of communication.

Finally, there is frequency modulation. As with PM, the broadcast wave has constant strength, but now the signal $f(t)$ is used to vary the frequency, i.e., the rate at which the phase is increasing. Thus, the broadcast wave is $\exp(i[bt + g(t)])$, where $g(t) = \int f(t)dt$. FM radio stations are assigned bands of size 200KHz. This greater bandwidth allows FM radio to carry much more information, hence higher quality sound than AM.

Exercises

1. Show that, for any **f**, $(XK - KX)\mathbf{f} = i\mathbf{f}$.

2. Find μ_x, V_x, and Δx for the function
$$f(x) = \begin{cases} 1 - x^2 & -1 < x < 1; \\ 0 & \text{otherwise.} \end{cases}$$

3. Show that the variance of any Hermitian operator M is the expectation of M^2 minus the square of the expectation of M.

Expectations and variances of the momentum operator can be computed without taking Fourier transforms:

4. For any (sufficiently well behaved) function **f**, show that $\langle \mathbf{f}|K|\mathbf{f}\rangle = -i\int_{-\infty}^{\infty} \overline{f(x)}f'(x)dx$.

5. For any (sufficiently well behaved) function **f**, show that $\langle \mathbf{f}|K^2|\mathbf{f}\rangle = \int_{-\infty}^{\infty} |f'(x)|^2 dx$.

6. Find μ_k, V_k, and Δk for the function of Exercise 2.

7. Suppose $g(x) = f(x - a)$ for all x, where a is a fixed number. Show that $\mu_x(\mathbf{g}) = \mu_x(\mathbf{f}) + a$, but $\sigma_x(\mathbf{g}) = \sigma_x(\mathbf{f})$, $\mu_k(\mathbf{g}) = \mu_k(\mathbf{f})$, and $\sigma_k(\mathbf{g}) = \sigma_k(\mathbf{f})$.

8. Suppose $h(x) = \exp(ibx)f(x)$, for all x, where b is a fixed number. Show that $\mu_k(\mathbf{h}) = \mu_k(\mathbf{f})+b$, but $\sigma_k(\mathbf{h}) = \sigma_k(\mathbf{f})$, $\mu_x(\mathbf{h}) = \mu_x(\mathbf{f})$, and $\sigma_x(\mathbf{h}) = \sigma_x(\mathbf{f})$.

9. Extend the proof of Theorem 10.5 to cover arbitrary functions $\mathbf{f}$, not just those with $\mu_x = \mu_k = 0$.

10. For any $\mathbf{f}$ with $\langle \mathbf{f}|\mathbf{f}\rangle = 1$, show that $V_x = \langle \mathbf{f}|X^2|\mathbf{f}\rangle - (\langle \mathbf{f}|X|\mathbf{f}\rangle)^2$.

11. Prove the generalized uncertainty principle: If A and B are Hermitian operators, and $[A, B] = iC$, then $(\Delta A)(\Delta B) \geq |\mu_C|/2$.

10.6 Fourier Transforms on the Half Line

We return now to analysis on the half line $[0, \infty)$. Problems on the half line come up in a variety of settings, in particular in systems with symmetry. A function on the whole line with a symmetry (say, $f(-x) = -f(x)$, or $f(-x) = f(x)$, or even $f(-x) = \pm\overline{f(x)}$) is determined by its values on the positive half line. Additionally, a function in physical 3-dimensional space (or in other dimensions) that is rotationally symmetric is a function of a single variable, the radius r, that ranges from 0 to ∞.

In this section we develop two forms of Fourier analysis that are applicable to the half line. The first, involving sine functions, is best suited for problems involving Dirichlet boundary conditions. The second, involving cosines, is designed for problems with Neumann boundary conditions.

The sine transform

Theorem 10.6 *Let $\mathbf{f}$ be a square-integrable function on $[0, \infty)$. Then*

$$f(x) = \sqrt{\frac{2}{\pi}} \int_0^\infty \hat{f}^s(k) \sin(kx)\,dk, \tag{10.84}$$

where

$$\hat{f}^s(k) = \sqrt{\frac{2}{\pi}} \int_0^\infty f(x) \sin(kx)\,dx. \tag{10.85}$$

We call $\hat{f}^s$ the *sine transform* of f. Notice that the two equations (10.84) and (10.85) are completely symmetrical, so f is also the sine transform of $\hat{f}^s$.

Proof: We use the method of images, extending the function $f(x)$ to the whole line by defining

$$f(-x) = -f(x). \tag{10.86}$$

Since **f** is square-integrable on the half line and is symmetric, it is also square-integrable on the whole line. By Theorem 10.1, we then have

$$f(x) = \frac{1}{\sqrt{2\pi}} \int_{-\infty}^{\infty} \hat{f}(k) e^{ikx} dk, \qquad (10.87)$$

where

$$\begin{aligned}
\hat{f}(k) &= \frac{1}{\sqrt{2\pi}} \int_{-\infty}^{\infty} f(x) e^{-ikx} dx \\
&= \frac{1}{\sqrt{2\pi}} \int_{-\infty}^{\infty} f(x)(\cos(kx) - i\sin(kx)) dx \\
&= -i\sqrt{\frac{2}{\pi}} \int_{0}^{\infty} f(x) \sin(kx) dx = -i \hat{f}^s(k), \qquad (10.88)
\end{aligned}$$

since $\sin(kx)$ and $f(x)$ are odd functions, while $\cos(kx)$ is even. This also implies that $\hat{f}(-k) = -\hat{f}(k)$. We then have

$$\begin{aligned}
f(x) &= \frac{1}{\sqrt{2\pi}} \int_{-\infty}^{\infty} \hat{f}(k) e^{ikx} dk, \\
&= \frac{1}{\sqrt{2\pi}} \int_{0}^{\infty} (\hat{f}(k) e^{ikx} + \hat{f}(-k) e^{-ikx}) dk \\
&= \frac{1}{\sqrt{2\pi}} \int_{0}^{\infty} \hat{f}(k)(e^{ikx} - e^{-ikx}) dk \\
&= \frac{2i}{\sqrt{2\pi}} \int_{0}^{\infty} \hat{f}(k) \sin(kx) dk \\
&= \sqrt{\frac{2}{\pi}} \int_{0}^{\infty} \hat{f}^s(k) \sin(kx) dk. \quad \blacksquare \qquad (10.89)
\end{aligned}$$

Let $\Delta = d^2/dx^2$ on $[0, \infty)$ with Dirichlet boundary conditions. The function $\sin(kx)$ is an eigenfunction of Δ with eigenvalue $-k^2$, and the sine transform is really just an eigenfunction expansion for the operator Δ. If for $k \geq 0$ we define

$$|\Delta, -k^2\rangle = \sqrt{\frac{2}{\pi}} \sin(kx), \qquad (10.90)$$

then Theorem 10.6 reduces to the statement that

$$|\mathbf{f}\rangle = \int_{0}^{\infty} \hat{f}^s(k) |\Delta, -k^2\rangle dk, \qquad (10.91)$$

where

$$\hat{f}^s(k) = \langle \Delta, -k^2 | \mathbf{f} \rangle. \qquad (10.92)$$

10.6. Fourier Transforms on the Half Line

This implies the decomposition

$$I = \int_0^\infty dk |\Delta, -k^2\rangle\langle\Delta, -k^2| \tag{10.93}$$

of the identity and the normalization

$$\langle\Delta, -k^2|\Delta, -k'^2\rangle = \delta(k - k'). \tag{10.94}$$

Furthermore, the fact that $|\Delta, -k^2\rangle$ is an eigenfunction of Δ implies that the sine transform of $f''(x)$ is $-k^2 \hat{f}^s(k)$. Since the sine transform and the inverse sine transform are the same, it also implies that the sine transform of $x^2 f(x)$ is $-d^2 f(k)/dk^2$.

We use the sine transform to revisit the wave equation

$$\frac{\partial^2 f(x,t)}{\partial x^2} - \frac{1}{v^2}\frac{\partial^2 f(x,t)}{\partial t^2} = 0 \tag{10.95}$$

on the half line with Dirichlet boundary conditions, and initial data $f(x,0) = f_0(x)$ and $\dot{f}(x,0) = g_0(x)$. The sine transform of (10.95) is

$$\frac{\partial^2 \hat{f}^s(k,t)}{\partial t^2} = -k^2 v^2 \hat{f}^s(k,t), \tag{10.96}$$

with initial conditions

$$\hat{f}^s(k,0) = \hat{f}_0^s(k), \qquad \frac{\partial \hat{f}^s}{\partial t}(k,0) = \hat{g}_0^s(k) \tag{10.97}$$

and solution

$$\hat{f}^s(k,t) = \hat{f}_0^s(k)\cos(kvt) + \frac{1}{kv}\hat{g}_0^s(k)\sin(kvt). \tag{10.98}$$

Applying the inverse transform, we get that $f(x,t)$ equals

$$\sqrt{\frac{2}{\pi}}\int_0^\infty \hat{f}_0^s(k)\cos(kvt)\sin(kx) + \frac{\hat{g}_0^s(k)}{kv}\sin(kvt)\sin(kx) dk$$

$$= \frac{1}{\sqrt{2\pi}}\int_0^\infty \Big[\hat{f}_0^s(k)\sin(k(x-vt)) + \hat{f}_0^s(k)\sin(k(x+vt))$$

$$+ \frac{\hat{g}_0^s(k)}{kv}\cos(k(x-vt)) - \frac{\hat{g}_0^s(k)}{kv}\cos(k(x+vt))\Big] dk. \tag{10.99}$$

The terms on the second line of (10.99) are forward and backward traveling waves, and integrate to $\frac{1}{2}f_0(x-vt)$ and $\frac{1}{2}f_0(x+vt)$, respectively. The remaining terms are also traveling waves, being functions of $x - vt$ and $x + vt$, but involve cosines instead of sines. To understand these terms we must first consider the cosine transform.

The cosine transform

The cosine transform is almost identical to the sine transform, only with cosines instead of sines.

Theorem 10.7 *Let* **f** *be a square-integrable function on* $[0, \infty)$. *Then*

$$f(x) = \sqrt{\frac{2}{\pi}} \int_0^\infty \hat{f}^c(k) \cos(kx)\,dk, \qquad (10.100)$$

where

$$\hat{f}^c(k) = \sqrt{\frac{2}{\pi}} \int_0^\infty f(x) \cos(kx)\,dx. \qquad (10.101)$$

Proof: As with the sine transform, we extend **f** to be a function on the whole line, only now we require it to be an even function: $f(-x) = f(x)$. The rest of the proof, applying Theorem 10.1 and using symmetry to cancel terms, is almost identical to the proof of Theorem 10.6, and is left to the exercises. ∎

While the sine transform is the eigenfunction expansion of the second derivative operator with Dirichlet boundary conditions, the cosine transform is the eigenfunction expansion of the second derivative operator with Neumann boundary conditions. The definition of the eigenfunctions, the decomposition of the identity, and the normalizations are entirely analagous to equations (10.90–10.94).

We obtain a relation between transforms and first derivatives by taking the derivative of both sides of (10.84) and (10.100) with respect to x:

$$\begin{aligned} f'(x) &= \sqrt{\frac{2}{\pi}} \int_0^\infty k\hat{f}^s(k) \cos(kx)\,dk, \\ &= -\sqrt{\frac{2}{\pi}} \int_0^\infty k\hat{f}^c(k) \sin(kx)\,dk. \qquad (10.102) \end{aligned}$$

In other words, the *cosine* transform of $f'(x)$ is k times the *sine* transform of $f(x)$, while the sine transform of $f'(x)$ is $-k$ times the cosine transform of $f(x)$.

With this we can identify the last two terms of (10.99). If we let $G_0(x) = \int_0^x g_0(x')\,dx'$ be the integral of $g_0(x)$, then the last two terms are $-G_0(k(x-vt))/2v$ and $G_0(k(x+vt))/2v$, respectively. This agrees with our previous decomposition (8.16–8.18) of a solution of the wave equation on the half line as a sum of traveling waves.

10.6. Fourier Transforms on the Half Line

Exercises

1. Compute the cosine transform of $e^{-x^2/2}$.
2. Compute the sine transform of $xe^{-x^2/2}$ directly, and compare to (10.102) and the results of Exercise 1.
3. Compute the cosine transform of $x^2 e^{-x^2/2}$.
4. For general n, derive a formula for the sine transform of $x^{2n+1} e^{-x^2/2}$ and the cosine transform of $x^{2n} e^{-x^2/2}$.
5. Use the sine transform to solve the heat equation on the half line $[0, \infty)$ with Dirichlet boundary conditions at the origin.
6. Use the sine transform to solve the Klein-Gordon equation on the half line $[0, \infty)$ with Dirichlet boundary conditions at the origin.
7. Use the cosine transform to solve the heat equation on the half line $[0, \infty)$ with Neumann boundary conditions at the origin.
8. Prove Theorem 10.7.

Chapter 11

Green's Functions

A recurring theme of this book is that, on an n-dimensional vector space, the choice of a basis converts vectors into n-tuples of numbers (i.e., elements of $\mathbb{R}^n$) and converts operators into $n \times n$ matrices, which can then be manipulated by ordinary matrix multiplication. Similarly, on an infinite dimensional vector space, the choice of an (ordinary) basis converts vectors into infinite sequences of numbers, and converts operators into matrices of infinite size, which can again be handled by ordinary matrix multiplication.

In this chapter we explore the analogous construction for a basis of generalized eigenfunctions, especially δ functions. This is a very simple idea, but the implications are profound. In Section 11.1 we develop the general formalism. In the remainder of the chapter we use the formalism to solve a variety of ordinary and partial differential equations.

11.1 Delta Functions and the Superposition Principle

In Chapter 9 we introduced the Dirac δ function. We now use δ functions to represent elements of $L^2(\mathbb{R})$ as functions of one variable (which they already are, of course) and to represent operators on L^2 as functions of two variables.

Definition *If L is an operator on $L^2(\mathbb{R})$, and K is a function of two variables such that, for every x,*

$$(Lf)(x) = \int_{-\infty}^{\infty} K(x,y)f(y)dy, \qquad (11.1)$$

then we say that K is the integral kernel *of the operator* L.[1] *We also say that L is an* integral operator.

Example: Let $g(x)$ be any smooth function, such as e^{-x^2}, and let Lf be the convolution $\mathbf{f} * \mathbf{g}$. Then L is an integral operator, and $K(x,y) = g(x-y)$.

Formally we can construct integral kernels for all operators, and we will do so. However, in many cases we will discover that the "function" K is really a distribution, not a function. In such cases we do not call L an integral operator, as the action of L does not really involve integration. For example, the simple fact that

$$f(x) = \int_{-\infty}^{\infty} \delta(x-y)f(y)dy \qquad (11.2)$$

shows that $\delta(x-y)$ is the integral kernel of the identity operator. Similarly, $\delta'(x-y)$ is the integral kernel of the derivative operator. However, the identity operator and the derivative operator are not considered to be integral operators.

To construct integral kernels, we note that every function $\mathbf{f}$ can be viewed as a linear combination of δ functions:

$$|\mathbf{f}\rangle = \int_{-\infty}^{\infty} f(y)|X,y\rangle dy. \qquad (11.3)$$

Applying L to each side we obtain

$$L|\mathbf{f}\rangle = \int_{-\infty}^{\infty} f(y)L|X,y\rangle dy. \qquad (11.4)$$

To get $Lf(x)$ we take the inner product of $\langle X, x|$ with $L|\mathbf{f}\rangle$:

$$\begin{aligned} Lf(x) &= \langle X,x|L|\mathbf{f}\rangle = \int_{-\infty}^{\infty} f(y)\langle X,x|L|X,y\rangle dy \\ &= \int_{-\infty}^{\infty} K(x,y)f(y)dy, \end{aligned} \qquad (11.5)$$

where

$$K(x,y) = \langle X,x|L|X,y\rangle. \qquad (11.6)$$

Note the close analogy between our derivation of the integral kernel of an operator on $L^2(\mathbb{R})$ and the derivation in Section 3.2 of

---

[1] Do not confuse the integral kernel, which is a function, with the kernel of L, which is the set of vectors $\mathbf{f}$ for which $L\mathbf{f} = 0$. It is unfortunate that these very different objects have almost identical names.

11.1. Delta Functions and the Superposition Principle

the matrix of a linear transformation. This is our yoga for converting operations from discrete expansions to continuous ones: In place of lists of numbers with one index (e.g., the coefficients v_i of a vector in $\mathbb{R}^n$), we have a function of one variable (e.g., $f(x)$ or $\hat{f}(k)$). In place of lists of numbers with two indices (e.g., matrix elements A_{ij}), we have functions of two variables (such as $K(x,y)$). In place of summation over an index

$$(A\mathbf{v})_i = \sum_j A_{ij} v_j, \tag{11.7}$$

we have integration over a variable as in (11.2).

This analogy extends to products of operators. Just as the product of two matrices A and B has matrix elements

$$(AB)_{ij} = \sum_k A_{ik} B_{kj}, \tag{11.8}$$

the product of two operators L_1 and L_2 has integral kernel

$$K(x,y) = \int_{-\infty}^{\infty} dz\, K_1(x,z) K_2(z,y), \tag{11.9}$$

where K_1 is the integral kernel of L_1, and K_2 is the integral kernel of K_2. To see this, just compute

$$\begin{aligned} L_1 \circ L_2 f(x) &= \int dz\, K_1(x,z) L_2 f(z) \\ &= \int dz\, K_1(x,z) \int dy\, K_2(z,y) f(y) \\ &= \int dy \left(\int dz\, K_1(x,z) K_2(z,y) \right) f(y). \end{aligned} \tag{11.10}$$

Although we have defined integral kernels and integral operators only for operators on $L^2(\mathbb{R})$, the same idea applies to $L^2(U)$, where U is any domain, such as the half line or an interval. U can even be multi-dimensional. For example, in $\mathbb{R}^3$, if $\rho(\mathbf{x})$ is the electric charge density at the point $\mathbf{x} \in \mathbb{R}^3$, then the electrostatic potential ϕ is given by

$$\phi(\mathbf{x}) = L\rho(\mathbf{x}) = \int_{\mathbb{R}^3} d^3y\, \frac{\rho(\mathbf{y})}{|\mathbf{x} - \mathbf{y}|}. \tag{11.11}$$

In this case, L is an integral operator with integral kernel $K(\mathbf{x},\mathbf{y}) = |\mathbf{x} - \mathbf{y}|^{-1}$.

Many theorems about matrices extend naturally to integral operators. As an example, we state (but do not prove) the following analog of Theorem 5.1:

Theorem 11.1 *Let L be an integral operator on $L^2(U)$ whose integral kernel $K(x,y)$ is real and non-negative for all x,y, and such that, for all y, $\int_U dx K(x,y) = 1$. (This is the analog of a probability matrix.) Then 1 is a (possibly generalized) eigenvalue of L, with an eigenfunction that is everywhere non-negative. All other (generalized) eigenvalues of L have magnitude at most 1, and the corresponding eigenfunctions integrate to zero. If $K(x,y)$ is strictly positive for all x,y, then these additional eigenvalues all have magnitude strictly less than 1.*

Exercises

Find the integral kernels of the following operators:

1. $Lf(x) = \int_0^x f(t)dt$.
2. $Lf(x) = f(x-1)$.
3. $Lf(x) = (x^2+1)f(x)$.
4. $Lf(x) = x\int_0^1 f(t)dt$.
5. $Lf(x) = f(2x)$.
6. $Lf(x) = \sin(x)\int_0^\infty f(xt)e^{-t}dt$.
7. Show that the integral kernel of d/dx is $\delta'(x-y)$, as previously claimed.
8. Let $L_1 = d/dx$ and let L_2 be the integration operator of Exercise 1. What is $L_1 \circ L_2$? Show explicitly that $\int dz K_1(x,z)K_2(z,y)$ is indeed the integral kernel of $L_1 \circ L_2$.

11.2 Inverting Operators

Integrating a function makes it smoother, as the indefinite integral of a function **f** is always once more differentiable than **f**. Similarly, for most integral operators L, Lf is at least as smooth as **f**. Even when **f** is a δ distribution $f(x) = \delta(x-a)$, $Lf(x) = K(x,a)$ is still an honest function. The flip side of this observation is that an operator that does not smooth things out (e.g., a differential operator) is rarely an integral operator. However, the inverse of such an operator often is:

Definition *The* Green's function *of a differential operator L is the integral kernel of L^{-1}.*

11.2. Inverting Operators

The main use for Green's functions is to solve differential equations with a variety of source terms. If $G(x, y)$ is the Green's function for a differential operator L, then the solution to the differential equation

$$L\mathbf{f} = \mathbf{g}, \tag{11.12}$$

where $\mathbf{g}$ is given and we have to solve for $\mathbf{f}$, is

$$f(x) = \int G(x,y)g(y)dy. \tag{11.13}$$

If we can find $G(x, y)$, we can solve equation (11.12) for every possible $\mathbf{g}$.

We have already seen an example of this in Section 10.3. The solution to the differential equation

$$f''(x) - f(x) = g(x) \tag{11.14}$$

on $L^2(\mathbb{R})$ is

$$f(x) = \frac{-1}{2} \int e^{-|x-y|} g(y) dy. \tag{11.15}$$

That is, the Green's function for the operator $L = d^2/dx^2 - 1$ is $G(x, y) = -e^{-|x-y|}/2$.

Computing Green's functions

Recall the superposition principle: If $\mathbf{f}_1$ is the solution to $L\mathbf{f} = \mathbf{g}_1$, and if $\mathbf{f}_2$ is the solution to $L\mathbf{f} = \mathbf{g}_2$, then $c_1\mathbf{f}_1 + c_2\mathbf{f}_2$ is the solution to $L\mathbf{f} = c_1\mathbf{g}_1 + c_2\mathbf{g}_2$. In other words, we can decompose the right-hand side of our differential equation into manageable pieces, solve the differential equation for each piece, and recombine to get a solution to our original differential equation. You have probably already guessed that these manageable pieces are δ functions.

Example 1: We illustrate the method by finding the Green's function for the differential operator $L_1 = \frac{d}{dx} + 1$. For every constant y, we solve the differential equation

$$L_1 f_y(x) = \delta(x - y). \tag{11.16}$$

That is, we want

$$\frac{df_y(x)}{dx} + f_y(x) = \delta(x - y). \tag{11.17}$$

Figure 11.1: The strategy for computing a Green's function

We then let $G(x, y) = f_y(x)$. Multiplying both sides of (11.17) by $g(y)$ and integrating over y then shows that $f(x) = \int G(x,y)g(y)dy$ is a solution to $Lf = g$.

We solve equation (11.17) in pieces, first in the region $x < y$, then in the region $x > y$, and finally match the solutions at $x = y$. See Figure 11.1. When $x < y$ we have $f_y'(x) + f_y(x) = 0$, so $f_y(x)$ is a multiple of e^{-x}. Exactly the same argument applies when $x > y$, so f_y must take the form

$$f_y(x) = \begin{cases} c_1 e^{-x} & \text{if } x < y; \\ c_2 e^{-x} & \text{if } x > y. \end{cases} \quad (11.18)$$

Since e^{-x} diverges as $x \to -\infty$, we must have $c_1 = 0$. The only remaining question is to determine c_2.

This is where the δ function comes in. Integrating equation (11.17) from $x = y - \epsilon$ to $x = y + \epsilon$ gives

$$f_y(y+\epsilon) - f_y(y-\epsilon) = 1 - \int_{y-\epsilon}^{y+\epsilon} f_y(x)dx, \quad (11.19)$$

which goes to 1 as $\epsilon \to 0$. In other words, $f_y(x)$ has a jump discontinuity of size 1 at $x = y$. Since $c_1 = 0$, this implies that $c_2 = e^y$, so our Green's function is

$$G(x, y) = f_y(x) = \begin{cases} e^{y-x} & \text{if } x > y; \\ 0 & \text{otherwise}, \end{cases} \quad (11.20)$$

and the solution to $L_1 \mathbf{f} = \mathbf{g}$, for any function $\mathbf{g}$, is

$$f(x) = \int_{-\infty}^{\infty} dy G(x,y)g(y)dy = \int_{-\infty}^{x} e^{y-x} g(y)dy. \quad (11.21)$$

See Figure 11.2.

Example 2: We compute the Green's function for the operator $L_2 = d^2/dx^2 - 1$ on the half line $[0, \infty)$, with Dirichlet boundary conditions at $x = 0$. We need to solve

$$L_2 f_y(x) = f_y''(x) - f_y(x) = \delta(x - y) \quad (11.22)$$

11.2. Inverting Operators

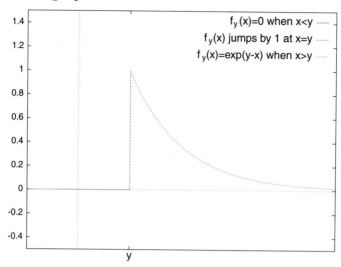

Figure 11.2: The Green's function for $L = d/dx + 1$

for each value of y. As before, we do this in two regions, $x < y$ and $x > y$, and then match the solutions appropriately at $x = y$.

When $x < y$ we have $f_y''(x) = f_y(x)$, so $f_y(x)$ must take the form $c_1 e^x + c_2 e^{-x}$. The Dirichlet condition $f_y(0) = 0$ then forces us to have $c_2 = -c_1$. Similarly, when $x > y$ we must have $f_y(x) = c_3 e^x + c_4 e^{-x}$. However, e^x blows up as $x \to \infty$, so we must have $c_3 = 0$. Thus our solution takes the form

$$f_y(x) = \begin{cases} c_1(e^x - e^{-x}) & \text{if } x < y; \\ c_4 e^{-x} & \text{if } x > y. \end{cases} \quad (11.23)$$

Next, we match the solutions at $x = y$. Integrating the differential equation (11.22) from $y - \epsilon$ to $y + \epsilon$ shows that f_y is continuous at $x = y$, but that f_y' has a jump discontinuity of size 1. Together this means that

$$\begin{aligned} c_4 e^{-y} + c_1(e^{-y} - e^y) &= 0 & (f_y \text{ is continuous}), \\ -c_4 e^{-y} - c_1(e^{-y} + e^y) &= 1 & (f_y' \text{ jumps by 1}), \end{aligned} \quad (11.24)$$

so

$$c_1 = -\frac{e^{-y}}{2}, \quad c_4 = \frac{e^y - e^{-y}}{2}, \quad (11.25)$$

and our Green's function is

$$G(x, y) = \begin{cases} (e^{-(x+y)} - e^{x-y})/2 & \text{if } x < y; \\ (e^{-(x+y)} - e^{y-x})/2 & \text{if } x \geq y \end{cases} = \frac{e^{-(x+y)} - e^{-|x-y|}}{2}, \quad (11.26)$$

as shown in Figure 11.3.

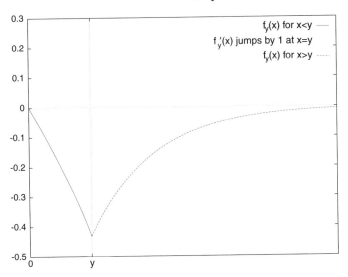

Figure 11.3: The Green's function for $L = d^2/dx^2 - 1$ on $[0, \infty)$

Matching solutions for more general operators

We have seen two examples of how to match solutions at $x = y$. When L was $d/dx + 1$, f_y was required to jump by 1 at $x = y$. When L was $d^2/dx - 1$, f_y was required to be continuous at $x = y$, but f'_y had a jump discontinuity. Here is the general rule:

If L is a k-th order differential operator,

$$L = a_k(x)\frac{d^k}{dx^k} + a_{k-1}(x)\frac{d^{k-1}}{dx^{k-1}} + \cdots + a_0(x), \tag{11.27}$$

where $a_0(x), \ldots a_k(x)$ are smooth at $x = y$, and $a_k(y) \neq 0$, then the $(k-1)$st derivative of f_y has a jump discontinuity of size $1/a_k(y)$ at $x = y$, while all lower derivatives of f_y are continuous.

To see this, rewrite the equation $Lf_y(x) = \delta(x-y)$ as

$$\frac{d^k f_y(x)}{dx^k} = \frac{\delta(x-y)}{a_k(y)} + \text{(terms with lower order derivatives of } f_y\text{)}. \tag{11.28}$$

Integrating this from $x = y - \epsilon$ to $y + \epsilon$ and taking the $\epsilon \to 0$ limit shows that the $(k-1)$st derivative of f_y must jump by $1/a_k(y)$. It also shows that the lower order derivatives do not jump, because if they did, the k-th derivative would involve derivatives of δ functions, which it does not.

To summarize, the procedure for finding the Green's function for a k-th order differential operator L is:

11.2. Inverting Operators

1. Solve $L\mathbf{f}_y = 0$ in the region $x < y$. The general solution will involve k arbitrary constants.

2. Solve $L\mathbf{f}_y = 0$ in the region $x > y$. This too will involve k arbitrary constants.

3. Apply boundary conditions or, if the domain is unbounded, apply growth conditions at $\pm\infty$. This will eliminate k of the $2k$ arbitrary constants.

4. Require that the function, and the first $k - 2$ derivatives be continuous at $x = y$, and that the $(k-1)$st derivative jump by $1/a_k(y)$. These k conditions will then determine the remaining constants.

Exercises

1. Find the Green's function for $d/dx - 1$ on the whole line.

2. If G_1 and G_2 are Green's functions for operators L_1 and L_2, find a formula for the Green's function of $L_1 \circ L_2$. What about $L_2 \circ L_1$?

3. Use the results of the last two exercises to derive the Green's function for $d^2/dx^2 - 1$ on the whole line from the Green's functions of $d/dx \pm 1$.

4. Find the Green's function for $d^2/dx^2 - 3d/dx + 2$ on the whole line.

5. Find the Green's function for $d^2/dx^1 - 1$ on the half line $[0, \infty)$ with Neumann boundary conditions.

6. Find the Green's function for $d^2/dx^2 - 1$ on the interval $[0, 1]$ with Dirichlet boundary conditions.

7. Find the Green's function for $x^2 d^2/dx^2 + x d/dx - n^2$ on the half line $[0, \infty)$ with Dirichlet boundary conditions, where n is a positive integer. [Hint: Rewrite the operator as $(x d/dx)^2 - n^2$.]

8. Find the Green's function for $d^2/dx^2 - n(n+1)/x^2$ on the half line $[0, \infty)$ with Dirichlet boundary conditions, where n is a positive integer.

9. Find the Green's function for $Lf(x) = df(x)/dx - x^2 f(x)$ on the whole line.

10. Find the Green's function for $Lf(x) = df(x)/dx + x^2 f(x)$ on the whole line. How does this differ from the solution to Exercise 9?

11. Let $h(x)$ be a function for which $\int_{-\infty}^{0} f(x)dx = \int_{0}^{\infty} f(x)dx = +\infty$. Find the Green's function for $Lf(x) = df(x)/dx + h(x)f(x)$. How would the situation change if the integrals were $-\infty$?

If an operator is not invertible, then its Green's functions either does not exist, or is not unique, depending on whether the operator has a left inverse, a right inverse, or neither. The following examples explore these possibilities:

12. What happens when you try to compute the Green's function of d/dx on the whole line?

13. What happens when you try to compute the Green's function of $d/dx + x$ on the whole line?

14. What happens when you try to compute the Green's function of $d/dx - x$ on the whole line?

11.3 The Method of Images

Green's functions are used to solve both ordinary and partial differential equations on a variety of domains. When the domain is highly symmetric, like a half line, a disk, or a ball, a differential equation can frequently be solved in two steps. First we solve the equation on all of $\mathbb{R}$ (or $\mathbb{R}^n$, if we are working in higher dimensions). Then we use symmetry to construct a solution on a domain, with appropriate boundary conditions, from the solution on $\mathbb{R}$ or $\mathbb{R}^n$. We already saw this approach in Chapter 8, where we first solved the wave equation on all of $\mathbb{R}$, then on the half line, and then on an interval.

For illustration, we consider the Green's function for $L = d^2/dx^2 - 1$ on four domains. First we construct the Green's function on the whole line, then on the half line, the interval $[0, 1]$ with Dirichlet boundary conditions, and the interval with periodic boundary conditions. (We denote the four Green's functions by G_w, G_h, G_d, and G_p, respectively.)

Example 1: We begin by computing G_w. See Figure 11.4. Away from $x = y$, we must solve $Lf_y = 0$, whose solutions are $e^{\pm x}$. To avoid unbounded growth as $x \to \pm\infty$ our solution must thus be of the form

$$f_y(x) = \begin{cases} c_1 e^x & \text{for } x < y; \\ c_2 e^{-x} & \text{for } x > y, \end{cases} \quad (11.29)$$

for some constants c_1 and c_2. Matching the values of the two func-

11.3. The Method of Images

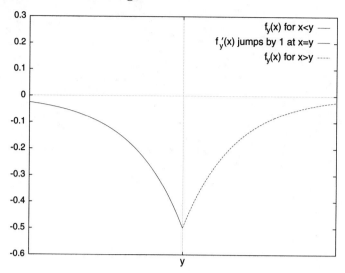

Figure 11.4: The Green's function for $d^2/dx^2 - 1$ on $(-\infty, \infty)$

tions at $x = y$ gives $c_2 = c_1 e^{2y}$, so our function takes the form

$$f_y(x) = c_1 e^y e^{-|x-y|}. \tag{11.30}$$

Finally, we compute the jump in the first derivative:

$$-2e^y c_1 = \lim_{\epsilon \to 0^+} f'_y(y + \epsilon) - f'_y(y - \epsilon) = 1, \tag{11.31}$$

so $c_1 = -e^{-y}/2$, and we have

$$G_w(x, y) = -e^{-|x-y|}/2. \tag{11.32}$$

Example 2: Next we compute G_h. To do this we solve

$$\frac{d^2 f_y(x)}{dx^2} - f_y(x) = \delta(x - y) \tag{11.33}$$

on the half line, with Dirichlet boundary condition by the method of images. As in Chapter 8, we extend f_y to a function on the whole line by defining

$$f_y(-x) = -f_y(x), \tag{11.34}$$

so that f_y satisfies

$$\frac{d^2 f_y(x)}{dx^2} - f_y(x) = \delta(x-y) - \delta(-x-y) = \delta(x-y) - \delta(x+y) \tag{11.35}$$

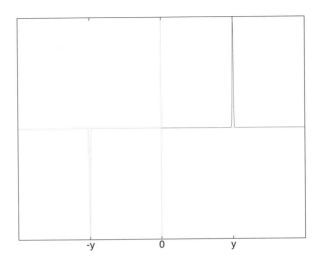

Figure 11.5: A source and its mirror image

on the whole line. We can solve this using G_w! The solution is

$$\begin{aligned}G_h(x,y) &= f_y(x) = \int G_w(x,y)(\delta(x-y) - \delta(x+y))dy \\ &= G_w(x,y) - G_w(x,-y) = \frac{e^{-(x+y)} - e^{-|x-y|}}{2},\end{aligned} \quad (11.36)$$

as we previously determined.

In other words, the response on the half line (with Dirichlet boundary conditions) generated by a source at y is equal to the response on the whole line generated by a source at y, plus the response on the whole line generated by a mirror image source, of opposite sign, at $-y$, as in Figure 11.5.

This argument does not depend on the details of the operator $d^2/dx^2 - 1$. It will work for any operator that is invariant under the symmetry $x \to -x$. In particular, if L is any differential operator of the form

$$L = a_2(x)d^2/dx^2 + a_1(x)d/dx + a_0(x), \quad (11.37)$$

with a_2 and a_0 being even functions of x, and with a_1 being an odd function of x, then we will have $G_h(x,y) = G_w(x,y) - G_w(x,-y)$. If we had worked on the half line with Neumann boundary conditions, the image charge would have been positive, and we would have had $G_h(x,y) = G_w(x,y) + G_w(x,-y)$.

Example 3: Next we consider the interval $[0,1]$ with Dirichlet boundary conditions. The reflections of the point y are at points

11.3. The Method of Images

$y + 2n$ (with positive sign) and $2n - y$ (with negative sign). Therefore

$$G_i(x,y) = \sum_{n=-\infty}^{\infty} G_w(x, y+2n) - \sum_{n=-\infty}^{\infty} G_w(x, 2n-y). \quad (11.38)$$

Example 4: Finally we consider the problem on $[0,1]$ with periodic boundary conditions. The images of the point y under the symmetry $x \to x + 1$ are at points $y + n$, so the Green's function G_p for the operator with periodic boundary conditions is

$$G_p(x,y) = \sum_{n=-\infty}^{\infty} G_w(x, y+n). \quad (11.39)$$

The formulas (11.38–11.39) apply to all operators that are invariant under the relevant symmetries, not just to $d^2/dx^2 - 1$.

Poisson's equation and electrostatics

In one dimension, the list of possible symmetries with which to apply the method of images is fairly short. One can reflect and translate, and that's about it. In two or more dimensions, however, there are many more possible symmetries, hence many more possible ways to apply the method of images.

In $\mathbb{R}^n$, the *Laplacian* operator is defined to be[2]

$$\Delta = \nabla \cdot \nabla = \frac{\partial^2}{\partial x_1^2} + \cdots + \frac{\partial^2}{\partial x_n^2}. \quad (11.40)$$

This operator is invariant under translations, rotations, and reflections. The differential equation

$$\Delta \mathbf{f} = \mathbf{g}, \quad (11.41)$$

where $\mathbf{g}$ is given, and we must solve for $\mathbf{f}$, comes up so frequently in math, physics, and engineering that it has a name: *Poisson's equation*. In particular, in 3-dimensional electrostatics, the electrostatic potential ϕ and charge density ρ are related by $\Delta \phi = -4\pi\rho$, and one frequently wants to find the potential induced by a particular charge density on a particular domain.

---

[2] Some authors define the Laplacian to be $-\sum \partial^2/\partial x_i^2$ rather than $\sum \partial^2/\partial x_i^2$. Both sign conventions are widely used.

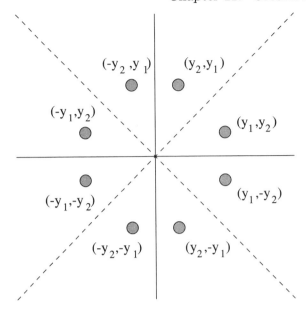

Figure 11.6: Reflections of the region $x_1 \geq x_2 \geq 0$

The Green's function for Δ on all of $\mathbb{R}^n$ is easy to write down:

$$G(\mathbf{x}, \mathbf{y}) = \begin{cases} \ln(|\mathbf{x}-\mathbf{y}|)/2\pi & \text{if } n = 2 \\ \dfrac{-|\mathbf{x}-\mathbf{y}|^{2-n}}{(n-2)S_n} & \text{if } n \neq 2, \end{cases} \qquad (11.42)$$

where S_n is the area of the unit sphere in $\mathbb{R}^n$. (In particular, $S_3 = 4\pi$.) To see that this is correct, one must check that $\Delta G(\mathbf{x}, \mathbf{y}) = 0$ when $\mathbf{x} \neq \mathbf{y}$, and that the flux of $\nabla G(\mathbf{x}, \mathbf{y})$ through a small sphere centered at $\mathbf{x} = \mathbf{y}$ equals 1. This is left as an exercise.

To illustrate the method of images, we compute the Green's function for Δ on the domain $\{x_1 \geq x_2 \geq 0\}$ in $\mathbb{R}^2$, with Dirichlet boundary conditions at the lines $x_2 = x_1$, and $x_2 = 0$. Reflecting the point $\mathbf{y} = (y_1, y_2)$ about the horizontal axis gives $(y_1, -y_2)$, while reflecting about the line $x_2 = x_1$ gives (y_2, y_1). See Figure 11.6. Applying these symmetries repeatedly, we find that the images of the point (y_1, y_2) are at (y_2, y_1), $(-y_2, y_1)$, $(-y_1, y_2)$, $(-y_1, -y_2)$, $(-y_2, -y_1)$, $(y_2, -y_1)$, and $(y_1, -y_2)$.

Summing over $\mathbf{y}$ and its images, we obtain our Green's function:

$$G(\mathbf{x}, \mathbf{y}) = G(x_1, x_2, y_1, y_2) =$$
$$\tfrac{1}{2\pi} \Big(\ln\left(|(x_1, x_2) - (y_1, y_2)|\right) - \ln\left(|(x_1, x_2) - (y_2, y_1)|\right)$$

$$+ \ln\left(||(x_1,x_2) - (-y_2,y_1)|\right) - \ln\left(||(x_1,x_2) - (-y_1,y_2)|\right)$$
$$+ \ln\left(||(x_1,x_2) - (-y_1,-y_2)|\right) - \ln\left(||(x_1,x_2) - (-y_2,y_1)|\right)$$
$$+ \ln\left(||(x_1,x_2) - (y_2,-y_1)|\right) - \ln\left(||(x_1,x_2) - (y_1,-y_2)|\right)\Big)$$
$$= \frac{1}{4\pi}\Big[\ln\left((x_1-y_1)^2 + (x_2-y_2)^2\right) - \ln\left((x_1-y_2)^2 + (x_2-y_1)^2\right)$$
$$+ \ln\left((x_1+y_2)^2 + (x_2-y_1)^2\right) - \ln\left((x_1+y_1)^2 + (x_2-y_2)^2\right)$$
$$+ \ln\left((x_1+y_1)^2 + (x_2+y_2)^2\right) - \ln\left((x_1+y_2)^2 + (x_2+y_1)^2\right)$$
$$+ \ln\left((x_1-y_2)^2 + (x_2+y_1)^2\right) - \ln\left((x_1-y_1)^2 + (x_2+y_2)^2\right)\Big].$$
(11.43)

This solution is not pretty, consisting of eight terms, but this very fact demonstrates the strength of the method. Even when the symmetry and the final answer are complicated, the method itself is straightforward.

Exercises

1. Use the method of images to find the Green's function for $d^2/dx^2 - 1$ on the half line $[0,\infty)$ with Neumann boundary conditions. Compare to Exercise 5 of Section 11.2.

2. Use the method of images to find the Green's function for $d^2/dx^2 - 1$ on the interval $[0,1]$ with Dirichlet boundary conditions. Compare to Exercise 6 of Section 11.2.

3. Find the Green's function for $d^2/dx^2 - 1$ on the interval $[0,1]$ with Neumann boundary conditions.

4. Find the Green's function for $d^2/dx^2 - 1$ on the interval $[0,1]$ with periodic boundary conditions.

5. Let $G(\mathbf{x},\mathbf{y})$ be as in equation (11.42). Show that $\nabla G(\mathbf{x},0)$ is a rotationally symmetric vector field centered at $\mathbf{x} = 0$. Compute the flux of ∇G over a sphere of radius R centered at $\mathbf{x} = \mathbf{y}$. Use the divergence theorem and rotational symmetry to conclude that $\Delta G(\mathbf{x},\mathbf{y}) = \delta^n(\mathbf{x}-\mathbf{y})$.

6. Compute the Green's function of the Laplacian on the first quadrant of $\mathbb{R}^2$, with Dirichlet boundary conditions on the x_1 and x_2 axes. Simplify your result as much as possible.

7. Compute the Green's function of the Laplacian on the first quadrant of $\mathbb{R}^2$, with Neumann boundary conditions (i.e., normal derivative equals zero) on the x_1 and x_2 axes. Simplify your result as much as possible.

8. Compute the Green's function of the Laplacian on the first quadrant of $\mathbb{R}^2$, with Dirichlet boundary conditions on the x_1 axis, and Neumann boundary conditions on the x_2 axis. Simplify as much as possible.

9. Suppose we wish to solve Poisson's equation on the region in $\mathbb{R}^2$ described in polar coordinates by $0 < \theta < 2\pi/n$, with Dirichlet boundary conditions. Show that the method of images works if n is an even integer. What happens if n is an odd integer?

10. Suppose we wish to solve Poisson's equation on the region in $\mathbb{R}^2$ described in polar coordinates by $0 < \theta < 2\pi/n$, with Neumann boundary conditions. Show that the method of images works if n is an integer. What happens if n is not an integer?

11.4 Initial Value Problems

We have seen the use of integral kernels to invert linear operators. In this section we use integral kernels to solve boundary value problems, and in particular initial value problems.

First-order equations

As an example, suppose the function $f(x,t)$ is governed by the heat equation

$$\frac{\partial f(x,t)}{\partial t} = D\frac{\partial^2 f(x,t)}{\partial x^2} \qquad (11.44)$$

where D is a fixed positive constant. We seek an integral operator that converts the initial function $f(x,0)$ into the function at time t. That is, we want to find a function $K_t(x,y)$ such that

$$f(x,t) = \int K_t(x,y) f(y,0) dy. \qquad (11.45)$$

Such a function is called the *Green's function* of the evolution equation (11.44).

We have already defined the term "Green's function" somewhat differently. Unfortunately, the term is generally used in both contexts, with somewhat different meanings. The Green's function of an *operator* is the integral kernel of the inverse of that operator. The Green's function of an *equation* is the integral kernel of the operator that describes time evolution.

11.4. Initial Value Problems

These two notions are closely related. If our differential equation takes the form

$$\frac{d\mathbf{f}}{dt} = L\mathbf{f}, \tag{11.46}$$

then $\mathbf{f}(t) = e^{Lt}\mathbf{f}(0)$, so the Green's function for the *equation* (11.46) is the integral kernel of the operator e^{Lt}, and so is the same thing as the Green's function of the *operator* e^{-Lt}.

To find the Green's function of a linear differential equation, we must find a solution $f_y(x,t)$ to the equation with initial condition

$$f_y(x, 0) = \delta(x - y) \tag{11.47}$$

for each value of y. More precisely, we need $f_y(x,t)$ to satisfy the differential equation for all $t > 0$ and, for any fixed test function $\phi(x)$, we need

$$\lim_{t \to 0} \int \phi(x) f_y(x,t) dx = \phi(y). \tag{11.48}$$

We then set $K_t(x,y) = f_y(x,t)$.

To see that this is our correct Green's function, we must check that

$$f(x,t) = \int K_t(x,y) f(y,0) dy = \int f_y(x,t) f(y,0) dy \tag{11.49}$$

is a solution to the differential equation with initial condition $f(x,0)$. Since each $f_y(x,t)$ is a solution, since $f(x,t)$ is a linear combination of the functions $f_y(x,t)$, and since the equation is linear, $f(x,t)$ is a solution. As for the initial condition,

$$\lim_{t \to 0} f(x,t) = \int \lim_{t \to 0} f_y(x,t) f(y,0) dy = \int \delta(x-y) f(y,0) dy = f(x,0). \tag{11.50}$$

Example: We compute the Green's function of the heat equation (11.44). One can directly check that the function

$$K_t(x,y) = f_y(x,t) = \frac{e^{-(x-y)^2/4Dt}}{\sqrt{4Dt\pi}} \tag{11.51}$$

satisfies the differential equation (11.44) for $t > 0$, and that the limit $\lim_{t \to 0} K_t(x,y)$ equals $\delta(x-y)$. Therefore, $K_t(x,y)$ is the Green's function for the heat equation. This function comes up frequently in math, physics and engineering, and is often called the *heat kernel*.

Although the preceding paragraph shows that (11.51) is indeed the Green's function for the heat equation, it does not explain how that function is derived. The easiest method I know is via Fourier transforms, and is essentially a repeat of the argument of Section 10.4. The Fourier transform (with respect to x) of the heat equation is

$$\frac{\partial \hat{f}(k,t)}{\partial t} = -k^2 D \hat{f}(k,t), \qquad (11.52)$$

with solution

$$\hat{f}(k,t) = e^{-Dk^2} \hat{f}(k,0). \qquad (11.53)$$

The initial condition $f(x,0) = \delta(x-y)$ implies $\hat{f}(k,0) = e^{-iky}/\sqrt{2\pi}$, so

$$\hat{f}_y(k,t) = e^{-Dk^2 - iky}/\sqrt{2\pi}. \qquad (11.54)$$

Taking the inverse Fourier transform then gives (11.51).

The exact same method gives the kernel

$$K_t^S(x,y) = \sqrt{\frac{m}{2\pi i \hbar t}} \exp(im(x-y)^2/2\hbar t), \qquad (11.55)$$

for the free Schrödinger equation, as previously determined in Section 10.4.

Second-order equations

To solve a second-order differential equation we need more information than just the initial value. We either need to know the initial value and its time derivative, or the initial value and the final value, or the initial value and a condition on the behavior of the solution as $t \to \infty$. In this section we consider the case where the initial value and its time derivative are given.

In general, we want a linear map from our initial data to our configuration at time t. Schematically,

$$\begin{aligned} \text{(Solution at } t) &= G_t(\text{initial data}) \\ &= \begin{pmatrix} G_t^{(1)} & G_t^{(2)} \end{pmatrix} \begin{pmatrix} \text{initial value} \\ \text{initial time derivative} \end{pmatrix} \\ &= G_t^{(1)}(\text{initial value}) \\ &\quad + G_t^{(2)}(\text{initial time derivative}), \end{aligned} \qquad (11.56)$$

where G_t is a linear map with a component $G_t^{(1)}$ that acts on the initial value, and a component $G_t^{(2)}$ that acts on the initial time derivative. We reconsider some of our past results in this light.

11.4. Initial Value Problems

The simplest example is a second-order ordinary differential equation

$$\frac{d^2x}{dt^2} = ax, \qquad (11.57)$$

with $a > 0$, whose solution we already know to be

$$x(t) = \cosh(\sqrt{a}t)x(0) + \frac{\sinh\sqrt{a}t}{\sqrt{a}}\dot{x}(0). \qquad (11.58)$$

This is of the form (11.56) with

$$G_t^{(1)} = \cosh(\sqrt{a}t) \qquad \text{and} \qquad G_t^{(2)} = \frac{\sinh\sqrt{a}t}{\sqrt{a}}. \qquad (11.59)$$

In general, for an ordinary differential equation $G_t^{(1)}$ is the trajectory that starts with initial position 1 and initial velocity zero, while $G_t^{(2)}$ is the trajectory that starts with initial position zero and initial velocity 1.

In Chapter 5 we solved coupled systems

$$\frac{d^2\mathbf{x}}{dt^2} = A\mathbf{x} \qquad (11.60)$$

of second-order ordinary differential equations. We obtained solutions by decomposing the initial data into eigenvectors of A, multiplying the coefficients by appropriate (possibly hyperbolic) sines and cosines, and recombining into $\mathbf{x}(t)$. Each step was linear, and the final result could have been written in the general form (11.56), with the operators $G_t^{(1)}$ and $G_t^{(2)}$ being $m \times m$ matrices. The k-th column of $G_t^{(1)}$ is the solution with $\mathbf{x}(0) = \mathbf{e}_k$ and $\dot{\mathbf{x}}(0) = 0$, while the k-th column of $G_t^{(2)}$ is the solution with $\mathbf{x}(0) = 0$ and $\dot{\mathbf{x}}(0) = \mathbf{e}_k$.

In Chapter 8 we solved the wave equation on the whole line with initial data

$$f(x,t) = f_0(x), \qquad \dot{f}(x,0) = g_0(x). \qquad (11.61)$$

Our solution was

$$f(x,t) = h_1(x - vt) + h_2(x + vt) \qquad (11.62)$$

with

$$h_1(x) = \frac{vf_0(x) - \int g_0(x)dx}{2v}; \qquad h_2(x) = \frac{vf_0(x) + \int g_0(x)dx}{2v}. \qquad (11.63)$$

This can be rewritten as

$$f(x,t) = \int_{-\infty}^{\infty} dy G_t^{(1)}(x,y) f_0(y) + G_t^{(2)}(x,y) g_0(y), \quad (11.64)$$

with

$$G_t^{(1)}(x,y) = \frac{\delta(x-y-vt) + \delta(x-y+vt)}{2}$$

$$G_t^{(2)}(x,y) = \begin{cases} 1/2v & \text{if } |x-y| < vt; \\ 0 & \text{otherwise.} \end{cases} \quad (11.65)$$

We recognize $G_t^{(1)}(x,y)$ as the solution to the wave equation with initial data $f(x,0) = \delta(x-y)$ and $\dot{f}(x,0) = 0$, while $G_t^{(2)}(x,y)$ is the solution to the wave equation with $f(x,0) = 0$ and $\dot{f}(x,0) = \delta(x-y)$, in complete analogy with (11.47).

Exercises

1. Find the Green's function of the heat equation on the half line with Dirichlet boundary conditions.

2. Repeat for Neumann boundary conditions.

3. Find the Green's function of the heat equation on interval $[0,1]$ with Dirichlet boundary conditions.

4. Repeat for Neumann boundary conditions.

5. Repeat for periodic boundary conditions.

6. Find Green's functions $G_t^{(1)}$ and $G_t^{(2)}$ for the (finite dimensional) evolution equation $d^2\mathbf{x}/dt^2 = \begin{pmatrix} -2 & 1 \\ 1 & -2 \end{pmatrix} \mathbf{x}$.

7. Find the Green's function for the equation

$$\frac{\partial f(x,t)}{\partial t} = D \frac{\partial^2 f(x,t)}{\partial x^2} - \alpha \frac{\partial f(x,t)}{\partial x},$$

where α is a constant. [This equation describes diffusion in a uniformly moving fluid.]

8. Show that

$$G_t(x,y) = \frac{e^{(x-e^{-t}y)^2/(2-2e^{-2t})}}{\sqrt{\pi(2-2e^{-2t})}}$$

is the Green's functions for the Ornstein-Uhlenbeck equation:

$$\frac{\partial f(x,t)}{\partial t} = \frac{\partial}{\partial x}\left(\frac{\partial}{\partial x} + x\right) f.$$

[This equation governs diffusion in a quadratic potential, and so is important in modeling thermal fluctuations away from equilibrium in a variety of systems. In particular, it describes velocities of particles in a fluid, and the integral of x over time describes the position of a realistic particle undergoing Brownian motion.]

9. Find the Green's functions for the Klein-Gordon equation:

$$\frac{\partial^2 f(x,t)}{\partial t^2} = \frac{\partial^2 f(x,t)}{\partial x^2} + f.$$

You may leave your answer in terms of Fourier transforms.

11.5 Laplace's Equation on $\mathbb{R}^2$

As a demonstration of Green's functions for boundary value problems, in this section we find solutions to Laplace's equation

$$\Delta \mathbf{f} = 0 \qquad (11.66)$$

(also known as harmonic functions) with specified boundary conditions on $\mathbb{R}^2$. This could describe an electrostatic potential in a region with no charge in which the potential on the boundary of the region was known. It also describes a steady-state distribution of temperature. There is an extensive literature on the subject of Laplace's equation in 2 dimensions, and almost any complex analysis book describes a variety of "conformal mapping" methods. Here we limit ourselves to a few simple examples where the Green's function for the equation (11.66) can be readily computed.

Problems with rectilinear symmetries

Example 1: Our first problem is to find a bounded solution to Laplace's equation on the half plane $x_2 \geq 0$, with specified values on the x_1 axis. See Figure 11.7. Specifically we assume that

$$f(x_1, 0) = f_0(x_1). \qquad (11.67)$$

As with the heat and wave equations, we take a Fourier transform of the equation

$$\frac{\partial^2 f(x_1, x_2)}{\partial x_1^2} + \frac{\partial^2 f(x_1, x_2)}{\partial x_2^2} = 0 \qquad (11.68)$$

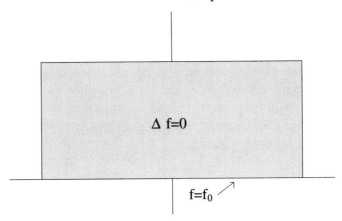

Figure 11.7: Laplace's equation on the upper half plane

with respect to x_1 to get

$$\frac{\partial^2 \hat{f}(k, x_2)}{\partial x_2^2} = k^2 \hat{f}(k, x_2). \quad (11.69)$$

This can be solved one value of k at a time, giving

$$\hat{f}(k, x_2) = c_1(k)e^{kx_2} + c_2(k)e^{-kx_2}, \quad (11.70)$$

where we still have to determine the functions c_1 and c_2. The boundary condition at $x_2 = 0$ gives half the required information:

$$\hat{f}_0(k) = c_1(k) + c_2(k). \quad (11.71)$$

The other half is given by the condition that our solution remain bounded for all x_2. Depending on whether k is positive or negative, one of the functions $e^{\pm kx_2}$ grows as $x_2 \to \infty$, so its coefficient must be zero. By (11.71), the other coefficient is then equal to $\hat{f}_0(k)$. In short,

$$\hat{f}(k, x_2) = \hat{f}_0(k)e^{-|k|x_2}. \quad (11.72)$$

Combining Table 10.2 with Theorem 10.3, we see that $e^{-|k|x_2}$ is the Fourier transform of

$$\tilde{G}_{x_2}(x_1) = \sqrt{\frac{2}{\pi}} \frac{1}{x_2(1 + (x_1/x_2)^2)} = \sqrt{\frac{2}{\pi}} \frac{x_2}{x_1^2 + x_2^2}. \quad (11.73)$$

By Theorem 10.4, $f(x_1, x_2)$ is then a convolution

$$f(x_1, x_2) = \int_{-\infty}^{\infty} \frac{\tilde{G}_{x_2}(x_1 - y_1) f_0(y_1) dy_1}{\sqrt{2\pi}}, \quad (11.74)$$

11.5. Laplace's Equation on $\mathbb{R}^2$

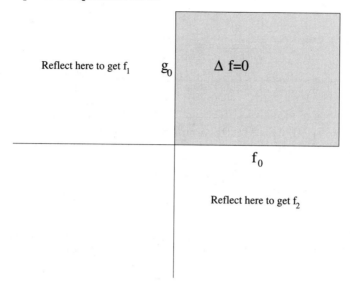

Figure 11.8: Laplace's equation on the first quadrant

which can we rewritten as

$$f(x_1, x_2) = \int_{-\infty}^{\infty} G_{x_2}(x_1, y_1) f_0(y_1) dy_1, \quad (11.75)$$

where

$$G_{x_2}(x_1, y_1) = \frac{1}{\pi} \frac{x_2}{(x_1 - y_1)^2 + x_2^2}. \quad (11.76)$$

is our Green's function.

Example 2: As a second problem, we derive a bounded solution to Laplace's equation on the first quadrant $x_1 \geq 0, x_2 \geq 0$, with specified values on the positive x_1 and x_2 axes. See Figure 11.8. In other words, given two functions $f_0(x_1)$ for $x_1 \geq 0$ and $g_0(x_2)$ for $x_2 \geq 0$, we seek a bounded function $f(x_1, x_2)$ that satisfies

$$\begin{array}{ll} \Delta f = 0 & \text{for all } x_1 > 0, x_2 > 0; \\ f(x_1, 0) = f_0(x_1) & \text{for all } x_1 > 0; \\ f(0, x_2) = g_0(x_2) & \text{for all } x_2 > 0. \end{array} \quad (11.77)$$

We do this by first solving the problem for $\mathbf{g}_0 = 0$, then by solving the problem for $\mathbf{f}_0 = 0$, and adding the two solutions.

We solve (11.77) with $\mathbf{g}_0 = 0$ by the method of images. We extend $\mathbf{f}_0$ and $\mathbf{g}_0$ to functions on the whole line by defining

$$f_0(-x_1) = -f_0(x_1), \quad g_0(-x_2) = -g_0(x_2). \quad (11.78)$$

For all x_1 and for $x_2 \geq 0$ we then set

$$f_1(x_1, x_2) = \int_{-\infty}^{\infty} G_{x_2}(x_1, y_1) f_0(y_1) dy_1. \tag{11.79}$$

This function satisfies $\Delta \mathbf{f}_1 = 0$, with $f_1(x_1, 0) = f_0(x_1)$ and, by symmetry, $f_1(0, x_2) = 0$. Reversing the roles of x_1 and x_2, we define, for $x_1 \geq 0$ and all x_2,

$$f_2(x_1, x_2) = \int_{-\infty}^{\infty} G_{x_1}(x_2, y_2) g_0(y_2) dy_2. \tag{11.80}$$

This satisfies $\Delta \mathbf{f}_2 = 0$, with $f_2(0, x_2) = g_0(x_2)$, and $f_2(x_1, 0) = 0$. Finally we add $\mathbf{f}_1$ and $\mathbf{f}_2$ to get our desired solution

$$\begin{aligned} f(x_1, x_2) &= f_1(x_1, x_2) + f_2(x_1, x_2) \\ &= \int_0^{\infty} \left(G_{x_2}(x_1, y_1) - G_{x_2}(x_1, -y_1) \right) f_0(y_1) dy_1 \\ &\quad + \int_0^{\infty} \left(G_{x_1}(x_2, y_2) - G_{x_1}(x_2, -y_2) \right) g_0(y_2) dy_2 \\ &= \frac{1}{\pi} \int_0^{\infty} \left(\frac{x_2}{(x_1-y_1)^2 + x_2^2} - \frac{x_2}{(x_1+y_1)^2 + x_2^2} \right) f_0(y_1) dy_1 \\ &\quad + \frac{1}{\pi} \int_0^{\infty} \left(\frac{x_1}{(x_2-y_2)^2 + x_1^2} - \frac{x_1}{(x_2+y_2)^2 + x_1^2} \right) g_0(y_2) dy_2. \end{aligned} \tag{11.81}$$

A problem with rotational symmetry

We close with a final application of Green's functions to reconstruct a harmonic function on the unit disk from its values on the unit circle. See Figure 11.9. In this case, we specify the value of the function on the entire boundary but do not specify derivatives. Since the region of interest is bounded, we do not have to worry about growth conditions at infinity.

Because of the rotational symmetry of the problem, we will use polar coordinates (r, θ). It is also convenient to consider the complex coordinates $z = re^{i\theta}$ and $\bar{z} = re^{-i\theta}$. Instead of deriving our Green's functions from Fourier transforms on $\mathbb{R}$, we use Fourier series on the unit circle. The basic result is

Theorem 11.2 *The harmonic function on the unit disk with boundary value* $f(1, \theta) = \sum a_n e^{in\theta}$ *is* $f(r, \theta) = \sum a_n r^{|n|} e^{in\theta}$.

11.5. Laplace's Equation on $\mathbb{R}^2$

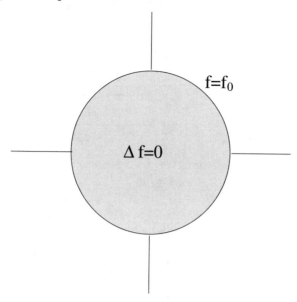

Figure 11.9: Laplace's equation on the unit disk

Proof: For $n \geq 0$, z^n is complex analytic, and therefore harmonic, and equals $e^{in\theta}$ on the unit circle. Similarly, $\bar{z}^n$ is the complex conjugate of an analytic function, and so is harmonic, and $\bar{z}^n = e^{-in\theta}$ on the unit circle. Therefore $\sum_{n=0}^{\infty} a_n z^n + \sum_{n=-1}^{\infty} a_{-n} \bar{z}^n$ is harmonic, and equals $\sum_{n=0}^{\infty} a_n e^{in\theta} + \sum_{n=-1}^{\infty} a_{-n} e^{-in\theta} = \sum_{n=-\infty}^{\infty} a_n e^{in\theta}$ on the unit circle. ∎

We now have the tool to construct a Green's function $G_r(\theta, \phi)$ such that, given f on the unit circle,

$$f(r, \theta) = \int_0^{2\pi} G_r(\theta, \phi) f(1, \phi) d\phi \tag{11.82}$$

is our desired extension to the interior. We first derive $G_r(\theta, 0)$. This (viewed as a function of (r, θ)) is a harmonic function that equals $\delta(\theta)$ when $r = 1$. However, we know how to find the Fourier series of a δ function:

$$\delta(\theta) = \frac{1}{2\pi} \sum_{n=-\infty}^{\infty} e^{in\theta}. \tag{11.83}$$

By Theorem 11.2 we then have

$$G_r(\theta, 0) = \frac{1}{2\pi} \left(\sum_{n=0}^{\infty} z^n + \sum_{n=1^{\infty}} \bar{z}^n \right)$$

$$= \frac{1}{2\pi}\left(\frac{1}{1-z} + \frac{\bar{z}}{1-\bar{z}}\right)$$

$$= \frac{1}{2\pi}\left(\frac{1-z\bar{z}}{1-z-\bar{z}+z\bar{z}}\right)$$

$$= \frac{1}{2\pi}\frac{1-r^2}{1+r^2-2r\cos(\theta)}. \tag{11.84}$$

Finally, by rotational symmetry, G_r can only depend on the difference between the angles θ and ϕ, so

$$G_r(\theta,\phi) = \frac{1}{2\pi}\frac{1-r^2}{1+r^2-2r\cos(\theta-\phi)}. \tag{11.85}$$

Rescaling we obtain a result, known as Poisson's formula, that applies to disks of arbitrary size:

Theorem 11.3 (Poisson's formula) *Let f be a harmonic function on a disk of radius R about the origin. Then for all $r < R$,*

$$f(r,\theta) = \frac{R^2-r^2}{2\pi}\int_0^{2\pi}\frac{f(\phi)}{R^2+r^2-2Rr\cos(\theta-\phi)}d\phi. \tag{11.86}$$

Exercises

1. Find a harmonic function on the upper half plane that equals 0 on the positive x_1 axis and equals 1 on the negative x_1 axis. Your answer should be a closed-form expression, not an integral. [You may find it easier to express your final answer in polar coordinates.]

2. For each ordered pair (a_1, a_2) with $a_1 < a_2$, find a harmonic function on the upper half plane that, on the x_1 axis equals 1 for $a_1 \leq x_1 < a_2$ and 0 otherwise.

3. More generally still, pick n points $a_1 < a_2 < \cdots < a_n$ on the x_1 axis and $n+1$ values $c_0, \ldots, c_n$. Find a harmonic function on the upper half plane that, on the x_1 axis, equals

$$\begin{cases} c_0 & \text{if } x_1 < a_1 \\ c_i & \text{if } a_i \leq x_1 < a_{i+1},\ i = 1,\ldots n-1 \\ c_n & \text{if } x_i \geq a_n. \end{cases}$$

4. Find a harmonic function on the first quadrant of $\mathbb{R}^2$ that equals 0 on the positive x_1 axis, and 1 on the positive x_2 axis.

11.5. Laplace's Equation on $\mathbb{R}^2$

5. Use Poisson's formula to show that a harmonic function cannot have a local maximum or minimum at the origin. [By translational invariance, this then implies that harmonic functions cannot have local maxima or minima anywhere.]

6. Find a harmonic function on the unit disk that equals 1 on the semicircle $r = 1$, $0 < \theta < \pi$, and equals -1 on the semicircle $r = 1$, $\pi < \theta < 2\pi$.

7. Consider Laplace's equation on the exterior of the unit disk, with boundary data on the unit circle. Show that the Green's function for this problem is

$$G_r(\theta, \phi) = \frac{1}{2\pi} \frac{r^2 - 1}{1 + r^2 - 2r\cos(\theta - \phi)}.$$

Index

$\mathbb{C}^n$, 10
C^0, 10
L^2, 179, 181, 250, 257
ℓ_2, 179, 250
M_{nm}, 10
$\mathbb{R}^n$, 10

adjoint, 192, 194
ADSL, 309
AM, 311

bandwidth, 307, 309
basis, 15, 17, 22, 59
    orthogonal, 159, 163, 199, 210, 214
    orthonormal, 159, 163, 167, 198, 205, 257
beats, 115
boundary conditions, 230, 234, 326
    Dirichlet, 230, 237, 314, 316
    Neumann, 230, 231, 234, 238, 316
    periodic, 244
bra, 153, 154, 156, 263
Brownian motion, 301

Cayley-Hamilton Theorem, 96
change of basis, 6, 25, 46
    matrix, 26
characteristic polynomial, 62
closure, 11
column space, 53
common mistakes, 168

commute, 82
complexification, 69, 152
computer graphics, 40
convolution, 299
coordinates, 22
    good, 97
    cosine transform, 316
coupled oscillators, 112
covector, 157

data fitting, 175
decoupling, 1, 6, 60
deficiency, 74
degeneracy, 74
delta function, 264, 265, 268–270, 273, 295, 319
determinant, 20, 78
diagonalizable, 72, 200, 205, 210
diagonalize, 60, 63, 64
    simultaneously, 82
difference equations, 2, 97, 119, 135, 136
diffusion, 301
dimension, 15, 24
dimension theorem, 54
direct sum, 30
    external, 30
    internal, 31, 170
discrete time, 2
distributions, 269, 274
doubling time, 104
dragons, 103
dual space, 153, 156

Index

eigenfunction, 226
   generalized, 264, 265, 268
   expansion, 263
      generalized, 271, 277
eigenspace, 59, 63
eigenvalues, 7, 57, 62, 63
   complex, 68, 210, 214
   dominant, 121
   real, 198, 205, 214
eigenvectors, 7, 57, 59, 63
   of rotations, 216
   orthogonal, 214
evaluation map, 38
evolution equations, 2
   continuous-time, 103, 108
   coupled, 4
   discrete-time, 97
expectation, 309
exponential
   complex, 87
   matrix, 88
   operator, 89
   real, 86

Fibonacci, 99
fixed points, 136
   stable, 138
   unstable, 138
FM, 311
Fourier
   coefficients, 248, 250, 273
      decay, 259
   integral, 283, 287, 288
   series, 184, 238, 246, 257, 283, 289
   transform, 287–289, 302, 308
      decay rates, 296
frequency, 241
   angular, 241, 288
frequency domain, 256
frog, 131

golden mean, 100
Gram-Schmidt process, 165, 167, 199
Green's function, 322, 324–326, 328, 334
group, 212

half line, 104, 313
Hamilton's equations of motion, 149
harmonic functions, 339
heat equation, 301
Heisenberg, 308
Hermitian conjugate, 194

infinite matrix, 50
initial value problems, 334
inner product, 145, 150
   complex, 150
   Lorentzian, 148
   real, 145, 146
   standard, 145
integral kernel, 320
integral operator, 320
isometry, 212
isomorphism, 24

Jordan form, 94

kernel, 52, 53
ket, 153, 154, 156, 263
Klein-Gordon equation, 305

Lagrange multipliers, 57, 202, 203, 208
Laplace's equation, 339, 341
least squares, 172, 173
   weighted, 176
Lie algebra, 221
Lie group, 221
linear combination, 15
linear independence, 15, 16

linear map, 37
linear transformation, 37
    matrix, 39, 42, 162
linearization, 135, 137, 139

Markov chains, 126
matrix
    block diagonal, 79
    block triangular, 79
    conjugate, 65, 78
    ill-conditioned, 73
    linear transformation, 39, 42
    orthogonal, 212
    Pauli, 223
    probability, 80, 126–128
    rotation, 68, 70, 216
    symmetric, 203
    unitary, 213
method of images, 232, 235, 328, 330, 332, 339
metric matrix, 154
mirror, 232
mode
    dominant, 121
    fundamental, 241
    neutrally stable, 121
    normal, 111, 113
    stable, 111, 121, 128
    unstable, 111, 121
momentum operator, 265, 266
multiplicity
    algebraic, 74
    geometric, 74
music, 242

null space, 53

ordinary differential equations
    first-order, 2, 103, 135, 139
    second-order, 3, 108, 135

operators
    Hermitian, 191, 197, 198, 219, 220, 239, 257, 264, 267
    orthogonal, 191, 219, 220
    symmetric, 191, 197, 202, 205, 219
    unitary, 191, 213, 219, 220
order reduction, 117
orthogonal, 198
orthogonal complements, 169
oscillations, 3

partial differential equations, 301
    first-order, 334
    second-order, 336
period, 241
periodic functions, 245, 246
Plancherel's theorem, 293
Poisson bracket, 149
Poisson's equation, 331
Poisson's formula, 344
position operator, 264, 279
power vector, 91
predator-prey, 4, 97
projection, 31, 32, 164, 278, 279
    orthogonal, 165, 169, 171

quadratic function, 202
quotient space, 30, 33, 52

rabbit, 100, 102
range, 52, 53
rank, 54
reflection, 230
Riemann-Stieljes integral, 280
rotation, 70
    symmetry, 342
rotations, 210, 216

Schrödinger equation, 302
Schwarz functions, 297

Index

Schwarz inequality, 147, 152
self-adjoint, 197
sine transform, 313
smoothing operator, 300
span, 15, 17
spectral theorem, 239, 277, 283
spectrum, 264, 265
stability, 120
stage model, 99
standard basis, 17
standard deviation, 309
standing waves, 238, 241, 243, 253, 256
string, 235
subspace, 11
symplectic form, 149

telephone, 308
Tennis Racket Theorem, 143
terminology, xvi
test function, 274
time domain, 255
time scale, 104
trace, 51, 77
transpose, 192
traveling waves, 227, 243, 253, 256

uncertainty principle, 308, 309, 311

variance, 309
vector space, 9, 10, 49
　　axioms, 10
　　complex, 9, 10
　　infinite dimensional, 17, 24, 49
　　real, 9, 10

wave equation, 225, 304
wave number, 241, 288
wave operator, 225

wavelength, 241